Genetic Engineering

Principles and Methods

Volume 17

GENETIC ENGINEERING
Principles and Methods

Genetic Engineering

Principles and Methods

Volume 17

Edited by
Jane K. Setlow
Brookhaven National Laboratory
Upton, New York

Plenum Press • New York and London

The Library of Congress cataloged the first volume of this title as follows:

Genetic engineering: principles and methods, v. 1–
New York, Plenum Press, [1979–
v. ill. 26 cm.
Editors: 1979– J. K. Setlow and A. Hollaender
Key title: Genetic engineering, ISSN 0196-3716

1. Genetic engineering—Collected works. I. Setlow, Jane K. II. Hollaender, Alexander, [DATE]

QH442.G454 575.1 79-644807
MARC-S

ISBN 0-306-45071-2

A Division of Plenum Publishing Corporation
233 Spring Street, New York, N.Y. 10013

10 9 8 7 6 5 4 3 2 1

Printed in the United States of America

CONTENTS OF EARLIER VOLUMES

VOLUME 12 (1990)

VOLUME 13 (1991)

VOLUME 14 (1992)

ACKNOWLEDGMENT

The final processing of the manuscripts has been well done by Sarah L. Zill. She has also eliminated some errors. The Editor salutes her efficiency.

CONTENTS

THE MOLECULAR BIOLOGY OF NUCLEOTIDE EXCISION REPAIR AND DOUBLE-STRAND BREAK REPAIR IN EUKARYOTES

A.R. Lehmann

MRC Cell Mutation Unit
University of Sussex, Falmer
Brighton BN1 9RR, United Kingdom

INTRODUCTION

In the thirty years since the discovery by Setlow and Carrier (1) of excision-repair of ultraviolet damage in *Escherichia coli*, it has become apparent that DNA repair is of fundamental importance for cells and organisms. Recent developments have shown that components of the DNA repair machinery are also involved at the cellular level in transcription, replication and recombination, and in man they are required for protection against cancer, development of the immune response and different aspects of differentiation and development. In all organisms a series of processes exist to remove almost all possible types of DNA damage, and to tolerate damage which has not yet been removed before replication takes place. Different types of chemical alteration are repaired by different mechanisms, namely by base excision repair or nucleotide excision repair. I shall only touch on the former while describing the latter in detail, concentrating on damage produced by ultraviolet light. Ionizing radiation, as well as producing altered bases in DNA, disrupts the physical structure of the phosphodiester backbone, by producing single- and double-strand breaks. The latter requires a special repair system, which I shall also discuss in this article. I shall for the most part confine my discussion to eukaryotic organisms.

BASE EXCISION REPAIR

Before discussing the complex nucleotide excision repair processes, I shall outline the mechanisms of the much simpler base excision repair (BER) (see 2,3 for reviews). Simple single subunit glycosylases are able to disrupt the glycosylic bonds between a variety of altered bases in DNA and the deoxyribose moiety. Most glycosylases can act on specific substrates, such as uracil, 3-methyladenine, pyrimidine oxidation products or 8-hydroxyguanine residues in DNA. Glycosylase action results in an abasic site in DNA (AP site), and a series of enzymes (AP endonuclease, deoxyribophosphodiesterase, DNA

Genetic Engineering, Vol. 17, Edited by J.K. Setlow
Plenum Press, New York, 1995

polymerase and ligase) remove the AP site, replace a single nucleotide and complete the repair process (4). Base excision repair has been reconstituted *in vitro* with the use of bacterial and mammalian enzymes (4) and many of the genes encoding the various enzymes have been cloned from prokaryotic and eukaryotic sources. Sequence conservation is high and transformation of an *E. coli* glycosylase or AP endonuclease mutant with the homologous eukaryotic gene often results in restoration of a wild-type phenotype (e.g., 5-8).

NUCLEOTIDE EXCISION REPAIR IN *E. coli*

Excision repair of ultraviolet damage in *E. coli* has been characterized in detail and reviewed recently (9,10). The products of six genes are involved. A dimer of the UvrA protein binds to DNA and loads the UvrB protein onto damaged DNA. This A_2B" complex tracks along the DNA and, on encountering a damaged site, the UvrB protein binds very tightly and UvrA dissociates from the DNA/UvrB complex. UvrC protein then binds and two nicks are inserted, 8 bases 5′ and 3 to 4 bases 3′ to the damaged nucleotides. DNA helicase II, the product of the *uvrD* gene, releases the 12- or 13-base oligonucleotide containing the damage and the gap is repaired by DNA polymerase I and DNA ligase.

NUCLEOTIDE EXCISION REPAIR (NER) IN EUKARYOTES

During the last 10 to 15 years many genes involved in NER in eukaryotes have been isolated from the budding yeast *Saccharomyces cerevisiae*, the distantly-related fission yeast *Schizosaccharomyces pombe* and from mammalian cells. A major finding has been the high level of structural and functional conservation of these genes between the three organisms, and by extrapolation, throughout eukaryotes. The implication, that the mechanism of NER is similar, if not identical, between species, means that findings obtained in one organism can be applied to others with confidence.

Nomenclature

A regrettable but unavoidable aspect of genetics is that mutants and their corresponding genes have to be named before their structure and function are understood. This often leads to confusion, and DNA repair genes are no exception in this respect. Mutants of *S. cerevisiae* sensitive to UV or ionizing radiation have been designated *rad* (e.g., *rad3*) and the corresponding genes as *RAD* (e.g., *RAD3*). In *S. pombe*, both mutants and their genes are designated *rad*, but the *S. pombe rad* numbers do not correspond to those of *S. cerevisiae*. Thus the *S. cerevisiae RAD3* and *S. pombe rad3* genes are quite different. The *S. pombe* homologue of *S. cerevisiae RAD3* is in fact *rad15* (see below).

Mammalian mutants and genes are derived from a variety of sources. Rodent mutants deficient in NER have been assigned to 11 complementation groups (cgs) and the human genes complementing these mutants have been designated *ERCC* (Excision Repair Cross Complementing) *1-11*. Another series of mutants is derived from individuals with the human genetic disorders, xeroderma pigmentosum (XP), Cockayne's syndrome (CS) and trichothiodystrophy (TTD). Several cgs of these cell strains have been identified and designated XP-A through G, CS-A, CS-B, TTD-A and their genes as *XPA*, etc. (formerly *XPAC*). As these genes were cloned and analyzed, it became apparent that some of the *ERCC* and *XP*, *CS* or *TTD* genes were identical. Thus for example the *ERCC2* and the *XPD* genes were found to be the same. It was recently agreed that in these cases the disease gene designation (e.g., *XPD*) should replace the *ERCC* designation (11).

Finally a number of genes isolated by other means have been found to be involved in NER. Their designations do not fall into any of the above schemes.

Genes involved in NER in *S. cerevisiae, S. pombe* and man are listed in Table 1, and to avoid excessive overexertion of his or her memory cells, the reader is advised to have a copy of Table 1 to hand while reading this section. The genes from both yeasts have in nearly all cases been cloned by their ability to complement the UV sensitivity of the corresponding mutants. *ERCC* genes were cloned in a similar manner with the use of human genomic DNA as donor to correct the UV sensitivity of rodent mutants, followed by recovery and cloning of the correcting DNA. The cloned *ERCC* genes were then transfected or microinjected into cells from the various human syndromes to establish possible equivalence between *ERCC* and human disease genes.

Human NER Diseases

The clinical and cellular defects in XP, CS and TTD have been reviewed extensively (see for example 12, 13) and they will be only briefly summarized here. XP patients have a greatly elevated incidence of sunlight-induced skin alterations, including freckling, pigmentation alterations, dryness, atrophy and multiple skin cancers including carcinomas and melanomas. A proportion of patients have neurological abnormalities which can be ascribed to premature neuronal death. The clinical and cellular features show a wide variation in severity. Cellular studies have revealed that cultured XP fibroblasts, lymphocytes and keratinocytes are hypersensitive to the lethal and mutagenic effects of UV light and chemical carcinogens which produce bulky adducts in DNA. They retain a normal response to ionizing radiation, alkylating agents and chemicals which produce damage normally repaired by BER. In most XP patients the cellular hypersensitivity can be ascribed to a deficiency in NER at an early step at or prior to incision of the DNA in the vicinity of the damaged site. Genetic studies with cell fusion have revealed the existence of 7 cgs (XP-A to G), indicating that defects in seven or more genes can result in the XP phenotype. A class of XP patients, comprising 10 to 20% of all XPs, are designated XP variants as they have normal NER and only mild cellular sensitivity to UV irradiation. They are however UV-hypermutable and this has been ascribed to a defect in the ability to replicate UV-damaged DNA.

The clinical features of CS are quite different. Although the skin is hypersensitive to sunlight, no pigmentation abnormalities result and there are no reports of skin cancer. CS is a multisystem disorder with severe mental and physical retardation, cachexia, retinal and neurological degeneration, skeletal abnormalities and various neurological defects (reviewed in 14). Cells are hypersensitive to killing by UV, but excision repair of damage in the bulk of the genome is unaffected. In normal cells, damage in the transcribed strand of active genes is repaired rapidly and preferentially, the non-transcribed strand somewhat more slowly and inactive genes more slowly still. In CS cells all damage is repaired at the relatively slow rate of untranscribed genes, i.e., the preferential rapid repair of active genes is defective (15). Two complementation groups (CS-A and CS-B) have been identified (16). Furthermore, patients in two out of the three identified families in the XP-B group (17,18), and two families in each of the XP-D (19,20) and XP-G (21) groups, have the clinical features of both XP and CS. Since these disorders are exceedingly rare, the existence of several patients having the features of both disorders cannot be coincidental. Rather it indicates some close causal connection between the molecular defects underlying the two disorders.

Table 1. NER Genes
(see references 29, 48, 49, 60, 61, 76, 82, 132-136)

S. cerevisiae	*S. pombe*	Human	Product
RAD1	*rad16* (includes *rad10, rad20, swi9*)	*XPF (ERCC4)*	Nuclease component
RAD2	*rad13*	*XPG (ERCC5)*	Nuclease
RAD3	*rad15*	*XPD (ERCC2)*	Helicase-TFIIH[a]
RAD4		*XPC*	
RAD23		*HHR23A,B*	
RAD10	*swi10*	*ERCC1*	Nuclease component
RAD14		*XPA*	
RAD25 (SSL2)	*ERCC3[sp]*	*XPB (ERCC3)*	Helicase-TFIIH[a]
SSL1		TFIIH-p44	
TFB1		TFIIH-p62	
		TFIIH-p34	
		TTDA	
		XPE	
		CSA	
		CSA (ERCC6)	
CDC9	*cdc17*	*LIGI*	DNA ligase
RFA1,2,3		*RPA* (p70, p32, - p14)	Single-strand binding protein
POL30	*pcn1*	*PCNA*	Sliding clamp
CDC44		*RFC1*	Replication factor

[a]Subunits of TFIIH (48-50, 136)

Trichothiodystrophy (TTD) is also a multisystem disorder, whose cardinal feature is sulphur-deficient brittle hair. There is associated mental and physical retardation, ichthyotic skin and unusual facial features (reviewed in 22,23). Some patients are sun-sensitive, but this does not result in pigmentation abnormalities or skin cancer. Sun-sensitive TTDs have cellular responses to UV which are indistinguishable from those of XPs with defects in NER of varying severity. Cell fusion has shown that cells from most TTD donors fall into the XPD cg (24). Recently, however, one patient has been found to represent a new cg (TTDA) (25) and two siblings in another family fall into the XPB group (26). It is important to note, however, that in no case reported to date do the clinical symptoms of TTD patients show any features characteristic of XP.

NER Proteins

The role of individual DNA repair proteins in the overall NER process is being deduced by a variety of approaches including (a) analysis of the predicted gene products from their amino-acid sequences, (b) investigation of activities of recombinant products expressed in *E. coli*, and (c) development of *in vitro* assays for NER. To date these studies have identified two nuclease activities, two helicases and two DNA binding proteins. In addition they have revealed the existence of several tightly bound multisubunit complexes.

Nucleases. The Rad1 and Rad10 proteins of *S. cerevisiae* (27,28), the homologous Rad16 and Swi10 products of *S. pombe* (29) and the human ERCC1 and ERCC4 (XPF?) proteins (30,31) have been shown to form a tight complex. The complex from *S. cerevisiae* has single-stranded (ss) endonuclease activity (28). Although this complex has no specificity for UV-irradiated substrates (32) its substrate specificity suggests a specific role in NER. In a study of different types of substrates, including fork-like structures (see Figure 1a), the Rad1/Rad10 nuclease specifically nicked the strand containing a free 3′ end close to the position of the fork (Friedberg, personal communication). The human complex may also contain the product of the *ERCC11* gene (31) and recent studies have shown that it interacts with the XPA protein (see below).

The *S. cerevisiae* Rad2 protein and its human homologue, the XPG protein, also have ss endonuclease activity (33,34). With the use of artificial substrates, the XPG nuclease was found to have the opposite polarity from that of the RAD1/RAD10 (ERCC1/4) nuclease (34a), i.e., nicks at forks, releasing the strand with a 5′ tail (Figure 1a). These two nucleases with specificities for strands of opposite polarity are the precise requirements for cutting an open structure containing damaged DNA residues on both sides of the damage, ERCC1/ERCC4 (Rad1/Rad10) cutting 5′ and XPG (Rad2) cutting 3′ to the damage (Figure 1b).

DNA-Binding Proteins. The XPA product (and its *S. cerevisiae* homologue Rad14) contains two zinc fingers and is a DNA-binding protein whose affinity for UV-irradiated DNA is 1000-fold increased over unirradiated DNA (35-37). The protein is localized in the nucleus and site-directed mutagenesis has shown that the N-terminus is not required for NER, whereas the C-terminus and one of the zinc fingers are required (38). Recent studies on protein-protein interactions with the yeast 2-hybrid system have demonstrated that the XPA protein interacts with the ERCC1 product (39,40). This suggests that the function of the XPA protein may be to direct the ERCC1/ERCC4 nuclease to damaged sites in the DNA.

Evidence suggests that the XPE protein may also be a damage-specific DNA-binding protein. A DNA-binding activity was found to be deficient in two XP-E cell strains, although it was normal in several other XP-E strains (41,42). Microinjection of this DNA-binding protein into the two deficient XP-E strains restored NER in these cells to normal levels (43). A candidate cDNA for this protein has been cloned recently (44). As yet there is no obvious yeast homologue.

Helicases: The Role of TFIIH. The predicted proteins of the *RAD3/rad15/XPD* and *RAD25/XPB* genes are of similar sizes and contain seven domains which are typical of DNA helicases (45). The RAD3 and XPD proteins have ATP-dependent DNA-DNA and DNA-RNA helicase activities and operate in a 5′-3′ direction with respect to the strand along which they are translocating (46-48). The RAD25/XPB helicase has been studied in less detail but the helicase appears to translocate in the 3′-5′ direction, opposite from that of RAD3/rad15/XPD (48).

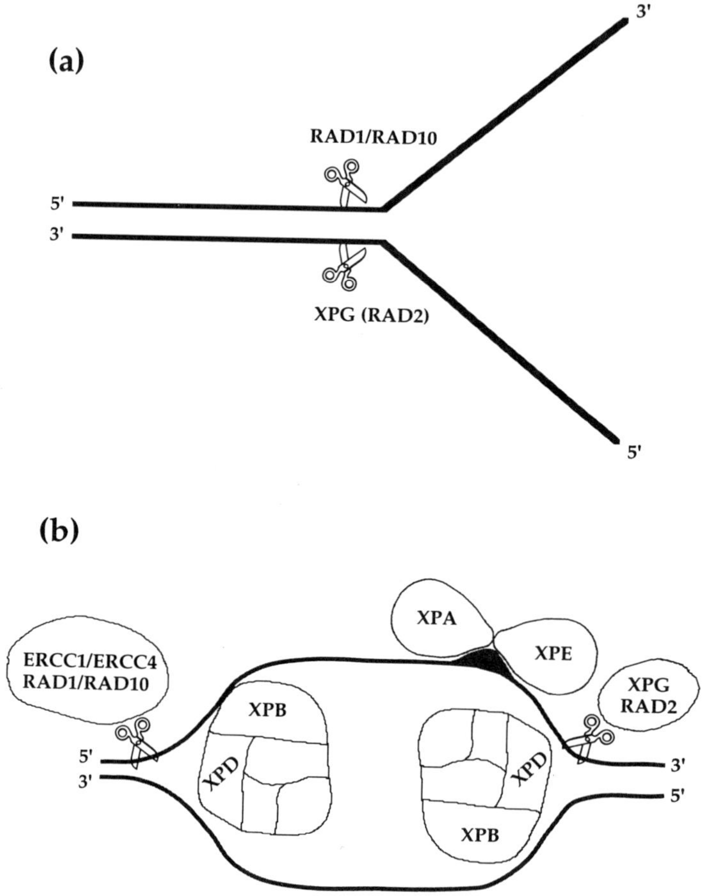

Figure 1(a). Site of nicking of a fork structure by the RAD1/RAD10 and the XPG (RAD2) proteins. The fork substrate is base-paired in the stem, but not in the arms.

(b). Model for the involvement of DNA repair gene products in excision repair of UV damage. XPA and XPE bind to the damaged site, XPG and ERCC1/ERCC4 cut the DNA on either side of the damage and TFIIH, containing the helicase activities of XPD and XPB, unwind the DNA structure.

Apart from their roles in NER the Rad3/rad15/XPD and Rad25/XPB products have essential functions, whose nature has been recently revealed from an unexpected direction. The basal transcription factor termed TFIIH in humans and factor b in yeast is one of a series of transcription factors whose role is the loading of RNA polymerase II onto the DNA at promoter sites to initiate transcription. TFIIH contains at least 5 and probably up to 9 different polypeptide subunits. The purified 89 kD subunit was found to have a number of activities including ATP-dependent helicase. The gene encoding this subunit was cloned and turned out to be identical to the *XPB* gene (49). Subsequent studies have shown that the *XPD* gene and the as yet uncloned *TTDA* gene also encode subunits of TFIIH (26,48,50). Analogous studies on factor b of yeast have demonstrated that Rad3 and the product of a gene termed *SSL1* are components of factor b and that Rad25 is closely associated with factor b (51,52). The involvement of these proteins in transcription provides a ready explanation for their essential functions and demonstrates that TFIIH has two separate roles, in basal transcription and in NER. Genetic analysis suggests however that its roles in these two processes are distinct. (a) Many alleles of *rad3* have been isolated or constructed, most of them were defective in NER, and the area in which mutations resulted in loss of viability was rather limited (53,54). (b) Likewise in the repair-deficient human syndromes, the XP-D group is relatively common (55), indicating that many mutations in the *XPD* gene are compatible with life. (c) The Rad3 protein contains the consensus sequence GCGKS/T for a nucleotide-binding site. Mutation of K48 to R at this site abolished DNA helicase activity, rendered the cells UV-sensitive but did not affect viability (56). Likewise analysis of a series of UV-sensitive *rad3* mutants showed that many of the mutations were located in helicase domains, presumably abolishing helicase activity, but growth and proliferation were unaffected (54). Turning to *RAD25/XPB* the situation is quite different. *rad25* mutants were not identified in screens for UV-sensitive mutants. The *RAD25(SSL2)* gene was only isolated coincidentally by an unexpected route (57) and subsequently by hybridization with the cloned *XPB* gene (58). This suggests that viable *rad25* mutants are relatively rare. Likewise only three families in the XP-B group are known. Genetic studies have shown that in contrast to *rad3*, mutations in *rad25* which abolish helicase activity are lethal (59). The picture that emerges from these studies is that RAD3/XPD is a protein which is principally involved in NER and whose role in transcription is relatively indirect, perhaps in maintaining the structural integrity of the TFIIH complex. In contrast RAD25/XPB has a major catalytic role in transcription which involves its helicase activity.

Other proteins. The *XPC* gene was cloned by Legerski and Peterson (60) and found to have some homology to *S. cerevisiae RAD4,* but there is no clue from the structure as to its function. The XPC protein was purified by its ability to complement NER activity in XP-C extracts and found to contain two subunits. The larger subunit was the product of the *XPC* gene and the smaller subunit was encoded by one of two human homologues of the *S. cerevisiae RAD23* gene (61). The function of the protein is as yet unknown.

A simple model for involvement of this multitude of proteins is shown in Figure 1b. *In vitro* studies have shown that nicks are made in the damaged strand, about 29 bases 5′ and 5 bases 3′ to the damaged site, releasing a fragment considerably larger than that in *E. coli* and consistent with a repair patch size of about 30 nucleotides determined in early cellular studies (62).

All of the proteins discussed so far appear to be required either prior to or at the incision step of NER. Work using cell-free extracts has shown that the subsequent polymerization and joining steps require the ss binding protein (RPA or HSSB in humans), PCNA and DNA polymerase δ and/or ϵ (63).

A Second ER Pathway in *S. pombe* - A Possible Link with Recombination and Replication

Deletion mutants of *RAD1,2,3,4,10,14* and *25* of *S. cerevisiae* abolish NER completely, and NER in most XP-A cell strains is barely detectable, as is the case for some XP-G cells. In *S. pombe*, however, deletion mutants of the *rad16* or *rad13* genes, though UV-sensitive, have only slightly reduced rates of removal of UV photoproducts (64). Mutants in at least 3 other genes, *rad2, rad18* and *rhp51* (Table 2) also are reduced in excision repair (64 and unpublished data), but no mutant has been found to date totally defective in repair of UV photoproducts. These results suggest that there is a second ER pathway in *S. pombe* involving the products of some or all of these genes. Structural homologues of these genes have been identified or isolated from *S. cerevisiae* (Table 2) and, in the case of *rad2* (65) and *rhp51* (66), also from human cells. The RAD51 product forms nucleoprotein filaments with DNA which are involved in recombination (see below). The *S. pombe rad2* product is a member of the *S. cerevisiae RAD2* family. It contains two conserved regions found in RAD2 homologues but a much smaller intervening region (65). Recent work has shown that the product of the mammalian homologue of the *S. pombe rad2* gene is like the *S. cerevisiae* Rad2 protein, a nuclease. A nuclease designated FEN-1 isolated from mouse cells nicks so-called Flap structures which contain single-stranded tails at the junction with the double-stranded stem. Purification of this nuclease and cloning of the gene showed it to be the mouse homologue of *S. pombe rad2* (67). Furthermore, peptide sequencing of the human protein revealed that it was identical to a ss exonuclease MF1 described by Waga and Stillman (68), which is an essential requirement for replication of SV40 DNA *in vitro*. The involvement of rhp51 in this second pathway suggests a role for recombination (see below).

Table 2. Genes Involved in Second Excision-Repair Pathway in *S. pombe*

S. pombe	*S. cerevisiae*	Human	Product
rad2	*YKL510*	*FENI = MF-I*	Nuclease
rad18	*rhc18*		
rhp51	*RAD51*	*Hhr51*	Nucleoprotein filament with DNA

Transcription-coupled Repair

Studies on excision repair of UV damage in specific genes showed that active genes were repaired more rapidly than the bulk of the DNA (69) and that the transcribed strand was repaired preferentially in both mammalian cells (70) and in *E. coli* (71). In *E. coli* transcription-coupled repair is brought about by TRCF (transcription-repair coupling factor), the product of the *mfd* gene which removes stalled RNA polymerase complexes from damaged regions of DNA and recruits the UvrA protein to the transcription blocking lesion in the transcribed strand (72).

In eukaryotes genetic studies have separated the transcription-coupled repair from repair in the bulk of the genome. XPC cells repair only the transcribed strands of active genes (73), implying that the XPC product is involved in repair in the inactive regions of the genome. The products of three genes in yeast (*RAD7,16* and *23*) are similarly involved only in repair of inactive DNA (74). Two human homologues of *RAD23* have recently been isolated, and, as described above, the products of one of these is tightly bound to the XPC protein (61). The functions and roles of these gene products are not yet understood.

In contrast to the above, CS cells strains are proficient in repairing damage in inactive regions of the genome, but they are unable to repair damage in active genes at an accelerated rate, all damage being repaired at the characteristic slow rate of bulk DNA (75). The *CSA* gene has not been cloned but *CSB* was found to be identical to the *ERCC6* gene (76). This gene has the seven domains characteristic of DNA helicases, but as yet no biochemistry has been carried out on the protein, and the precise roles of the *CSA* and *CSB* genes in transcription-coupled repair in active genes remains a matter of speculation.

Genotype-phenotype Relationships in NER Diseases

It is likely that over the next few years the role of the various proteins involved in NER will be elucidated, and the major challenge will then be to determine the relationships of the mutations in human subjects to their clinical features. Two approaches are currently being used to address this problem: (a) determination of the nature and location of the mutations in different donors, and (b) generation of transgenic mice with mutations in DNA repair genes.

To date mice deleted for the *ERCC1* and *XPA* gene have been generated. The *ERCC1* mice (77) are very small at birth and die during weaning. Their principal defects are in the liver. The *XPA* mice are viable but also very small. As expected, cultured fibroblasts from these mice are hypersensitive to UV-irradiation and deficient in NER.

Mutations in XP Genes

XPA. A large number of mutations have been identified in XP-A patients by Tanaka and colleagues (78-80). In Japan, 80% of XP-A donors have the same mutation at the splice acceptor site of intron 3. This results in an alteration of the reading frame and it is likely that the protein is completely inactive. Individuals homozygous for this mutation have severe symptoms. Other mutations found in Japan include both nonsense and missense mutations, the former resulting in severe symptoms, whereas the latter give rise to mild symptoms. In XP-A therefore, genotype-phenotype relationships appear relatively straightforward, and XP-A fulfills the paradigm that defective NER results in sunlight-induced hypermutation in the skin, which in time gives rise to an elevated incidence of skin abnormalities including skin cancers.

XPC. The mutations in 5 XP-C individuals have been reported recently by Li et al. (81). Four out of five were homo- or hemizygous mutations. The mutations detected include two dinucleotide deletions, a trinucleotide insertion, an 83 bp insertion and two single-base changes. This limited spectrum is quite different from that of *XPA* mutations, which are almost entirely base substitutions. The existence of frameshift mutations provides strong support to the idea that the *XPC* gene is not essential for viability.

XPB and XPD. The *XPB* and *XPD* mutations give much more complex phenotypes. In the rare XP-B group, in one patient with severe features of XP and CS, there is a splicing abnormality in the last intron in the only expressed allele, so that the C-terminus is out of frame (82). In two brothers in the second family, who have only mild features of both XP and CS, there is a simple missense T→C mutation resulting in a Phe to Ser change at aa 99 in the middle of the gene product (18). In the third family, however, the features are not of XP and CS but of TTD (26). The mutation(s) has not yet been identified.

In the XP-D group the problem is more marked. Patients in this group can have the features of XP alone, TTD alone, or in two individuals, XP and CS. Can these differing features be attributed to different locations of the mutations? So far, out of about 10 different mutations identified, most but not all are clustered in the C-terminal 20% of the protein (83 and unpublished data). These mutations are in XP,XP/CS and TTD individuals. Thus with the limited data available so far, it is not possible to delineate clearly "XP domains" or "TTD domains." On the other hand the same mutation has not, so far, been found in an XP and a TTD individual.

XP has long been thought of as resulting from the consequences of a defect in NER. The totally different features of TTD have been a mystery. The recent discovery that the products of the three genes in which mutations can result in TTD, namely *XPD, XPB* and *TTDA*, are all subunits of the transcription factor TFIIH (see above) has led to the proposal that the features of TTD result not from the defect in NER, but rather from a subtle defect in transcription, which has rather specific effects on development. This idea has been expounded in some detail by Vermeulen et al. (26). It remains unclear, however, why TTD patients do not develop the clinical features of XP in addition to those of TTD. To date no patient with the features of both TTD and XP has been described in the literature. Evidently there is much about these complex disorders still to be learned.

IONIZING RADIATION

Whereas the products of UV irradiation in DNA are limited and well characterized, ionizing radiation (IR) produces a very diverse set of lesions including base damage, single-strand breaks (ssbs) with different ends and double-strand breaks (dsbs). Most of the base damage and ssbs are repaired by BER, but other mechanisms are required to repair some lesions, especially dsbs, which are potentially highly deleterious. Our understanding of the biochemistry and molecular biology of repair of IR damage is rather rudimentary when compared with repair of UV damage.

The *RAD52* Group of *S. cerevisiae*

The *RAD50-57* genes of *S. cerevisiae* are sensitive to killing by ionizing radiation whereas their response to UV is either normal or shows very mild sensitivity. The role that these genes play in the rejoining of dsbs has been recently reviewed by Game (84). Direct measurements have shown that the Rad51, Rad52 and Rad54 proteins of *S. cerevisiae* are absolutely required for rejoining of dsbs. Cells carrying mutations at these loci also show a reduced level of both spontaneous and induced mitotic recombination. The Rad51 and Rad52 products are also required in meiosis. Furthermore, a large body of evidence shows that cells with two copies of the genome are much more resistant to radiation than cells with one copy. Taken together these data provide compelling evidence that the rejoining of dsbs in *S. cerevisiae* involves a recombinational exchange process mediated by the Rad51, 52 and 54 products. The Rad50 product is partially required for dsb repair and the Rad53, 55, 56, 57 proteins are probably also involved, although the role of these genes has not yet been studied in sufficient detail.

The *RAD50, 51, 52* and *54* genes and their products have been subjected to detailed molecular and biochemical analysis, but their precise roles in repair of dsbs remain fairly speculative. The Rad51 protein shows significant structural homology to the RecA protein of *E. coli* and recently it has been shown to form nucleoprotein filaments with double-stranded DNA with a configuration strikingly similar to that of RecA nucleoprotein filaments (85). The phenotypes of *rad51* and *rad52* mutants are very similar and the products of the two genes have been shown to associate with each other (86). During meiosis *rad51* and *rad52* mutants accumulate dsbs, and the Rad51 and Rad52 products are therefore thought to be involved in converting dsbs into the next intermediate in the processes of meiosis or repair. This early event probably involves bringing together the broken DNA molecule with its homologous sister chromatid to initiate an exchange event.

The Rad50 protein is absolutely required for meiosis but only partially required for repair of dsbs. It is a 153 kD protein with an ATP-binding site and several heptad repeats similar to those found in myosin (87). The protein forms a homodimer and has ATP-dependent DNA-binding activity (88). It is thought to be involved in an early event in recombination/dsb repair, possibly by recruiting other proteins to their site of action.

The *RAD54* gene is inducible by DNA damage (89). It contains a nucleotide binding site and the seven domains typical of DNA helicases (90). It is a member of the SNF2 subfamily of DNA helicases with a high level of sequence identity to several other proteins, namely, RAD5, RAD16, SNF2, MOT1, FUN30 from *S. cerevisiae*, brm, lodestar from *Drosophila*, human ERCC6 and *S. pombe* rad8, which are variously implicated in DNA repair, transcriptional activation and other undefined functions.

Both the *RAD55* (91) and *RAD57* genes (92) show significant amount of sequence identity to the *RAD51* gene, but their roles in recombination and repair have not been investigated in any detail.

Although we currently have some understanding of the activities of several of the RAD50-RAD57 proteins, it does not yet permit a detailed model to be put forward for their precise involvement in repair of dsbs.

Other Organisms. In *S. pombe* the *rhp51, rad22* and *rhp54* genes are structurally homologous to *RAD51,52* and *54* of *S. cerevisiae* (93,94 and Muris, unpublished data). Like the *S. cerevisiae* mutants, all three *S. pombe* homologues are very sensitive to ionizing radiation, and indirect evidence suggests a role of *S. cerevisiae rad22* in repair of dsbs. Unlike the *S. cerevisiae* mutants, however, the *S. pombe* homologues also show significant sensitivity to UV-irradiation, and the *rhp51* strain is deficient in excision of UV photoproducts and in combination with *rad13* shows extreme UV sensitivity (Lehmann et al., unpublished data). These data suggest that in *S. pombe* there may be a second excision repair pathway for UV damage which involves recombination (see above).

Chicken and human homologues of the *RAD51* and *RAD52* genes have been isolated with the use of homology techniques (66,95-97). The sequence of the *RAD51* gene has been highly conserved in its homologues, the *RAD52* to a lesser extent. Both human genes are highly expressed in the thymus and testis, suggesting an involvement in recombination. As yet, however, there is no evidence concerning their involvement in rejoining of dsbs. It is implied by extrapolation from the involvement of yeast genes.

End-Joining

Apart from the rejoining pathway(s) involving homologous recombination, eukaryotic cells are also able to rejoin dsbs by an end-joining activity. In *S. cerevisiae* this latter activity is a quantitatively minor pathway and is independent of *rad52*. It is specifically important for rejoining dsbs between repeated regions and appears to involve deletion of the sequences between repeats. This latter process requires the Rad1/Rad10 endonuclease to remove protruding ss tails (98,99). In mammalian studies evidence suggests that end-joining may be the major pathway for rejoining dsbs. Most of this evidence comes from *in vitro* or transfection studies with DNAs cleaved with restriction endonucleases. In contrast to yeast, integration of transfecting DNA by non-homologous recombination is very much more frequent (1000 x) than by homologous recombination.

Several studies have investigated the ability of whole mammalian cells or extracts from *Xenopus* or human cells to join linearized transfecting plasmids with different types of ends. Perhaps surprisingly all types of ends whether 5′ protruding, 3′ protruding, matched or mismatched can be joined with high efficiency. A common feature of the end-joining in both yeast and animal systems is the importance of repeated sequences in determining the nature of the joining reaction. The nature of the joined products is broadly similar in yeast, human and *Xenopus*, and is highly influenced by microhomologies which determine the precise nature of the joined products. In a proportion of rejoined molecules a region spanning the break flanked by repeats of 1 to 4 nucleotides is deleted from around the break. In other circumstances, blunt-end joining occurs, after fill-in reactions, if necessary. In yet further cases, protruding single strands can form short overlaps whose extent is determined by single- or two-base matches. Details of these reactions are outside the scope of this chapter but the reader is referred to papers by Roth and Wilson (100), Thacker et al. (101), Pfeiffer et al. (102), Kramer et al. (99) and references cited therein.

The importance of this kind of end-joining as opposed to recombination between homologous pairs of DNA molecules in rejoining of dsbs induced by ionizing radiation remains to be established by genetic, molecular and biochemical studies.

Human Genes Involved in the Repair of dsbs

As with UV-sensitive mutants, mammalian cells sensitive to ionizing radiation have been derived from two sources, namely from cultured rodent cells and from humans, in the latter case from individuals with the genetic disorder ataxia-telangiectasia (A-T) and the related Nijmegen Breakage Syndrome (NBS). Ten rodent cgs (103) and 6 human cgs (104) have been identified by cellular studies.

A-T and NBS. The clinical features of A-T comprise progressive cerebellar ataxia, oculocutaneous telangiectases, deficiencies in both cellular and humoral immunity, an increased frequency of lymphoreticular tumors and also carcinomas, and hypersensitivity of the individuals to radiation. A-T patients have a high frequency of chromosomal rearrangements with breakpoints on chromosomes 7 and 14 at the sites of T-cell receptor genes. In addition, there is an elevated level of general chromosome instability in A-T lymphocytes (105). NBS patients have similar features but without ataxia and telangiectasia but including microcephaly (see 104). A-T and NBS cells are sensitive to the lethal effects of irradiation. Extensive celluar biochemical studies have been relatively unsuccessful in helping to understand the basis of the radiosensitivity in A-T. All A-T cells fail to shut down DNA synthesis following irradiation, but this phenotype can be at least partially separated from the hypersensitivity (106,107). A bewildering array of abnormalities have been reported for some A-T cells, but the relationship to the basic defect remains unclear (108). It is possible that the A-T gene has a regulatory function, but there is also compelling evidence for a DNA repair defect. The radiosensitivity can be observed both in non-dividing cells (109) and within one hour of irradiation in G1 (110), thus ruling out an involvement of abnormal cell cycle control in these particular experiments. The possibility that a minor proportion of strand breaks remains unrepaired or that dsbs are repaired with decreased fidelity remains. Evidence for the latter has come from a series of studies in which plasmids digested with restriction endonucleases were introduced into normal and A-T cells and the nature of the rejoined breaks was examined. Normal cells rejoined these breaks accurately, with most reconstituted plasmids regaining the original structure, whereas in A-T cells the rejoined plasmids contained deletions around the breakpoint (111,112). Although the results of these experiments were clear-cut, the generality of the observations remains to be established.

Rodent Mutants. In the rodent mutants, direct evidence for a deficiency in the repair of dsbs has been obtained for mutants in three of the cgs (113-115), whereas other mutants do not show such a deficiency. The best characterized mutants defective in dsb rejoining compromise the *xrs* mutants of cg5, the *XR-1* mutant of cg4 and cells from *scid* mice. Microcell-mediated chromosome transfer has localized the human *XRCC5, XRCC4* and *scid* genes to chromosomes 2, 5 and 8 (116-118). In addition to their involvement in rejoining of irradiation-induced dsbs, mutants in these three cgs are also deficient in V(D)J rejoining of immunoglobulin genes (119,120). During development of the immune response, immunoglobin and T-cell receptor genes are rearranged such that the variable and constant regions are joined by a process termed V(D)J rejoining. V(D)J rejoining involves a series of breakage and rejoining steps. The detailed mechanism of this process is beyond the scope of this review, but the fact that mutants are deficient in both the rejoining of radiation-induced dsbs and in V(D)J joining once again demonstrates the dual function of DNA repair proteins.

Ku Protein and the *XRCC5* Gene. Cell-free extracts of *xrs* mutants were recently shown to be deficient in a DNA end-binding activity (121). Such an activity is associated with a heterodimeric protein known as Ku (122). The gene encoding the 80 kD subunit of Ku and the *XRCC5* gene both map to the long arm of chromosome 2 at 2q33-35 (123,124). Furthermore, Ku activity and radiation resistance co-segregate in a series of hamster-human hybrids containing different fragments of chromosome 2 (125). The Ku80 cDNA corrected both the V(D)J joining defect and the radiation sensitivity of *xrs* cells, demonstrating that the 80 kD subunit of Ku is the product of the *XRCC5* gene (125). The end-binding activity of Ku provides an immediately evident role for it in an early step of dsb rejoining as part of a recognition complex. The 70 kD and 80 kD subunits of Ku are very tightly bound to each other, and their binding to DNA ends activates a third subunit. This is a 350 kD protein with protein kinase (PK) activity. The PK activity is absolutely dependent on the binding of Ku to DNA ends and the whole complex is known as DNA-PK.

The role of Ku and DNA-PK in rejoining of dsbs remains speculative. Suggestions proposed by Taccioli et al. (125) include a protective role against nuclease attack or holding of DNA ends in an appropriate conformation for rejoining. The gene encoding the 70 kD subunit of Ku maps to chromosome 22, which does not correspond to the chromosomal location of any other genes involved in dsb repair. The very large p350 subunit could be the product of the *XR-1* or *scid* genes, and, among other possibilities, it could provide a framework around which repair enzymes are recruited to the complex. A possible role for the kinase activity of DNA-PK in rejoining of dsbs or in the response to dsbs remains a matter for some speculation. *In vitro* this activity is capable of phosphorylating a wide variety of substrates including SP1 and p53, but its substrates *in vivo* are unknown. One suggestion is that phosphorylation of transcription factors in the vicinity of dsbs might inhibit transcription in the environment of the dsb. Another possibility is that activation of the kinase by dsbs might provide a signal to the cell that the DNA is damaged. There is, however, no evidence that the various cellular responses to DNA damage, such as delays in cell cycle progression, are deficient in *xrs* mutants.

A-T Genes. Linkage analysis has localized the *ATA/B* and *ATC* genes to a 1 to 5 Mb region on chromosome 11q21-22 (126). The gene(s) itself has not yet been cloned. Transfection experiments have led to the cloning of a candidate for the *ATD* gene, also mapping on chromosome 11q. The gene encodes several species of mRNA and a 3 kb cDNA has been recently cloned and sequenced. The sequence does not give any clue to its function and as yet no mutation has been found in this gene in an AT-D patient (127). At the time of writing it remains therefore a candidate gene of unknown status.

Other Genes Involved in Repair of dsbs. Hoekstra et al. (128) identified a mutant of *S. cerevisiae* designated *hrr25* which was unable to rejoin the site-specific dsbs introduced into DNA by the HO endonuclease. The *HRR25* gene was found to be a homologue of casein kinase I. Two homologues of this gene (*hhp1* and *hhp2*) were isolated from *S. pombe* (129). A deletion of *hhp1* was sensitive to methyl methanesulfonate, whereas a disruptant of *hhp2* was not. Only the double mutant was sensitive to γ-irradiation. These sensitivities correlated with deficiencies in dsb rejoining. The authors suggested that the putative kinase activities of the hhp1 and hhp2 products were involved in the regulation of dsb repair.

In *S. pombe* the *rad21* gene (130) and the *rad32* gene (131) are deficient in dsb repair. The *rad21* gene is essential for cell proliferation, whereas *rad32* is not essential but is deficient in mitotic recombination.

CONCLUDING REMARKS

After many years of intensive research, molecular biological studies have recently made dramatic progress in elucidating the mechanism of NER. This work has in addition revealed unexpected links with transcription, replication and recombination, and the challenge for the future will be to relate the molecular deficiencies to the clinical features of the DNA repair disorders. With ionizing radiation, the first human gene involved in dsb repair has been identified only very recently (125), and it is likely that further progress will be made over the next few years. Again, unexpected links with other cellular processes are already becoming apparent. When DNA repair was discovered in the early 1960s it was regarded as a somewhat esoteric process likely to be mainly of academic interest. Research since then continues to demonstrate that it is a fundamental process of central importance to all organisms.

ACKNOWLEDGMENT

I am indebted to E. Grenfell for excellent secretarial assistance.

REFERENCES

1 Setlow, R.B. and Carrier, W.L. (1964) Proc. Nat. Acad. Sci. U.S.A. 51, 226-231.
2 Wallace, S.S. (1988) Environ. Mol. Mutagen. 12, 431-477.
3 Doetsch, P.W. and Cunningham,, R.P. (1990) Mutat. Res. 236, 173-201.
4 Dianov, G., Price, A. and Lindahl, T. (1992) Molec. Cell. Biol. 12, 1605-1612.
5 Demple, B., Herman, T. and Chen, D.S. (1991) Proc. Nat. Acad. Sci. U.S.A. 88, 11450-11454.
6 Robson, C.N. and Hickson, I.D. (1991) Nucl. Acids Res. 19, 5519-5523.
7 O'Connor, T.R. and Laval, J. (1991) Biochem. Biophys. Res. Commun. 176, 1170-1177.
8 Chen, J., Derfler, B., Maskati, A. and Samson, L. (1989) Proc. Nat. Acad. Sci. U.S.A. 86, 7961-7965.
9 Lin, J.-J. and Sancar, A. (1992) Mol. Microbiol. 6, 2219-2224.
10 Hoeijmakers, J.H.J. (1993) Trends Genet. 9, 173-177.
11 Lehmann, A.R., Bootsma, D., Clarkson, S.G., Cleaver, J.E., McAlpine, P.J., Tanaka, K., Thompson, L.H. and Wood, R.D. (1994) Mutat. Res. 315, 41-42.
12 Cleaver, J.E. and Kraemer, K.H. (1989) in The Metabolic Basis of Inherited Disease (C.R. Scriver, A.L. Beaudet, W.S. Sly and D. Valle, eds.) pp. 2949-2971, McGraw-Hill, Inc., New York, NY.
13 Lehmann, A.R. and Dean, S.W. (1989) in Handbook of Experimental Pharmacology, Vol 94(II) (C.S. Cooper and P.L. Grover, eds.) pp. 71-101, Springer, Berlin.
14 Nance, M.A. and Berry, S.A. (1992) Amer. J. Med. Genet. 42, 68-84.

15 van Hoffen, A., Natarajan, A.T., Mayne, L.V., van Zeeland, A.A., Mullenders, L.H.F. and Venema, J. (1993) Nucl. Acids Res. 21, 5890-5895.
16 Lehmann, A.R. (1982) Mutat. Res. 106, 347-356.
17 Robbins, J.H., Kraemer, K.H., Lutzner, M.A., Festoff, B.W. and Coon, H.G. (1974) Annals Internal Med. 80, 221-248.
18 Vermeulen, W., Scott, R.J., Potger, S., Muller, H.J., Cole, J. Arlett, C.F., Kleijer, W.J., Bootsma, D., Hoeijmakers, J.H.J. and Weeda, G. (1994) Amer. J. Hum. Genet. 54, 191-200.
19 Johnson, R.T., Elliot, G.C., Squires, S. and Joysey, V.C. (1989) Hum. Genet. 81, 203-210.
20 Broughton, B.C., Thompson, A.F., Harcourt, S.A., Vermeulen, W., Hoeijmakers, J.H.J., Botta, E., Stefanini, M., King, M., Weber, C., Cole, J., Arlett, C.F. and Lehmann, A.R., (1995) Amer. J. Hum. Genet. (in press).
21 Vermeulen, W., Jaeken, J., Jaspers, N.G.J., Bootsma, D. and Hoeijmakers, J.H.J. (1993) Amer. J. Hum. Genet. 53, 185-192.
22 Itin, P.H. and Pittelkow, M.R. (1990) J. Amer. Acad. Dermatol. 20, 705-717.
23 Lehmann, A.R. (1987) Cancer Rev. 7, 82-103.
24 Stefanini, M., Lagomarsini, P., Giliani, S., Nardo, T., Botta, E., Peserico, A., Kleijer, W.J., Lehmann, A.R. and Sarasin, A. (1993) Carcinogenesis 14, 1101-1105.
25 Stefanini, M., Vermeulen, W., Weeda, G., Giliani, S., Nardo, T., Mezzina, M., Sarasin, A., Harper, J.I., Arlett, C.F., Hoeijmakers, J.H.J. and Lehmann, A.R. (1993) Amer. J. Hum. Genet. 53, 817-821.
26 Vermeulen, W., van Vuuren, A.J., Chipoulet, M., Schaeffer, L., Appeldoorn, E., Weeda, G., Jaspers, N.G.J., Priestley, A., Arlett, C.F., Lehmann, A.R., Bootsma, D., Egly, J.-M. and Hoeijmakers, J.H.J. (1994) Cold Spring Harbor Symp. Quant. Biol. (in press).
27 Bailly, V., Sommers, C.H., Sung, P., Prakash, L. and Prakash, S. (1992) Proc. Nat. Acad. Sci. U.S.A. 89, 8273-8277.
28 Tomkinson, A.E., Bardwell, A.J., Bardwell, L., Tappe, N.J. and Friedberg, E.C. (1993) Nature 362, 860-862.
29 Carr, A.M., Schmidt, H., Kirchhoff, S., Muriel, W.J., Sheldrick, K.S., Griffiths, D.J., Basmacioglu, C.N., Subramani, S., Clegg, M., Nasim, A. and Lehmann, A.R. (1994) Molec. Cell. Biol. 14, 2029-2040.
30 Biggerstaff, M., Szymkowski, D.E. and Wood, R.D. (1993) EMBO J. 12, 3685-3692.
31 van Vuuren, A.J., Appeldoorn, E., Odijk, H., Yasui, A., Jaspers, N.G.J., Bootsma, D. and Hoeijmakers, J.H.J. (1993) EMBO J. 12, 3693-3701.
32 Tomkinson, A.E., Bardwell, A.J., Tappe, N., Ramos, W. and Friedberg, E.C. (1994) Biochemistry 33, 5305-5311.
33 Habraken, Y., Sung, P., Prakash, L. and Prakash, S. (1993) Nature 366, 365-368.
34 O'Donovan, A., Scherly, D., Clarkson, S.G. and Wood, R.D. (1994) J. Biol. Chem. 269, 15965-15968.
34a O'Donovan, A., Davies, A.A., Moggs, J.G., West, S.C. and Wood, R.D. (1994) Nature 371, 432-435.
35 Robins, P., Jones, C.J., Biggerstaff, M., Lindahl, T. and Wood, R.D. (1991) EMBO J. 10, 3913-3921.
36 Jones, C.J. and Wood, R.D. (1993) Biochemistry 32, 12096-12104.
37 Guzder, S.N., Sung, P., Prakash, L. and Prakash, S. (1993) Proc. Nat. Acad. Sci. U.S.A. 90, 5433-5437.

38 Tanaka, K. (1993) in Frontiers of Photobiology, (A. Shima, M. Ichahashi, Y. Fujiwara and H. Takebe, eds.) pp. 293-302, Elsevier, Amsterdam.
39 Li, L., Elledge, S.J., Peterson, C.A., Bales, E.S. and Legerski, R.J. (1994) Proc. Nat. Acad. Sci. U.S.A. 91, 5012-5016.
40 Park, C.-H. and Sancar, A. (1994) Proc. Nat. Acad. Sci. U.S.A. 91, 5017-5021.
41 Kataoka, H. and Fujiwara, Y. (1991) Biochem. Biophys. Res. Commun. 175, 1139-1143.
42 Keeney, S., Wein, H. and Linn, S. (1992) Mutat. Res. 273, 49-56.
43 Keeney, S., Ekers, A.P.M., Brody, T., Vermeulen, W., Bootsma, D., Hoeijmakers, J.H.J. and Linn, S. (1994) Proc. Nat. Acad. Sci. U.S.A. 91, 4053-4056.
44 Takao, M., Abramic, M., Moos, M., Otrin, V.R., Wootton, J.C., McLenigan, M., Levine, A.S. and Protic, M. (1993) Nucl. Acids Res. 21, 4111-4118.
45 Hoeijmakers, J.H.J. (1993) Trends Genet. 9, 211-217.
46 Sung, P., Prakash, L., Matson, S.W. and Prakash, S. (1987) Proc. Nat. Acad. Sci. U.S.A. 84, 8951-8955.
47 Naegeli, H., Bardwell, L., Harosh, I. and Friedberg, E.C. (1992) J. Biol. Chem. 267, 7839-7844.
48 Schaeffer, L., Moncollin, V., Roy, R., Staub, A., Mezzina, M., Sarasin, A., Weeda, G., Hoeijmakers, J.H.J. and Egly, J.M. (1994) EMBO J. 13, 2388-2392.
49 Schaeffer, L., Roy, R. Humbert, S., Moncollin, V., Vermeulen, W., Hoeijmakers, J.H.J., Chambon, P. and Egly, J.-M. (1993) Science 260, 58-63.
50 Drapkin, R., Reardon, J.T., Ansari, A., Huang, J.-C., Zawel, L., Ahn, K., Sancar, A. and Reinberg, D. (1994) Nature 368, 769-772.
51 Feaver, W.J., Svejstrup, J.Q., Bardwell, L., Bardwell, A.J., Buratowski, S., Gulyas, K.D., Donahue, T.F., Friedberg, E.C. and Kornberg, R.D. (1993) Cell 75, 1379-1387.
52 Bardwell, L., Bardwell, A.J., Feaver, W.J., Svejstrup, J.W., Kornberg, R.D. and Friedberg, E.C. (1994) Proc. Nat. Acad. Sci. U.S.A. 91, 3926-3930.
53 Naumovski, L. and Friedberg, E.C. (1986) Molec. Cell. Biol. 6, 1218-1227.
54 Song, J.M., Montelone, B.A., Siede, W. and Friedberg, E.C. (1990) J. Bacteriol. 172, 6620-6630.
55 Johnson, R.T. and Squires, S. (1992) Mutat. Res. 273, 97-118.
56 Sung, P., Higgins, D., Prakash, L. and Prakash, S. (1988) EMBO J. 7, 3263-3269.
57 Gulyas, K.D. and Donahue, T.F. (1992) Cell 69, 1031-1042.
58 Park, E., Guzder, S.N., Koken, M.H.M., Jaspers-Dekker, I., Weeda, G., Hoeijmakers, J.H.J., Prakash, S. and Prakash, L. (1992) Proc. Nat. Acad. Sci. U.S.A. 89, 11416-11420.
59 Guzder, S.N., Sung, P., Bailly, V., Prakash, L. and Prakash, S. (1994) Nature 369, 578-581.
60 Legerski, R. and Peterson, C. (1992) Nature 359, 70-73.
61 Masutani, C., Sugasawa, K., Yanagisawa, J., Sonoyama, T., Ui, M., Enomoto, T., Takio, K., Tanaka, K., van der Spek, P.J., Bootsma, D., Hoeijmakers, J.H.J. and Hanaoka, F. (1994) EMBO J. 13, 1831-1843.
62 Huang, J.C., Svoboda, D.L., Reardon, J.T. and Sancar, A. (1992) Proc. Nat. Acad. Sci. U.S.A. 89, 3664-3668.
63 Shivji, M.K.K., Kenny, M.K. and Wood, R.D. (1992) Cell 69, 367-374.
64 McCready, S., Carr, A.M. and Lehmann, A.R. (1993) Mol. Microbiol. 10, 885-890.
65 Murray, J.M., Tavassoli, M., Al-Harithy, R., Sheldrick, K.S., Lehmann, A.R., Carr, A.M. and Watts, F.Z. (1994) Molec. Cell. Biol. 14, 4878-4888.

66 Shinohara, A., Ogawa, H., Matsuda, Y., Ushio, N., Ikeo, K. and Ogawa, T. (1993) Nature Genetics 4, 239-243.
67 Harrington, J.J. and Lieber, M.R. (1994) Genes Dev. 8, 1344-1355.
68 Waga, S. and Stillman, B. (1994) Nature 369, 207-212.
69 Bohr, V.A., Smith, C.A., Okumoto, D.S. and Hanawalt, P.C. (1985) Cell 40, 359-369.
70 Mellon, I., Spivak, G. and Hanawalt, P.C. (1987) Cell 51, 241-249.
71 Mellon, I. and Hanawalt, P.C. (1989) Nature 342, 95-98.
72 Selby, C.P. and Sancar, A. (1993) Science 260, 53-58.
73 Venema, J., van Hoffen, A., Karcagi, V., Natarajan, A.T., van Zeeland, A.A. and Mullenders, L.H.F. (1991) Molec. Cell. Biol. 11, 4128-4134.
74 Terleth, C., Schenck, P., Poot, R., Brouwer, J. and Van de Putte, P. (1990) Molec. Cell. Biol. 10, 4678-4684.
75 Venema, J., Mullenders, L.H.F., Natarajan, A.T., van Zeeland, A.A. and Mayne, L.V. (1990) Proc. Nat. Acad. Sci. U.S.A. 87, 4707-4711.
76 Troelstra, C., van Gool, A., de Wit, J., Vermeulen, W., Bootsma, D. and Hoeijmakers, J.H.J. (1992) Cell 71, 939-953.
77 McWhir, J., Selfridge, J., Harrison, D.J., Squires, S. and Melton, D.W. (1993) Nature Genetics 5, 217-223.
78 Satokata, I., Tanaka, K., Miura, N., Miyamoto, I., Satoh, Y., Kondo, S. and Okada, Y. (1990) Proc. Nat. Acad. Sci. U.S.A. 87, 9908-9912.
79 Satokata, I., Tanaka, K., Yuba, S. and Okada, Y. (1992) Mutat. Res. 273, 203-212.
80 Satokata, I., Tanaka, K., Miura, N., Narita, M., Mimaki, T., Satoh, Y., Kondo, S. and Okada, Y. (1992) Mutat. Res. 273, 193-202.
81 Li, L., Bales, E.S., Peterson, C.A. and Legerski, R.J. (1993) Nature Genetics 5, 413-417.
82 Weeda, G., van Ham, R.C.A., Vermuelen, W., Bootsma, D., van der Eb, A.J. and Hoeijmakers, J.H.J. (1990) Cell 62, 777-791.
83 Broughton, B.C., Steingrimsdottir, H., Weber, C. and Lehmann, A.R. (1994) Nature Genetics 7, 189-194.
84 Game, J.C. (1993) Seminars in Cancer Biology 4, 73-83.
85 Ogawa, T., Yu, X., Shinohara, A. and Egelman, E.H. (1993) Science 259, 1896-1899.
86 Shinohara, A., Ogawa, H. and Ogawa, T. (1992) Cell 69, 457-470.
87 Alani, E., Subbiah, S. and Kleckner, N. (1989) Genetics 122, 47-57.
88 Raymond, W.E. and Kleckner, N. (1993) Nucl. Acids Res. 21, 3851-3856.
89 Cole, G.M., Schild, D., Lovett, S.T. and Mortimer, R.K. (1987) Molec. Cell. Biol. 7, 1078-1084.
90 Emery, H.S., Schild, D., Kellogg, D.E. and Mortimer, R.K. (1991) Gene 104, 103-106.
91 Lovett, S.T. (1994) Gene 142, 103-106.
92 Kans, J.A. and Mortimer, R.K. (1991) Gene 105, 139-140.
93 Muris, D.F.R., Vreeden, K., Carr, A.M., Broughton, B.C., Lehmann, A.R., Lohman, P.H.M. and Pastink, A. (1993) Nucl. Acids Res. 21, 4586-4591.
94 Ostermann, K., Lorentz, A. and Schmidt, H. (1993) Nucl. Acids Res. 21, 5940-5944.
95 Bezzubova, O., Shinohara, A., Mueller, R.G., Ogawa, H. and Buerstedde, J.-M. (1993) Nucl. Acids Res. 21, 1577-1580.
96 Bezzubova, O.Y., Schmidt, H., Ostermann, K., Heyer, W.-D. and Buerstedde, J.-M. (1993) Nucl. Acids Res. 21, 5945-5949.

97 Yashimura, Y., Morita, T., Yamamoto, A. and Matsushiro, A. (1993) Nucl. Acid Res. 21, 1655.
98 Fishman-Lobell, J. and Haber, J.E. (1992) Science 258, 480-484.
99 Kramer, K.M., Brock, J.A., Bloom, K., Moore, J.K. and Haber, J.E. (1994) Molec. Cell. Biol. 14, 1293-1301.
100 Roth, D.B. and Wilson, J.H. (1986) Molec. Cell Biol. 6, 4295-4304.
101 Thacker, J., Chalk, J., Ganesh, A. and North, P. (1992) Nucl. Acids Res. 20, 6183-6188.
102 Pfeiffer, P., Thode, S., Hancke, J. and Vielmetter, W. (1994) Molec. Cell. Biol. 14, 888-895.
103 Jeggo, P.A., Tesmer, J. and Chen, D.J. (1991) Mutat. Res. 254, 125-133.
104 Jaspers, N.G.J., Gatti, R.A., Bann, C., Linssen, P.C.M.L. and Bootsma, D. (1988) Cytogenet. Cell Genet. 49, 259-263.
105 Gatti, R.A. and Swift, M. (1985) Ataxia-telangiectasia. Genetics, Neuropathology, and Immunology of a Degenerative Disease of Childhood. Alan R. Liss, Inc., New York, NY.
106 Lehmann, A.R., Arlett, C.F., Burke, J.F., Green, M.H.L., James, M.R. and Lowe, J.E. (1986) Int. J. Radiat. Biol. 49, 639-643.
107 Kapp, L.N. and Painter, R.B. (1989) Int. J. Radiat. Biol. 56, 667-675.
108 Murnane, J.P. and Kapp, L.N. (1993) Seminars in Cancer Biology 4, 93-104.
109 Cox, R., Masson, W.K., Weichselbaum, R.R., Nove, J. and Little, J.B. (1981) Int. J. Radiat. Biol. 39, 357-369.
110 Cornforth, M.N. and Bedford, J.S. (1985) Science 227, 1589-1591.
111 Cox, R., Debenham, P.G., Masson, W.K. and Webb, M.B.T. (1986) Mol. Biol. and Medicine 3, 229-244.
112 Thacker, J. (1989) Mutat. Res. 220, 187-204.
113 Kemp, L.M., Sedgwick, S.G. and Jeggo, P.A. (1984) Mutat. Res. 132, 189-196.
114 Giaccia, A., Weinstein, R., Hu, J. and Stamato, T.D. (1985) Somatic Cell Molec. Genet. 11, 485-491.
115 Biedermann, K.A., Sun, J., Giaccia, A.J., Tosto, L.M. and Brown, J.M. (1991) Proc. Nat. Acad. Sci. U.S.A. 88, 1394-1397.
116 Jeggo, P.A., Hafezparast, M., Thompson, A.F., Broughton, B.C. Kaur, G.P., Zdzienicka, M.Z. and Athwal, R.S. (1992) Proc. Nat. Acad. Sci. U.S.A. 89, 6423-6427.
117 Giaccia, A.J., Denko, N., MacLaren, R.A., Mirmen, D., Waldren, C., Hart, I. and Stamato, T.D. (1990) Amer. J. Hum. Genet. 47, 459-469.
118 Itoh, M., Hamatani, K., Komatsu, K., Araki, R., Takayama, K. and Abe, M. (1993) Rad. Res. 134, 364-368.
119 Taccioli, G.E., Rathbun, G., Oltz, E., Stamato, T., Jeggo, P.A. and Alt, F.W. (1993) Science 260, 207-210.
120 Pergola, F., Zdzienicka, M.Z. and Lieber, M.R. (1993) Molec. Cell. Biol. 13, 3464-3471.
121 Rathmell, W.K. and Chu, G. (1994) Molec. Cell. Biol. 14, 4741-4748.
122 de Vries, E., van Driel, W., Bergsma, W.G., Arnberg, A.C. and van der Vliet, P.C. (1989) J. Mol. Biol. 208, 65-78.
123 Cai, Q.-Q., Plet, A., Imbert, J., Lafage-Pochitaloff, M., Cerdan, C. and Blanchard, J.-M. (1994) Cytogenet. Cell Genet. 65, 221-227.
124 Hafezparast, M., Kaur, G.P., Zdzienicka, M., Athwal, R.S., Lehmann, A.R. and Jeggo, P.A. (1993) Somat. Cell Molec. Genet. 19, 413-421.

125 Taccioli, G.E., Gottlieb, T.M., Blunt, T., Priestley, A., Demengeot, J., Mizuta, R., Lehmann, A.R., Alt, F.W., Jackson, S.P. and Jeggo, P.A. (1994) Science 265, 1442-1445.
126 Foroud, T., Wei, S., Ziv, Y., Sobel, E., Lange, E., Chao, A., Goradia, T., Huo, Y., Tolun, A., Chessa, L., Charmley, P., Sanal, O., Salman, N., Julier, C., Concannon, P., McConville, C., Taylor, A.M.R., Shiloh, Y., Lange, K. and Gatti, R. (1991) Amer. J. Hum. Genet. 49, 1263-1279.
127 Leonhardt, E.A., Kapp, L.N., Young, B.R. and Murnane, J.P. (1994) Genomics 19, 130-136.
128 Hoekstra, M.F., Liskay, R.M., Ou, A.C., DeMaggio, A.J., Burbee, D.G. and Heffron, F. (1991) Science 253, 1031-1034.
129 Dhillon, N. and Hoekstra, M. (1994) EMBO J. 12, 2777-2788.
130 Birkenbihl, R.P. and Subramani, S. (1992) Nucl. Acids Res. 20, 6605-6611.
131 Tavassoli, M., Shayeghi, M., Nasim, A. and Watts, F.Z. (1994) (unpublished data).
132 Carr, A.M., Sheldrick, K.S., Murray, J.M., Al-Harithy, R., Watts, F.Z. and Lehmann, A.R. (1993) Nucl. Acids Res. 21, 1345-1349.
133 Scherly, D., Nouspikel, T., Corlet, J., Ucla, C., Bairoch, A. and Clarkson, S.G. (1993) Nature 363, 182-185.
134 Weber, C.A., Salazar, E.P., Stewart, S.A. and Thompson, L.H. (1990) EMBO J. 9, 1437-1448.
135 van Duin, M., de Wit, J., Odijk, H., Westerveld, A., Yasui, A., Koken, M.H.M., Hoeijmakers, J.H.J. and Bootsma, D. (1986) Cell 44, 913-923.
136 Humbert, S., van Vuuren, H., Lutz, Y., Hoeijmakers, J.H.J., Egly, J.M. and Moncollin, V. (1994) EMBO J. 13, 2393-2398.

MANIPULATING AND MAPPING DNA WITH RecA-ASSISTED RESTRICTION ENDONUCLEASE (RARE) CLEAVAGE

Lance J. Ferrin

Genetics and Biochemistry Branch
National Institute of Diabetes
and Digestive and Kidney Disease
National Institutes of Health
Bethesda, MD 20892

INTRODUCTION

RARE cleavage is a general method to perform sequence-specific cleavage of DNA. The method is possible because of some very useful properties of the RecA protein from *E. coli*. RecA protein is a versatile enzyme that plays a key role in repair and general recombination pathways. It has also proven to be a versatile enzyme *in vitro*, and can catalyze a diverse set of pairing reactions using a variety of DNA substrates. For example, it can promote the reannealing of two complementary single strands of DNA, and can pair a single strand with a homologous duplex to form a stable three-stranded joint molecule. Joint molecules are probably stabilized by hydrogen bonds between the bases in the single strand and donors and acceptors in the major groove of the homologous duplex (1). Joint molecules formed by RecA protein have several useful biochemical properties: they can be formed with oligonucleotides, they are stable for hours but can be dissociated under gentle conditions, they form only at regions of homology between a single strand and a duplex, and they can block the action of several enzymes which utilize duplex DNA as a substrate. In a sense, a filament of RecA protein and an oligonucleotide can be used as a sequence-specific "masking tape" to protect DNA transiently from modifying enzymes.

PRINCIPLES OF THE METHOD

An outline of the RARE cleavage strategy is given in Figure 1. The first step is to select a particular restriction enzyme and site for cleavage (2-4). Cleavage at an *Eco*RI site is especially convenient since both the restriction enzyme and methylase are available free of nonspecific nucleases, and are active under conditions optimal for RecA protein. A homologous oligonucleotide is synthesized such that the *Eco*RI site is centered in the oligonucleotide. Slightly less efficient protection is obtained if the *Eco*RI site is near one end of the oligonucleotide. The oligonucleotide and RecA protein are incubated with duplex DNA and a complex is formed at the site of homology. In the RecA literature, this ternary structure is called a "synaptic complex." *Eco*RI methylase and S-adenosylmethionine are then added and allowed to methylate all available sites, but sparing the site involved in the synaptic complex. The complex is then dissociated, the methylase is inactivated and *Eco*RI restriction enzyme is added to cleave at the specific *Eco*RI site now exposed.

In a simple variation of the above strategy, two oligonucleotides can be added to cleave a fragment out of DNA. In addition, the synaptic complex can be used to prevent restriction enzyme cleavage at a protected site.

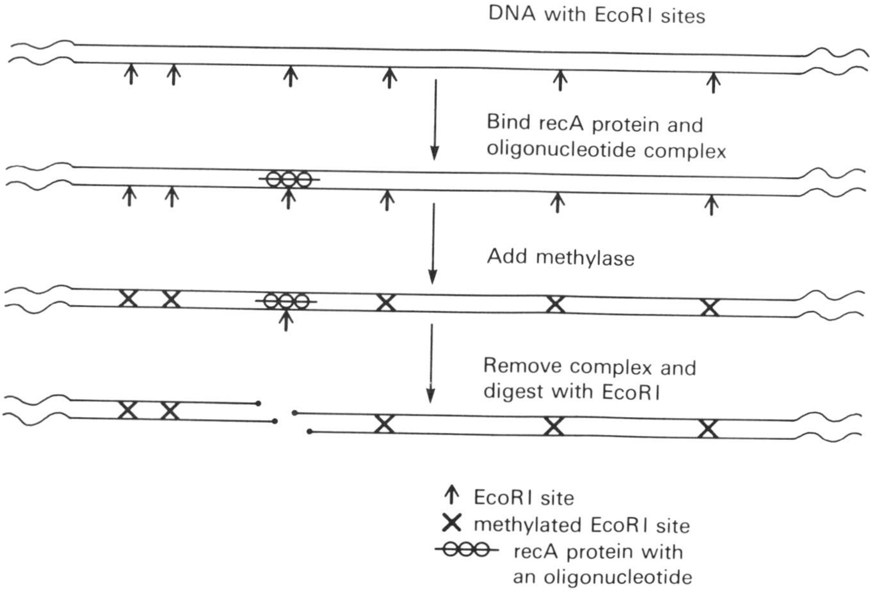

Figure 1. Schematic of the strategy of RARE cleavage.

REAGENTS

Preparation of DNA

When performing RARE cleavage on DNA in solution, one should dissolve the DNA in a low ionic strength buffer such as TE, as RecA protein will not readily form complexes if additional salts in the reaction exceed about 25 mM.

When RARE cleavage is performed for the purpose of preparing large fragments, it is necessary to embed the DNA in agarose to prevent shearing. RARE cleavage has been performed on bacterial, yeast and human DNA, and methods for preparing the DNA in agarose have been compiled (5-7). The following method for preparing DNA from mammalian cells works well with the RARE protocol given. If a different type or concentration of agarose is used, the optimal time between RecA protein addition and methylase addition should be determined.

Mammalian cells (1 X 10^8, 150 mg wet weight) are washed twice with phosphate-buffered isotonic saline and suspended in 1 ml of 10 mM Tris-HCl at pH 7.2, 20 mM NaCl, and 100 mM EDTA. The cell suspension is incubated for 1 to 2 min at 65°C and mixed with 1 ml of 1.6% InCert agarose (FMC Bioproducts). The solution is poured into molds or 1 ml syringes and allowed to set. The agarose blocks or rods are placed in 6 ml of proteinase K solution (1 mg/ml proteinase K, 100 mM EDTA at pH 8.0, 1% sodium N-laurylsarcosine, and 0.2% sodium deoxycholate) for 18 to 20 hr at 42°C with gentle shaking. The proteinase K solution is removed, and the agarose is washed twice with 10 ml of wash solution (20 mM Tris-HCl at pH 8.0, 50 mM EDTA) for 15 min at room temperature. The agarose is treated for 1 hr with 10 ml of 1 mM phenylmethylsulfonyl fluoride in wash solution, washed twice with wash solution alone for 15 min, and twice with a tenfold dilution of wash solution. The DNA is stable for several months at 4°C.

Immediately prior to use, a sample of the DNA in agarose is chopped with a coverslip to the consistency of a grainy syrup (8), incubated in 50 mM EDTA at pH 8.0 for 30 min at room temperature, and exchanged into 1 X RARE cleavage buffer. Five changes of buffer with a 10 min equilibration time per change are used. The chopped agarose can be pelleted by spinning for a few seconds in a microcentrifuge. The chopped agarose can be pipetted with a wide bore tip, and is stirred when mixing with reagents to avoid splashing and sticking to the vessel wall. The chopping is done to increase the rate of diffusion of proteins to the DNA.

Chemicals and Proteins

RARE cleavage buffer (5 X stock): 125 mM Tris acetate at pH 7.85, 20 mM Mg acetate, 2.5 mM spermidine trihydrochloride (Sigma), 2 mM DTT.

ADP/ATPγS (10 X stock): 11 mM ADP, 3 mM ATPγS (Fluka). Each concentration should be determined assuming a 1 M solution at pH 7.0 has an absorbance at 259 nm of 15,400.

EGTA: 5 mM.

S-adenosylmethionine (New England Biolabs): freshly diluted to 2.4 mM with water.

BSA (Sigma #A-7638): dilute to 2 mg/ml with water.

RecA protein: recently, several companies have begun to sell RecA protein. We have used protein from United States Biochemical and Boehringer Mannheim. The efficiency of synaptic complex formation varies slightly between batches.

Methylases: New England Biolabs has the largest selection. A list of which restriction enzymes are sensitive to methylases has been compiled (9). *Eco*RI methylase may be diluted into 10 mM Tris-HCl at pH 7.5, 100 μg/ml BSA immediately prior to use.

Oligonucleotides: 30-60 bases in length. The best efficiencies are obtained with 60-mers that have been purified on an acrylamide gel. Concentrations are determined assuming 33 μg per A_{260} unit.

*Eco*RI restriction enzyme buffer: star activity is very low in a buffer of 100 mM Tris-HCl at pH 8.0, 50 mM NaCl, 20 mM $MgCl_2$, 1.5 μM DTT, and 200 μg/ml BSA.

Stop solution: 6% SDS, 90 μM EDTA, 50 mM Tris-HCl at pH 7.5, 0.1% bromophenol blue.

PROTOCOLS

RARE Cleavage of DNA in Solution

1. In this order at room temperature, mix sufficient water to make the final volume 40 μl (including the methylase and S-adenosylmethionine), 8 μl of RARE buffer (5x stock), 4 μl of ADP/ATPγS, 0.18 μg of oligonucleotide (or 0.09 μg each of two oligonucleotides), 10 μg of RecA protein, and up to 1.0 μg of duplex DNA.
2. Incubate at 37°C for 10 min.
3. Add 2μl of BSA methylase (generally 4-8 units) and 2 μl of S-adenosylmethionine.
4. Incubate at 37°C for 20 min. This time may be increased to 1 hr for less active methylases.
5. Incubate at 65°C for 15 min to inactivate the methylase and RecA protein.
6. If desired, ethanol precipitate the DNA and redissolve into a specific restriction enzyme buffer.
7. Add 20 units of restriction enzyme, and digest to completion.
8. Add 5 μl of stop solution, and examine the products by gel electrophoresis.

RARE Cleavage of DNA Embedded in Agarose

1. Take a freshly chopped and washed aliquot of DNA in agarose and suspend about 1 volume of packed agarose in 3 volumes of 1 X RARE buffer. Set aside.
2. In this order at room temperature, mix sufficient water to make the final volume 200 μl (including the methylase and S-adenosylmethionine), 20 μl of RARE buffer, 4 μl of EGTA, 20 μl of ADP/ATPγS, 0.360 to 1.2 μg of oligonucleotide (or 0.18 to 0.6 μg each of two oligonucleotides), 30 μg of RecA protein, and 100 μl of the agarose suspension. Routinely, three or four individual reactions with different oligonucleotide concentrations are performed.
3. Incubate at 37°C for 15 min. Some users report higher efficiency of cleavage if an incubation time of 20-30 min is used.
4. Add 10 μl of BSA methylase (generally 80 units) and 10 μl of S-adenosylmethionine.
5. Incubate at 37°C for 1 hr. Some users report more efficient methylation if an incubation time of 1.5-2 hrs is used.
6. Add 20 μl of 10% SDS.
7. Incubate at 37°C for 15 min.
8. Remove the SDS by washing the agarose five times with a restriction enzyme buffer. Leave 150-200 μl of buffer in the tubes after the final exchange.
9. Add 80 units of a restriction enzyme for 1 hr.
10. Pellet the agarose and remove all but about 20 μl of the buffer.
11. Add 15 μl of stop solution and examine the products by gel electrophoresis. Southern blotting will be necessary to detect a band if complex genomic DNA was used.

Protecting a Restriction Site from Cleavage "Synaptic Complex Protection"

1. In this order at room temperature, mix sufficient water to make the final volume 40 μl, 8 μl of RARE buffer (5x stock), 4 μl of ADP/ATPγS, 0.18 μg of oligonucleotide, 10 μg of RecA protein,, and up to 1.0 μg of duplex DNA.
2. Incubate at 37°C for 10 min.
3. Add 20 units of restriction enzyme, and incubate at 37°C for 20 min.

4. Add 5 μl of stop solution and examine the products by gel electrophoresis. RARE buffer can be replaced by 100mM Tris-HCl at pH7.5, 62.5mM $MgCl_2$, 2mM DTT (5 x stock) if necessary to increase restriction enzyme activity. The digestion time may also be increased to 1 hr.

USING RARE CLEAVAGE TO MAP GENOMIC DNA

General Mapping Projects

Several methods are currently used for physical mapping of genomic DNA (reviewed in 10,11). The methods differ by the amount of effort required to obtain the data, and by the precision of the data obtained. RARE cleavage can provide distance information between two cleavage sites. Typically, this information is obtained by cleaving at two sites, running the cleavage products on a gel, and probing a Southern blot of the gel with a probe known to be between the two cleavage sites. The size of the band detected is the distance between the two cleavage sites, and with the use of pulsed field gel electrophoresis, distances of up to 1.7 megabases have been measured (4).

After the initial report, the first reported use of RARE cleavage was to measure DNA from mouse cells that had been injected with a yeast artificial chromosome (YAC) (12). The YAC contained an insert of 210 kb and vector DNA with a neomycin resistance gene. The sequences of the insert termini were determined by direct sequencing of the two vector-insert junctions, and oligonucleotides used for RARE cleavage were made with this information. After injection, three G418-resistant colonies were analyzed by RARE cleavage. A 210 kb band was detected, providing evidence that the entire YAC insert had integrated without rearrangement.

In several areas of the human genome, restriction maps have been constructed. These maps usually show the position of infrequently-cutting restriction enzyme sites. For example, the position of most of the *Not*I sites on chromosome 21 have been determined (13). The distance from a RARE cleavage site to the nearest of such restriction enzyme landmarks can be determined by performing one RARE cleavage along with a *Not*I (or other infrequently cutting restriction enzyme) digest (4).

Recently, the laboratory of David C. Schwartz has developed a method for the rapid physical mapping of DNA by optical imaging of DNA molecules (14), and has used the method to measure the size of RARE cleavage fragments (15).

Mapping Repetitive DNA

Long regions of repetitive DNA are especially difficult to map. These regions are unstable in YACs or other vectors, unique probes may not be available, and individual restriction enzymes usually cut either frequently or not at all. Thus, even the size of repetitive regions may be difficult to determine. Many such regions are known throughout the human genome, including all telomeric and centromeric regions. Telomeric regions have been mapped by RARE cleavage with the following strategy (4,16). A clone containing a unique region near the telomeric and subtelomeric repeats was isolated. An *Eco*RI site from the clone was sequenced and a unique probe telomeric to the site was obtained. RARE cleavage was then performed at that site and the fragment size detected with the probe gave the distance from the site to the end of the telomere. In the Riethman laboratory (16,17), several telomeres have been analyzed in this manner.

Confirming That a Clone is a Faithful Reproduction of Genomic DNA

Several cloning systems work poorly with certain types of DNA such as direct or inverted repeats. The problem is especially severe with YAC vectors, since yeast are efficient at rearranging clones *via* homologous recombination. The YAC systems also generally have a high frequency of chimeric clones, that is, DNA from two totally different regions of the genome may be spliced together into one clone. YAC clones are very useful, however, because of the large size of the inserts. It is possible with RARE cleavage to determine which clones are free of rearrangements (at least large rearrangements) and chimerism.

The strategy that has been used to analyze YAC clones is as follows. First, the two sequences at the insert-vector junction are determined by standard methods such as vectorette PCR (18). Oligonucleotides with the insert sequences are used to direct RARE cleavage of genomic DNA. Most YAC libraries have been constructed by cloning partially *Eco*RI-digested genomic DNA into the *Eco*RI sites of the vector, so that the convenience of the *Eco*RI/*Eco*RI methylase system can be used. If the genomic fragment generated by RARE cleavage is the same size as the YAC insert, then it is virtually certain that the YAC clone is a faithful reproduction of genomic DNA. This strategy has been used to analyze YAC clones containing human telomeres (16,17), and as part of a larger effort to generate a complete contig of chromosome 21 prior to initiating sequencing efforts (19).

YAC contigs are sets of overlapping YACs, but the extent of overlap is usually unknown. With RARE cleavage and oligonucleotides from the insert-vector junction, the extent and position of overlap can be readily determined.

THE USE OF RARE CLEAVAGE TO SUBCLONE OR ALTER DNA

RARE cleavage or synaptic complex protection are very flexible tools for the manipulation of DNA. Most common restriction enzymes, including many with four-base recognition sequences, can be blocked by commercially available methylases. In addition, most restriction enzymes are sufficiently active in the buffer optimal for RecA protein, and none have been found that can cleave at a site blocked by a synaptic complex.

The simplest way RecA protein is used in subcloning DNA is to block a site from restriction enzyme cleavage. Conventionally in molecular biology, when one wishes to avoid cleaving a given restriction site, a partial digestion is carried out, and the desired fragment is purified on a gel. It is generally simpler to block a site by forming a synaptic complex than by performing a partial digest. This is especially true if the restriction enzyme cleaves at multiple sites, and only one site needs to be blocked. If three or more sites are to be blocked, or if the restriction enzyme is inactive in the RecA buffers, it is generally more efficient to prepare a fragment with the appropriate methylase and RARE cleavage.

In a more complicated subcloning procedure in this laboratory, the goal was to place a chloramphenicol resistance gene from one plasmid into the middle of a polylinker region of a second plasmid. A convenient unique *Xmn*I site was just downstream of the gene, but none upstream of the gene. A *Taq*I site was just upstream of the gene, but the gene also contained three other *Taq*I sites. RARE cleavage of the plasmid was performed at the upstream *Taq*I site, and the *Xmn*I/*Taq*I fragment was purified on a gel and ligated into cohesive ends created in the polylinker of the second plasmid. The overwhelming majority of the clones obtained (even though chloramphenicol selection was not used) had the gene in an unrearranged and functional form. Notably, even though the *Taq*I methylase had to be used at 37°C instead of its optimal temperature of 65°C, it was still sufficiently active for the procedure.

RARE cleavage is especially useful for manipulating large DNA constructs, and has been used by groups seeking to make large altered constructs for transgenic mouse studies. One group wished to delete a 478 bp putative enhancer region from a cosmid containing the entire 35 kb chicken β-globin cluster (20). The region was deleted by performing RARE cleavage at two *Hha*I sites bordering the region, re-ligating the product without further purification, and transforming bacteria with the re-ligated cosmid. About half of the colonies contained the desired altered cosmid, which was used to make transgenic mice. The group also made a small insertion in the cosmid by performing RARE cleavage at a single *Aci*I site, adding a small duplex linker, and re-ligating.

For a similar reason, another group wished to make a point mutation in a 90 kb P1 clone containing the entire 43 kb human apolipoprotein B gene (21). The first attempt to use RARE cleavage involved deleting a 1550 bp fragment from the P1 clone (which worked well) and then inserting a mutated fragment back into the site. All the colonies screened, however, contained rearranged clones. To aid in screening for unrearranged clones, a multi-step strategy was used (Figure 2). The unmodified P1 clone (Figure 2, left), was subjected to RARE cleavage at sites A and B, and the product was re-ligated to generate a clone with a 1550 bp deletion. In a separate procedure, the 1550 bp fragment was cloned into M13, the point mutation was introduced, the fragment was transferred to another plasmid and the chloramphenicol resistance and *sacB* genes were spliced to the fragment. This cassette was then removed with RARE cleavage at sites A and C, the P1 clone with the deletion was opened by RARE cleavage, and the two were ligated together to generate a composite clone (Figure 2, middle). Selection with chloramphenicol aided in the isolation of an unrearranged clone. RARE cleavage was then performed at sites B and C, and the product was re-ligated to generate the final clone (Figure 2, right). Selection on sucrose plates (the *sac*B gene is lethal to bacteria in the presence of sucrose) aided in selection of the correct clone. 50 clones were screened and one was found that contained the entire desired P1 clone with the point mutation.

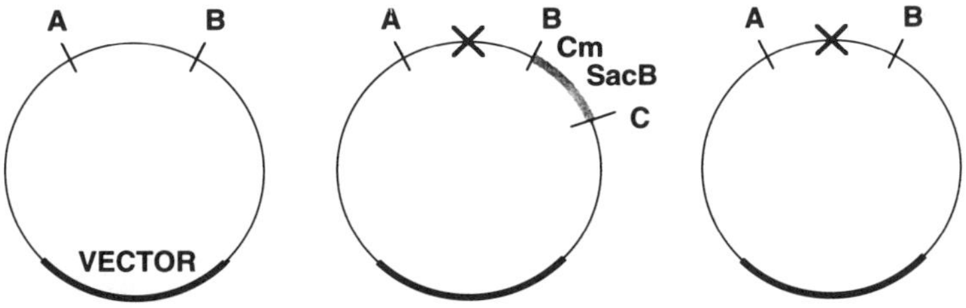

Figure 2. Introducing a point mutation into a 90 kb P1 clone. The position of vector sequences, and the chloramphenicol resistance and *sacB* genes are shown. The cross shows the position of the point mutation, and the letters show the position of *Eco*RI sites used for RARE cleavage.

Other Uses for RARE Cleavage and Synaptic Complex Protection

Often it is useful to know whether a given sequence is present in a clone. While this can be determined by sequencing or a Southern blot, if one has the necessary homologous oligonucleotide, it is usually easier to confirm the identity of a clone by performing a synaptic complex protection with the oligonucleotide, and comparing the products to the restriction pattern generated with a nonhomologous control oligonucleotide.

Additional uses for RARE cleavage have been mentioned in the literature such as using it to generate convenient substrates for oligonucleotide-directed site-specific mutagenesis (22), or as an adjunct to sequencing DNA by capillary electrophoresis (23). It should also soon be possible to increase the specificity of RARE cleavage to the point where selected fragments can be cloned directly from complex genomes.

TROUBLESHOOTING AND OPTIMIZING RARE CLEAVAGE

Performing RARE cleavage on DNA less than 50 kb in size is a relatively simple procedure, but for several reasons, cleaving intact fragments several hundred kb or longer can be difficult. On long fragments, there can be hundreds of individual restriction sites which must be methylated to keep the fragment intact during restriction enzyme digestion. The methylase must be very active, and nonspecific binding by free RecA protein must be kept to a minimum. In addition, all of the reagents must be almost entirely free of nucleases, and the restriction enzyme must be used under conditions where it does not exhibit star activity.

Before attempting to cleave large fragments from genomic DNA, it can be useful to perform three simple control reactions to assess the activity and purity of the reagents. In the first reaction, a simple DNA such as λ DNA is subjected to RARE cleavage at a single site (as in ref. 2). This tests the efficiency of the RecA protein pairing reaction. In the second reaction, all reagents except for the restriction enzyme are incubated together. This tests for the presence of nucleases. In the third reaction, the RARE cleavage protocol is followed except that the RecA protein and oligonucleotide are omitted. This tests for complete methylation or for the presence of star activity in the restriction enzyme. These reactions are qualitatively evaluated on gels stained with ethidium bromide and can be compared to lanes with the starting and the completely digested genomic DNA sample. Intact genomic DNA will remain in the well or upper compression zone of the gel. With complete RARE cleavage experiments, there is a significant amount of nonspecific degradation due to free RecA protein binding (and subsequent inhibition of the methylase), but a large proportion of the DNA should remain in a large molecular weight form (see (4) for examples of gels stained with ethidium bromide and autoradiographs of RARE cleavage products).

Individual Reagents

Oligonucleotides. Generally, oligonucleotides 30 to 60 bases in length are used. Cleavage with an efficiency of about 80% can be obtained when a 60-mer is used, and about 60% with a 30-mer. The best yield is obtained if the restriction site is centered in the oligonucleotide. If sequence is available on only one side of a restriction site, an efficiency of 60% can be obtained with an oligonucleotide that has 10 bases of any sequence at the 5' end, followed by the first 50 bases of homologous sequence at the 3' end. The sequence and polarity of the oligonucleotide can be identical to either strand of the target duplex, but oligonucleotides with a high pyrimidine content should be avoided. Although oligonucleotides with marked secondary structure have not been studied, they should probably also be avoided.

Methylases and Restriction Enzymes. *Eco*RI methylase has performed well in demanding applications of RARE cleavage. *Alu*I methylase performed poorly in such cases, but was satisfactory with DNA 50 kb in size. *Bam*HI, *Taq*I, *Sss*I, *Hpa*II and *Hha*I methylases (but not *Dam* methylase) worked well on 50 kb DNA samples, but have not been tested on larger DNA samples. Restriction enzymes that cleave infrequently are generally better choices than those that cleave frequently (2).

Buffers. When attempts are made to cleave large fragments of DNA, trace metals should be kept to a minimum by including EGTA or by passing the RARE buffer reagents (excluding the Mg acetate) through small Chelex 100 (Bio-Rad) columns.

Bacterial hosts. Several cloning hosts that are tolerant of most methylation patterns are commercially available. To test a given strain and methylase, one may transform the strain with methylated and unmethylated DNA, compare the transformation efficiencies, and analyze the clones for rearrangements.

CONCLUSIONS

RARE cleavage is a versatile tool for the precise manipulation of DNA. After an initial amount of effort in preparing and testing the reagents, the technique is fairly rapid and straightforward to use. It can be applied to a large number of mapping and subcloning projects that cannot be done efficiently by conventional procedures. With mapping projects, exact physical distances are obtained. With subcloning projects, even large clones can be manipulated efficiently, and most common restriction enzyme sites, even those with four base recognition sequences, can be turned into unique RARE cleavage sites.

REFERENCES

1 Camerini-Otero, R.D. and Hsieh, P. (1993) Cell 73, 217-223.
2 Ferrin, L.J. and Camerini-Otero, R.D. (1991) Science 254, 1494-1497.
3 Ferrin, L.J. and Camerini-Otero, R.D. (1993) in Methods in Molecular Genetics (Adolph, K.W., ed.), Vol. 2B, pp. 57-66, Academic Press, San Diego, CA.
4 Ferrin, L.J. and Camerini-Otero, R.D. (1994) Nature Genet. 6, 379-383.
5 McClelland, M. (1987) in Meth. Enzymol. (Wu, R., ed.), Vol. 155, pp. 22-32, Academic Press, San Diego, CA.
6 Carle, G.F. and Olson, M.V. (1987) in Meth. Enzymol. (Wu, R., ed.), Vol. 155, pp. 468-482, Academic Press, San Diego, CA.
7 Smith, C.L., Klco, S., Zhang, T., Fang, H., Oliva, R., Wang, D., Bremer, M and Lawrance, S. (1993) in Methods in Molecular Genetics (Adolph, K. W., ed.), Vol. 2B, pp. 155-194, Academic Press, San Diego, CA.
8 Wang, Y. and Schwartz, D. C. (1993) Nucl. Acids Res. 21, p. 2528.
9 McClelland, M. and Nelson, M. (1992) Nucl. Acids Res. 20, Suppl. pp. 2145-2157.
10 Patterson, D. (1993) Nature Genet. 4, 323-324.
11 Evans, G.A. and McElligott, D.L. (1992) in Genetic Engineering: Principles and Methods (Setlow, J. K., ed.) Vol. 14, pp. 269-278, Plenum Press, New York, NY.
12 Gnirke, A., Huxley, C., Peterson, K. and Olson, M.V. (1993) Genomics 15, 659-667.
13 Ichikawa, H., Hosoda, F., Arai, Y., Shimizu, K., Ohira, M. and Ohki, M.(1993) Nature Genet. 4, 361-366.
14 Schwartz, D.C., Li, X., Hernandez, L.I., Ramnarain, S.P., Huff, E.J. and Wang, Y. (1993) Science 262, 110-114.
15 Wang, Y., Huff, E.J. and Schwartz, D.C., Proc. Nat. Acad. Sci. U.S.A. (in press).

16 Negorev, D.G., Macina, R.A., Spais, C., Ruthig, L.A., Hu, X.-L. and Riethman, H.C. (1994) Genomics 22, 569-578.
17 Macina, R.A., Negorev, D.G., Spais, C., Ruthig, L.A., Hu, X. and Riethman, H.C. (1994) Hum. Molec. Genet. 3, 1847-1853.
18 Riley, J., Butler, R., Ogilvie, D., Finniear, R., Jenner, D., Powell, S., Anand, R., Smith, J.C. and Markham, A.F. (1990) Nucl. Acids Res. 18, 2887-2890.
19 Patterson, D. (personal communication).
20 Mason, M.M., Lee, E., Westphal, H. and Reitman, M., Molec. Cell. Biol. (in press).
21 Callow, M.J., Ferrin, L.J. and Rubin, E.M., Nucl. Acids Res. (in press).
22 Weiner, M.P., Felts, K.A., Simcox, T.G. and Braman, J.C. (1993) Gene 126, 35-41.
23 Noolandi, J. (1992) Electrophoresis 13, 394-395.

MOLECULAR STUDIES ON THE VIRULENCE OF *Listeria monocytogenes*

Michael Kuhn and Werner Goebel

Lehrstuhl für Mikrobiologie
Theodor-Boveri-Institut für Biowissenschaften
Universität Würzburg, 97074 Würzburg, Germany

INTRODUCTION

Listeria monocytogenes, a gram-positive, rod-shaped, non-spore-forming, facultative anaerobic bacillus (1), is the best characterized member of the genus *Listeria*, which comprises five additional species, *L. ivanovii*, *L. seeligeri*, *L. innocua*, *L. welschimeri*, and *L. grayi* (2). *L. monocytogenes* causes serious infections in immunocompromised individuals, newborns, pregnant women and older people. The major symptomatic manifestations of Listeria infections are stillbirth, septicemia, encephalitis and meningitis (3,4, reviewed in 5). The tracing of recent epidemics to food contaminated with *L. monocytogenes* suggests that the natural route of infection by bacteria is the gastrointestinal tract (6,7, reviewed in 8).

For many years, *L. monocytogenes* received considerable attention as a model pathogen for studying cell-mediated immunity (9,10). The murine model of listeriosis essentially mimics human infections by *L. monocytogenes* and provides a well-reproducible system for the determination of *L. monocytogenes* virulence. After intraperitoneal injection into the mouse, virulent *L. monocytogenes* bacteria are taken up within minutes by resident macrophages. Rapid killing of the ingested bacteria by the macrophage follows and the number of viable listeriae is greatly reduced within a few hours. Surviving bacteria start replicating inside macrophages and the number of viable bacteria increases again over 2 to 3 days in the liver and spleen. This step is followed by the rapid inactivation of the bacteria involving mainly cytotoxic and helper T-cells over the next days leading to the recovery of the infected animal (11,12). Experimental oral infection models are currently developed in the mouse, which probably resemble more the normal route of infection (13,14).

Compared to the large body of immunological work performed with *L. monocytogenes* as a model of an intracellular microorganism, the studies aimed to unravel the pathogenicity of *L. monocytogenes* and its interaction with host cells on the cellular, molecular and genetic levels were initiated only some years ago. The early studies used transposon mutagenesis (15-17) and infection of primary and established cell lines (epithelial) fibroblast and macrophage) (17-21) to obtain insights into the interaction of *L. monocytogenes* with eukaryotic host cells. Meanwhile, the development of new tools allowed the genetic manipulation of *L. monocytogenes*. The use of mammalian primary and established cells as cellular models have greatly broadened our understanding of the molecular and cell biology of *L. monocytogenes* infections.

THE INTRACELLULAR LIFE CYCLE OF *L. monocytogenes* IN CELL CULTURE MODELS

Epithelia-like and macrophage-like cell lines have been used to study the cell biology of *L. monocytogenes* infection (19,22,23). Whereas macrophages actively ingest *L. monocytogenes*, the internalization of the bacteria by normally nonphagocytic cells is triggered by *L. monocytogenes*-specific products. Besides the internalization step, there are no obvious differences in the intracellular life cycle of the listeriae in phagocytes or normally nonphagocytic mammalian cells. Upon uptake, the bacteria are first seen in a vacuole, which is subsequently lysed allowing the bacteria to escape into the cytoplasm. There they start to replicate, whereas those which remain in the phagosome are killed and digested. Shortly after the onset of intracellular replication, *L. monocytogenes* induces the nucleation of host actin filaments which form a cloud around the bacterial cell. The actin filaments are then rearranged to a polar tail which consists of short actin filaments and other actin-binding proteins which stabilize this structure. It is assumed that the formation of the actin tails at one pole of the bacterial cell is the propulsive force which moves the bacteria through the cytoplasm of the host cell. This requires continuous *de novo* actin polymerization. Some of the bacteria reach the surface of the infected host cell and induce the formation of pseudopod-like structures with the bacterium at the tip and the actin tail behind. These pseudopods are apparently recognized by the neighbouring cell and phagocytosed, resulting in bacteria surrounded by a double membrane vacuole which is subsequently lysed to release the bacteria into the cytoplasm of the new host cell.

In the following, we will summarize the major work performed to dissect the intracellular life cycle of *L. monocytogenes* at a molecular level.

MOLECULAR ASPECTS OF ADHESION AND INVASION

The growth of *L. monocytogenes* in macrophages of different origin is well documented (9,17,20). The invasion by *L. monocytogenes* of different, normally nonphagocytic mammalian cell types including murine and human fibroblasts (17,18,20, 24), murine and human epithelial cells (17,19,24,25), murine hepatocytes (26) and human endothelial cells (27,28) has also been described. Invasion and survival of *L. monocytogenes* was also reported (29) in protozoa of the genera *Acanthamoeba* and *Tetrahymena*.

Spontaneously occurring rough mutants of *L. monocytogenes* show significantly reduced uptake by 3T6 fibroblast cells and reduced expression of a 60 kD extracellular protein, termed p60 (21). These "p60 mutants" (also referred to as R-mutants because of their rough colony appearance) form long cell chains which possess double septa between the individual cells. Treatment with partially purified p60 protein deaggregates the cell chains to normal-sized single bacteria which are again able to invade the fibroblasts. Physical disruption of the cell chains by ultrasonication produces similar single cells which are, however, noninvasive. Treatment of these ultrasonicated mutant cells with wild-type p60 restores their ability to invade the fibroblasts (21,30). The reduced invasiveness of the p60 mutants is only observed with certain mammalian host cells. Cell chains of p60 mutants adhere normally to Caco-2 epithelial cells and are perfectly invasive upon disruption of the bacterial cell chains by ultrasonication without the addition of p60 (30). *Salmonella typhimurium*, expressing and secreting p60, seems to be also more invasive for phagocytic cells but not for enterocytes (31). In line with this assumption is that pretreatment of *L. monocytogenes* with a polyclonal anti-p60 antiserum inhibits the uptake of the listeriae by a macrophage-like cell line (31,32). Protein p60 is a major secreted protein of all *L. monocytogenes* isolates (21,30) but is also found on the cell surface of *L. monocytogenes* (33). In contrast to other virulence factors p60 is also an essential enzyme for the cell metabolism of *L. monocytogenes*. It possesses murein hydrolase activity which appears to be involved in a late step of cell division (34). The gene coding for this obviously bifunctional protein, called *iap* (*i*nvasion *a*ssociated *p*rotein), was cloned and sequenced (35). Its expression seems to be controlled on the post-transcriptional level by a yet unknown mechanism (36). The amino acid sequence predicts an extremely basic protein of 484 amino acids. The protein has a 27 amino acid signal sequence, and an extended repeat domain consisting of 19 threonine-asparagine units which are separated by a proline-serine-lysine motif. A single cysteine in its C-terminal part is probably essential for its enzymatic activity (34,35). The significance of the repeat domain for either the enzymatic activity of p60 or its function in invasion is still unknown.

By transposon mutagenesis, internalin, a surface protein of *L. monocytogenes*, was identified as a mediator of bacterial invasion for epithelial cells (37). These mutants exhibit a 100-fold lower invasive capacity than the wild-type strain when tested on different epithelial cells. In addition, the mutants have also lost the ability to adhere to eukaryotic cells. The transposon insertion occurred into a chromosomal region, which represents an operon consisting of the *inlA* and *inlB* genes. Expression of *inlA* in *L. innocua*, a noninvasive *Listeria* species closely related to *L. monocytogenes*, renders this species invasive. This shows that the *inlA* gene product is crucial for invasion. Internalin is an acidic protein (pI 4.4) (37,38) which possesses two extended repeat domains. Domain A consists of 15 repeats of 22 amino acids each, whereas domain B consists of 2.5 repeats of about 70 amino acids each. The InlA protein has a typical N-terminal transport signal sequence and in the C-terminal part a cell wall-spanning region followed by a hydrophobic sequence which represents a putative membrane anchor (38).

The mechanism by which internalin mediates adhesion and invasion is unknown and no specific receptor on the target cell membrane has yet been identified. Internalin shows a structural organization strikingly similar to streptococcal M proteins, the protein A of *Staphylococcus aureus* and the fibronectin-binding protein of Streptococci (38). It has been postulated that these two structurally related proteins (protein M and internalin) can cause adherence to target cells followed by extracellular colonization or trigger intracellular uptake by the target cells depending on the receptor of the target cells to which they bind and/or on the receptor density of the infected cells (39).

As mentioned above, *inlA* forms an operon with *inlB*. The InlB protein carries also an N-terminal transport signal sequence and repeat domains, but, in contrast to InlA, has no obvious membrane anchor and no cell wall-spanning region (38). The high sequence similiarity of *inlA* and *inlB* indicates that the two genes originate from a common ancestral gene. Recently it was shown that *inlB* mutants still expressing *inlA* are invasive for Caco-2 cells. On the other hand, a hepatocyte-like cell line (HepG-2) is invaded by an *inlA* mutant but not by an *inlB* mutant (P. Cossart, personal communication). Invasion of 3T3 fibroblasts by *L. monocytogenes* seems to be independent of either *inlA* or *inlB* and even double mutants are still invasive for these fibroblasts suggesting that different, cell type-specific adhesion and invasion systems are present in *L. monocytogenes* (38). Since *inlA* and *inlB* belong to a gene family of at least seven genes in *L. monocytogenes* (P. Cossart, personal communication), their gene products, possibly together with p60 and/or other yet unknown proteins, may be involved in cell type-specific interactions between *L. monocytogenes* and various host cells.

The mechanism of *L. monocytogenes* invasion is still largely unknown. Uptake of *L. monocytogenes* by macrophages and other mammalian cells is dependent on functional actin microfilaments, since invasion is blocked by treatment with actin depolymerizing drugs such as cytochalasins (19,20). *L. monocytogenes* adheres and enters polarized Caco-2 cells from the apical surface (40). Attachment of wild-type *L. monocytogenes* to these host cells induces structural modification(s) of microvilli (40). These modifications are not observed with a prfA mutant of *L. monocytogenes* (see below) which shows a reduced invasiveness (41). Since no receptors for *L. monocytogenes* are known, the signal pathway triggering phagocytosis by the adherent bacteria is unkown. Phosphorylation of MAP-kinase is observed 10 min post infection by *L. monocytogenes* in Caco-2 cells. Internalin mutants defective in adhesion do not trigger MAP-kinase phosphorylation (42), suggesting that MAP-kinase may be involved in triggering the uptake. Recent data indicate, however, that listeriolysin rather than internalin is responsible for the activation of MAP-kinase (B. B. Finlay, personal communication).

The kinetics of uptake and intracellular killing of *L. monocytogenes* by macrophages were determined (43-45). The macrophages ingest *L. monocytogenes* rapidly and intracellular killing starts within minutes after phagocytosis and leads to the destruction of most of the ingested bacteria. In a single macrophage, both killed bacteria inside acidified phagosomes and phagolysosomes and growing bacteria escaped into the cytoplasm are detected. This suggests a competition between phagosome-lysosome fusion followed by killing of the bacteria and escape of the bacteria from the acidified phagosome before phagosome-lysosome fusion occurs. This competition results in a population of bacteria able to grow inside the cytoplasm of the macrophage.

THE VIRULENCE GENE CLUSTER AND THE SIGNIFICANCE OF THE GENE PRODUCTS FOR THE INTRACELLULAR LIFE CYCLE

Most of the known virulence genes of *L. monocytogenes* are clustered on the chromosome in the so-called *prfA*-dependent virulence gene cluster. The cluster comprises six well-characterized genes, *prfA, plcA, hly, mpl, actA* and *plcB* (Table 1) and three small ORFs of unknown functions downstream of plcB, called ORF X, Y and Z. Both ends of the gene cluster are defined by genes coding for housekeeping enzymes. Distal from *prfA*, defining the "left" border of the gene cluster, the *prs* gene encoding a phosphoribosyl-pyrophosphate synthetase is located (46,47). The *ldh* gene coding for lactate dehydrogenase together with the ORFs A and B (47,48) mark the "right" border of the gene cluster downstream from *plcB* and the small ORFs X, Y and Z. The products of these virulence genes are: listeriolysin (encoded by *hly*), two phospholipases (*plcA* and *plcB*), a metalloprotease (*mpl*), ActA, a protein involved in actin polymerization (*actA*) and the positive transcription factor PrfA (*prfA*). The functions of these gene products will be discussed in the following.

Listeriolysin (LLO)

All clinical isolates of *L. monocytogenes* are hemolytic due to the action of a cytolysin, called listeriolysin (often called listeriolysin O and therefore abbreviated as LLO). In experimental infections all virulent strains were found to be hemolytic, whereas nonhemolytic strains were avirulent. Hemolysis was therefore always suspected to be essential for virulence. Nonhemolytic mutants ere obtained by several groups after trans-

Table 1. Features of the *L. monocytogenes* virulence determinants[a]

Gene	Protein	ORF bp	mRNA[e]	AA[f]	MWcal[g]	MW[h]	pI[g]	Sig. Seq.[i]	PrfA Reg.[j]	Temp. Reg.[k]
prfA[b]	PrfA	705	0.9/2.1	235	27.1	27	7.3	-	+	+
plcA[b]	PI-PLC	951	1.1/2.1	317	36.3	34	10.1	+	+	+
hly[c]	LLO	1617	1.8	529	58.6	58		+	+	+
mpl[b]	Mpl	1533	1.6/5.7	510	57.4	67/35	6.6	+	+	+
actA[b]	ActA	1917	2.9/5.7	639	67.0	92/72		-	+	+
plcB[c]	PC-PLC	867	2.9/5.7	289		32/29		+	+	+
inlA[c]	Internalin	2400	2.9/5	800	86	95	4.4	+	+	+
inlB[c]	InlB	1890	5	630	71.1		10.1	+	+	+
iap[b]	p60	1452	1.5	484	50.7	60	10.0	+	-	
lmaA[b]	LmaA	510		170	18.0	21	4.2	-		
lmaB[b]	LmaB	372		124	14.4					
lmsod[b]	SOD[l]	606		202	22.6	24		-		
cat[d]	Catalase	1464		488	55.9	67		-		

[a]see text for references
[b]from strain *L. monocytogenes* EGD
[c]from strain *L. monocytogenes* LO28
[d]from *L. seeligeri*
[e]given in kb
[f]amino acids (including signal sequence)
[g]molecular weight (MW) in kD (including signal sequence) and isoelectric point (pI) as calculated from sequence data
[h]molecular weight in kD as calculated from SDS-PAGE
[i]presence of signal sequence
[j]PrfA regulated expression
[k]temperature-regulated expression
[l]SOD=Superoxide dismutase

poson mutagenesis using the conjugative transposons Tn1545 or Tn916 (15-17). The nonhemolytic transposon mutants always proved to be avirulent in the mouse model. Virulence is restored in hemolytic revertants, which have lost the transposon insertion or by the introduction of the cloned *hly* gene into a nonhemolytic *L. monocytogenes* transposon mutant (49). Despite the clear correlation between hemolysis and virulence, the level of hemolysin production *in vitro* is not directly proportional to the virulence of the producing strains in the mouse (50), suggesting that synthesis of listeriolysin under intracellular conditions is different from that observed under the extracellular growth conditions.

Listeriolysin is a secreted protein of 58 to 60 kD. It is sulfhydryl-activated and belongs to a family of pore-forming cytolysins for which streptolysin O is the prototype (51). These cytolysins are inhibited by low concentrations of cholesterol and oxygen and activated by reducing agents. Listeriolysin cross-reacts with anti-streptolysin O antibodies. Cholesterol is considered as a receptor for these cytolysins since this component inhibits pore formation and toxicity. The toxin monomers oligomerize in the membrane of target cells to form stable pores which can be visualised by electron microscopy (52). Listeriolysin has been purified and its toxicity, determined by intraperitoneal injection in the mouse, shows a LD_{50} of $1.7\mu g$ per mouse. Optimal activity is found at pH 5.5, a pH value which is lower than that determined for the other SH-activated cytolysins (53), a property which is in agreement with the function of listeriolysin in the acidified phagosome.

To determine the role of listeriolysin in virulence, wild-type and nonhemolytic mutants of *L. monocytogenes* were injected intravenously and intraperitoneally in mice and the fate of the listeriae was followed in liver and spleen. In contrast to the wild-type strain, the nonhemolytic mutants are eliminated from these organs within a few hours (15-17,49). The role of listeriolysin in the intracellular survival was determined using different mouse and human cell lines. In the human enterocyte-like cell line Caco-2 (19), the mouse 3T6 fibroblasts (20), and the mouse CL.7 fibroblasts (17), it was shown that nonhemolytic *L. monocytogenes* mutants are as invasive as the isogenic wild-type strains. This indicates that listeriolysin is not required for the uptake into nonprofessional phagocytes. The nonhemolytic mutants are, however, incapable of intracellular growth and survival within these host cells and also in mouse peritoneal macrophages (20), mouse bone-marrow-derived macrophages (17), and the mouse macrophage-like cell line J774 (17). Interestingly nonhemolytic mutants are able to replicate in the human epithelial cell line Henle 407 (17), suggesting that hemolysin is probably not essential for intracellular growth in all (human) cells. Electron microscopy of infected macrophages and epithelial cells reveals that the nonhemolytic *L. monocytogenes* mutants are unable to escape from the phagosome into the cytoplasm of the host cells (19,22). These results suggest that listeriolysin is involved in the lysis of the phagosomal membrane. Further evidence for listeriolysin being essential for lysis of the phagosomal membrane and the intracellular growth was obtained by infection of macrophages with a *Bacillus subtilis* strain expressing listeriolysin (54). The listeriolysin-producing *Bacillus* strain escapes from the phagosome into the cytoplasm and can even replicate in this compartment of the host cell, whereas the nonhemolytic *B. subtilis* parental strain stays in the phagosome.

The listeriolysin gene, *hly*, was cloned from strains of different serovars of *L. monocytogenes* and sequenced (55-57). The deduced amino acid sequence for listeriolysin yielded 529 amino acids including a N-terminal signal sequence of 25 amino acids. The sequence shows extended homologies with other SH-activated cytolysins. The highest homology is observed in the C-terminal part and includes a highly conserved undecapeptide. This sequence contains the unique cysteine which was thought to be essential for cytolytic activity. Site-directed mutagenesis revealed, however, that the cysteine is not essential for hemolytic activity. In contrast, a tryptophan residue, in close vicinity to the cysteine, appears to be required for both hemolytic activity and virulence (58).

Phosphatidylinositol-specific Phospholipase C (PI-PLC)

Phospholipase activity of *L. monocytogenes* cultures was first discovered as a zone of opacity surrounding colonies on egg yolk agar (59). *L. monocytogenes* transposon mutants lacking phospholipase activity were identified by the formation of small plaques on fibroblast cell lawns (60) and by the reduced hemolysis on blood-agar plates (61). One transposon insertion was mapped in an ORF located adjacent to the *hly* gene on the chromosome of *L. monocytogenes* (60,62). The gene was cloned (63-65) and shown to encode a protein of 34 kD with a typical transport signal sequence which exhibits high homology to several gram-positive phospholipases. The mature enzyme possesses phosphatidylinositol-specific phospholipase C (PI-PLC) activity. The enzyme, purified from culture supernatants of an overexpressing *L. monocytogenes* strain (66), is highly specific for phosphatidylinositol with no detectable activity on phosphatidylethanolamine, phosphatidylcholine, or phosphatidylserine. It also does not cleave phosphatidylinositol-4-phosphate or phosphatidylinositol-4,5-bisphosphate, but is active, albeit with low specific activity, on glycosylated phosphatidylinositol-anchored proteins (67). The broad pH optimum, ranging from pH 5.5 to 7.0, is consistent with its postulated function in the acidified phagocytic vacuole of infected cells.

Early studies on the role of PI-PLC in the intracellular life cycle of *L. monocytogenes* suffered from the polar effect which the transposon insertion in *plcA* exerted on the transcription of the *prfA* gene with which *plcA* forms an operon. The *prfA* gene encodes a transcriptional activator necessary for the expression of all genes in the virulence gene cluster (see below for details). With the use of in-frame deletions in the plcA gene it could be convincingly shown that PI-PLC is required for the efficient escape of *L. monocytogenes* from the phagosome of mouse bone marrow-derived macrophages. However, the mutation in *plcA* has only a slight effect on virulence (68). It is assumed that PI-PLC acts in concert with listeriolysin inside the acidified phagosomal vacuole of the host cell to mediate lysis of the vacuolar membrane. To further assess the role of PI-PLC, the *plcA* gene was cloned and expressed in *L. innocua*, which lacks the *prfA*-dependent virulence gene cluster and is, therefore, unable to escape from the host cell vacuole. The PI-PLC-expressing *L. innocua* strain cannot escape from the phagosome of J774 macrophages, but shows limited intracellular growth inside the vacuoles which appear to be structurally intact (69). These data suggest that PI-PLC alone is unable to open the phagosomal membrane but affects the vacuole in a way which alters its function but not its structure.

Phosphaditylcholine-specific phospholipase C (Lecithinase, PC-PLC)

Besides the highly specific PI-PLC, *L. monocytogenes* produces a second phospholipase C which hydrolyzes phosphatidylcholine (lecithin), and is thus a phosphatidylcholine-specific phospholipase C (PC-PLC) or lecithinase (70,71). A 32 kD protein was detected in the supernatant of a *L. monocytogenes* culture which shows phospholipase activity on egg yolk overlays (61). The protein was purified to homogeneity (71,72) and shown to be a zinc-dependent phospholipase C of 29 kD (in SDS-PAGE). The pH optimum of the enzyme is between pH 6 and 7 and its activity is stimulated by 0.5 M NaCl and 0.05 mM $ZnSO_4$. Besides phosphatidylcholine it also hydrolyzes phosphatidylethanolamine, phosphatidylserine and with lower efficiency sphingomyelin. Phosphatidylinositol is not accepted as substrate. The purified protein exhibits a weak hemolytic activity but is not toxic in mice (71).

The gene, *plcB*, encoding PC-PLC, is part of an operon which consists of *mpl, actA, plcB* and the three small ORFs X, Y and Z. The gene was cloned and sequenced (48). The deduced amino acid sequence yields a protein of 289 amino acids with a 25 amino acids N-terminal transport signal and a putative propeptide of 26 amino acids. The maturation of the 32 kD precursor of PC-PLC takes place after secretion and is obviously accomplished by the metalloprotease of *L. monocytogenes* (see below) (73,74). *PlcB* mutants constructed to analyze the function of PC-PLC in the intracellular life cycle of *L. monocytogenes* are fully invasive and replicate and move inside host cells similar to the wild type strain (73). On monolayers of 3T3 fibroblasts the *plcB* mutants form, however, only small plaques suggesting that the cell-to-cell spread is impaired in these mutants. Electron micrographs of *plcB* mutants inside mammalian cells (48) show the bacteria with actin tails being trapped in vacoules surrounded by a double membrane. This indicates that the *plcB* mutants are unable to lyse the double membrane of the vacuole which is formed when listeriae spread from cell to cell. Other factors like listeriolysin or PI-PLC may be also required for this lysis step. Whether lecithinase contributes also to lysis of the primary phagosome after uptake of the bacteria from the medium remains to be analyzed. PC-PLC is involved in virulence in mice since *plcB* mutants are impaired in growth in the liver and spleen of infected mice and the LD_{50} is two orders of magnitude higher than the wild-type (73).

Metalloprotease

The gene, *mpl*, located immediately downstram of the *hly* gene (75,76) encodes a zinc-dependent metalloprotease. It is the first gene of the "lecithinase operon" (48). The deduced amino acid sequence of this protease shows high homology to a number of metalloproteases from *Bacillus* species (75,76) and yields 510 amino acids with a typical N-terminal signal sequence. Similar to other metalloproteases, there is a putative cleavage site for proteolytic maturation resulting in a 35 kD mature protein (75,76). A 60 kD protein is detected with an antiserum raised against *B. stearothermophilus* thermolysin which probably represents the proform of the metalloprotease. The postulated 35 kD mature form of the protein is detectable only in small quantities in the supernatant of a *L. monocytogenes* culture (75).

The role of the metalloprotease in pathogenicity and intracellular replication of *L. monocytogenes* is still unclear. Transposon mutants in the *mpl* gene are reduced in virulence but grow normally inside mammalian cell lines (73). The reduced virulence was attributed to the lack of proteolytic processing of the 32 kD PC-PLC proform (74,77). However, mutants with in-frame deletions in *mpl*, also impaired in PC-PLC maturation and *actA* degradation (see below), are as virulent as the isogenic wild-type strain when injected intraperitoneally in the mouse (D. A. Portnoy, personal communication and 78).

ActA and Intracellular Movement

The initial reports on the intra- and intercellular movement of *L. monocytogenes* (22,23) mediated by actin polymerization were followed by a series of detailed analyses which showed that *L. monocytogenes* moves rapidly through the cytoplasm with the help of the formed actin tails with a speed of up to 1.5 μm/sec (79). The rate of actin assembly equals the rate of actin-based motility (80) and most likely, the actin polymerization provides the propulsive force for intracellular movement (81). The mechanism of actin nucleation and initiation of tail formation were investigated in detail (82-84). Since the cell biology of this intracellular movement has been reviewed recently (85) we will focus here on the molecular aspect of this process. *L. monocytogenes* mutants defective in the intracellular motility were obtained by transposon mutagenesis. These mutants have either lost the ability to initiate actin polymerization (60) or still induce actin polymerization but are unable to rearrange the actin filaments to actin tails (86). The gene downstream of *mpl*, called *actA*, was shown to be essential for actin polymerization. It codes for a proline-rich protein (ActA) of 639 amino acids. Its apparent molecular weight determined by SDS-PAGE is 92 kD (48,87). The ActA protein has a transport signal sequence (88) and represents a surface protein anchored to the bacterial membrane by a putative C-terminal membrane anchor (87,88). In its central part, the protein carries several repeats with homology to vinculin (87). Mutations in the *actA* gene resulted in the loss of virulence in mice (87), in the lack of intracellular actin polymerization around the bacteria and in the inability of intracellular movement (87,88). Inside the host cells, the *actA* mutant forms microcolonies which are located near the nucleus (87). The ActA protein seems to be distributed asymmetrically on the bacterial surface. After cell division, it is not present at the new bacterial pole but is concentrated at the old pole. Since ActA is present at the site of actin assembly, it is likely that ActA is involved either directly or indirectly in actin nucleation on the bacterial surface (89). The ActA protein is, however, not found within the actin tail (90).

The possible role of the ActA protein as nucleator for actin polymerization was also demonstrated in mammalian cells transfected with actA (91). Expression of the complete ActA protein (including the membrane anchor) results in targeting of the protein to mitochondria, which subsequently recruits actin to these organelles, suggesting that ActA alone is sufficient to polymerize actin. ActA protein without the internal proline-rich repeats is also localized in the mitochondrial membrane but does not induce actin polymerization, suggesting that this domain is essential for actin nucleation (91). This assumption is supported by the finding that microinjection of a peptide representing one internal repeat unit of ActA inhibits actin-mediated motility of *L. monocytogenes* (92). The injected peptide probably binds profilin which, as recently shown in a cell-free system (93), is directly involved in actin-mediated motility of *L. monocytogenes*. Similar to ActA, profilin is localized to the site of actin filament assembly which suggests that profilin binds directly to ActA. Profilin appears to promote rapid filament elongation on the bacterial surface (reviewed in 94,95). Other host cell proteins like α-actinin, tropomyosin, vinculin and talin seem to be also associated with the actin tails (79,96). While tropomyosin may protect actin filaments from cleavage by severing proteins, α-actinin obviously stabilizes the actin filaments by crosslinking. Vinculin and talin, proteins located usually in the attachment plaques, probably enable the actin tails to bind to the membrane of the pseudopodia which are formed during cell-to-cell spread.

As recently shown by SDS-PAGE, the 92 kD ActA protein is cleaved by the listerial metalloprotease (see above) resulting in a major 72 kD degradation product and, dependent on the strains, additional smaller degradation products (41,90). Probably more important is the phosphorylation of the ActA protein inside the host cells which yields three distinct forms of this protein with slightly different sizes in SDS-PAGE (97). The significance of these modifications which may modulate the intracellular activity of ActA is presently unknown.

L. ivanovii, pathogenic to animals but not to humans, is also invasive to most mammalian cells tested and the intracellular life cycle of these bacteria is similar to that of *L. monocytogenes*. Inside the host cells, *L. ivanovii* polymerizes F-actin as does *L. monocytogenes* albeit at a reduced efficiency. Actin tail formation and cell-to-cell spread are also observed (98). However, only a weak signal was obtained when chromosomal DNA of *L. ivanovii* was hybridized with *actA* gene probes of *L. monocytogenes* (47). The recent cloning and sequencing of the *actA*-related gene from *L. ivanovii* (99) shows surprisingly little sequence homology with the *actA* gene of *L. monocytogenes*. On the protein level there is high homology at the N- and C-termini and in the proline-rich repeat sequences between the two proteins both of which are active in actin polymerization (99). The ActA-related protein of *L. ivanovii* is larger in size (1021 amino acids) than ActA of *L. monocytogenes* (639 amino acids) due to two insertions which are missing in ActA of *L. monocytogenes* (99).

PRFA AND THE REGULATION OF VIRULENCE GENE EXPRESSION IN *L. MONOCYTOGENES*

Spontaneously-occurring nonhemolytic mutants of *L. monocytogenes* which carried deletions in a region upstream of the *hly* gene were the first indication for the regulation of virulence genes in *L. monocytogenes* by a *trans*-acting factor (100,101). Cloning and sequencing of the locus affected by the deletion led to the identification of the *prfA* (*p*ositive *r*egulating *f*actor A) gene. Its product, PrfA, a cytoplasmic protein of 27 kD (102,103), regulates all virulence genes of the above-described gene cluster. The *prfA* deletion mutants are complemented *in trans* by the introduction of the cloned *prfA* gene to yield again a wild-type phenotype (102). Site-specific mutations or transposon insertions in the *prfA* promotor or in the *prfA* coding region block the transcription of the entire gene cluster, i.e. *plcA, hly, mpl, actA,* and *plcB* (87,103,104) indicating that the *prfA* gene encodes a transcriptional activator required for the expression of the *L. monocytogenes* virulence gene cluster. Additional evidence for this presumptive role of PrfA was provided by the transcriptional activation of the cloned *hly* gene by PrfA in *B. subtilis* (105).

A *prfA*-like gene with high sequence similarity to *prfA* from *L. monocytogenes* is also present in the closely related species *L. ivanovii* where it also controls a set of virulence genes similar to those of *L. monocytogenes* (46). The amino acid sequence of PrfA suggests that it is a member of the Crp/Fnr family of transcriptional activators which have been identified so far mainly in gram-negative bacteria (106). Like all members of this family, PrfA contains a conserved helix-turn-helix motif in its C-terminal part. In addition, adjacent to this motif PrfA carries a sequence with features of a leucine zipper (107) and a second helix-turn-helix motif at its N-terminus. The PrfA protein seems to recognize and bind to a 14 bp palindromic sequence which is located about 40 bp upstream of the transcriptional start site in front of each PrfA-regulated gene or operon (62).

The transcriptional organization of the virulence gene cluster is complex. Three transcripts of the *prfA* gene were identified: a long (2.1 kb) transcript, which is cotranscribed with the *plcA* gene and autoregulated by PrfA, and two shorter ones (0.8 and 0.9 kb) transcribed from two distinct promotors located in front of the *prfA* gene (108). Transcription of *prfA* from the latter two promoters seems to be negatively regulated by PrfA. Synthesis of the bicistronic *plcA-prfA* transcript depends on the activity of at least one of the *prfA* promotors (109), and is necessary for full expression of the PrfA-dependent genes (68). Translation of the monocistronic *prfA* transcript(s) appears to be very inefficient since PrfA-dependent virulence genes are poorly expressed even in the presence of large amounts of this transcript when the synthesis of the PrfA-regulated bicistronic transcript is blocked (110). The listeriolysin gene, *hly*, is the only gene transcribed in a monocistronic mRNA from two PrfA-dependent promotors, P1 and P2, located in the intragenic region between *hly* and *plcA* (62). A third *hly* promotor P3 downstream from P1 and P2, recently identified and shown to be PrfA-independent, results in low level transcription of the *hly* gene (111). The other three genes of the virulence gene cluster are transcribed from at least two PrfA-regulated promotors: one, located in front of the *mpl* gene, yields a 5,4 to 5,7 kb transcript comprising *mpl, actA,* and *plcB* and an additional mRNA of 1,8 kb comprising *mpl* alone (112). A second promotor, located directly in front of the actA gene, leads to a bicistronic transcript comprising *actA* and *plcB* (112).

It has been suggested that in an initial step, transcription of *prfA via* the *prfA* promotors results in the synthesis of a limited amount of PrfA sufficient to activate the high-affinity PrfA-dependent *hly* and *plcA* promotors. This in turn results in the synthesis of *plcA-prfA* transcripts which leads to enhanced PrfA synthesis. The higher cellular level of PrfA activates the *mpl* and *actA* promotors which seem to have a lower affinity to PrfA due to base mismatches in their palindromic "PrfA-boxes." The high cellular level of PrfA also leads to down-regulation of the monocistronic prfA transcripts (68,108-110).

There is, however, experimental evidence that regulation of the virulence genes of *L. monocytogenes* is more complex and may involve also other more global regulation systems. Temperature influences the expression of the virulence genes in two ways: synthesis of listeriolysin (and possibly other PrfA-controlled proteins) is greatly enhanced under heat-shock conditions, especially in *L. monocytogenes* strains which express only low amounts of this cytolysin in culture media at 37°C under normal culture conditions (113,114).

At lower temperature (20°C) which allows efficient growth of *L. monocytogenes*, transcription of the virulence genes is turned off although synthesis of the monocistronic *prfA* transcript(s) is still detectable. Shift to 37°C results in the synthesis of the transcripts of all PrfA-controlled virulence genes of *L. monocytogenes* (115). In Minimal Essential Medium (MEM) where only limited growth of *L. monocytogenes* is observed, the PrfA-dependent proteins encoded by the virulence gene cluster are preferentially synthesized. In addition to the known proteins of the virulence gene cluster, at least five PrfA-dependent cell surface proteins of unknown function are also induced under these conditions suggesting that PrfA is not only responsible for the regulation of the virulence gene cluster (116). The pattern of induction of the known PrfA-regulated transcripts in MEM indicates that the PrfA-controlled genes are differentially regulated in the presence of apparently constant levels of PrfA. In addition, the stability of these transcripts varies greatly and is unusually high for some of these transcripts. This suggests that in addition to PrfA, other transcription factors and/or modification(s) of the PrfA protein may be involved in the differential expression of the virulence genes (112).

Two sugars, glucose and cellubiose, were shown to repress expression of virulence genes. The reduced synthesis of listeriolysin in the presence of glucose is most likely caused by the acidification of the culture medium due to the glucose metabolism (117,118). A more direct repression of listeriolysin and PI-PLC synthesis by the disaccharide cellobiose seems to occur at the transcriptional level (119).

The only known gene which does not belong to the virulence gene cluster but seems to be also regulated by PrfA is the above-described gene *inlA* encoding internalin (120). However, substantial transcription of this gene still occurs in a *prfA* mutant (110) and only a rather incomplete palindromic "PrfA-box" has been identified upstream of the mapped internalin promotor (120).

OTHER *L. MONOCYTOGENES* GENES POSSIBLY INVOLVED IN VIRULENCE

Superoxide Dismutase and Catalase

Superoxide dismutase and catalase act in concert to detoxify potentially harmful superoxide radicals. Superoxide radicals generated by the oxidative burst in a phagocytic cell are converted into hydrogen peroxide by the action of superoxide dismutase. Hydrogen peroxide is then cleaved by catalase into water and molecular oxygen (121). Since macrophages react upon ingestion of bacteria with greatly enhanced production of reactive oxygen intermediates, bacterial catalases and superoxide dismutases were long suspected to be important virulence factors of intracellular bacteria. However, pathogens have also evolved other strategies to prevent oxygen-dependent killing by phagocytes (122). Expression of superoxide dismutase was analyzed in *L. monocytogenes* and shown to be dependent on different growth conditions (123,124), but no correlation with virulence was found (125). The genes for superoxide dismutase from *L. ivanovii* (126) and *L. monocytogenes* (127) were recently cloned in *E. coli*. The nucleotide sequence of the *L. monocytogenes* gene, *lmsod*, reveals an ORF coding for a protein of 202 amino acids with high homology to the manganese-containing superoxide dismutase from other organisms. The gene, which is conserved in all other *Listeria* species, was mutagenized and virulence of the mutant was tested in mice. Again no difference in the LD_{50} or in the survival of the bacteria in the spleen and liver of the infected animals is observed between the *lmsod* mutant and the wild-type strain (128).

Catalase activity was detected in all *L. monocytogenes* isolates tested albeit at different levels (129,130). Expression of catalase is modulated by temperature and sodium chloride concentration (131), but again there is no correlation between virulence and catalase activity (125) and catalase mutants obtained by transposon mutagenesis show wild-type virulence in infected mice (132). The isolation of a catalase-negative strain of *L. monocytogenes* from a human meningitis patient further supports the view that catalase does not play a critical role in the virulence of *L. monocytogenes* (133). On the other hand, catalase-negative *L. monocytogenes* mutants are killed by mouse resident macrophages already at low serum concentrations, whereas killing of the wild-type bacteria requires high serum concentrations suggesting that resistance to fully activated macrophages is partially mediated by catalase activity (134). The catalase gene (*cat*) of *L. seeligeri* was the first to be cloned and characterized (135). Recently, the corresponding gene of *L. monocytogenes* has also been isolated (128). The construction of defined catalase mutants and of catalase/superoxide dismutase double mutants are under way and should further elucidate the role of both enzymes in *L. monocytogenes* virulence.

LmaBA

The *lmaA* (*Listeria monocytogenes antigen A*) gene was isolated by screening a *L. monocytogenes* expression library with an antiserum raised against soluble antigens of the same strain (136). The product of the *lmaA* gene codes for a 21 kD protein which is able to induce a specific delayed-type hypersensitivity reaction in *L. monocytogenes*-immune mice. The amino acid sequence of LmaA reveals a strongly hydrophobic protein of a predicted molecular mass of 18 kD. A putative transmembrane region located at the N-terminal end of the protein suggests that LmaA may represent an integral membrane protein. The *lmaA* gene is part of an operon and is preceded by the *lmaB* gene encoding a 14 kD protein. Hybridization with *lmaA* gene probes is only observed with DNA of the pathogenic species *L. monocytogenes* and *L. ivanovii*. Since *lmaA* and *lmaB* mutants are not yet available, the role of *lmaBA* in virulence of *L. monocytogenes* is currently unknown.

RESPONSE TO *L. MONOCYTOGENES* INFECTION BY MAMMALIAN HOST CELLS

Since the pioneering work of G. B. Mackaness (9), a large number of publications on the concept of cell-mediated immunity against *L. monocytogenes* infections has appeared (reviewed in 137). The importance of cytokines, especially interferon-γ (IFN-γ) and tumor necrosis factor (TNF), in nonspecific and in T-cell mediated immunity during experimental listeriosis is well documented and has been reviewed in detail (138,139). In this chapter, we therefore will not discuss these immunological aspects but rather concentrate on what is known on the response of mammalian host cells to *L. monocytogenes* infection on the molecular and genetic level. We will focus on host genes involved in signal transduction and stress responses which are altered in their expression during infection by *L. monocytogenes*.

The first studies on host response performed with primary mammalian cells and established cell lines infected with *L. monocytogenes* assayed mainly cytokine activities in the culture supernatants of the infected cells. These studies could not discriminate between the release of preformed proteins and *de novo* synthesized proteins. More recent studies determined expression of the affected host genes more specifically by measuring expression on the transcriptional and translational level in an at least semiquantitative way.

To summarize the early reports: primary mouse embryonic fibroblasts infected with *L. monocytogenes* release IFN α/β into the culture medium (18). Interleukin-1 (IL-1) production by mouse peritoneal macrophages is observed with viable, virulent *L. monocytogenes* strains, but not with killed or avirulent *Listeria*. Northern blot analysis further shows that the increase in IL-1 secretion correlates with an increase of IL-1α specific mRNA after infection of macrophages with *L. monocytogenes* (140). Secretion of significant amounts of TNF and interleukin-6 (IL-6) is detected in alveolar macrophages after infection with viable *L. monocytogenes* (141); IL-6 is induced in embryonic fibroblasts even by heat-killed *L. monocytogenes* albeit to a lesser extent than by infection with viable bacteria. TNF is secreted only after treatment with killed but not with viable *L. monocytogenes* (141). Release of the proinflammatory cytokines, IL-1ß, IL-6 and TNF-α, takes place in human polymorphonuclear granulocytes and in the human epithelial cell line HEp-2 after *L. monocytogenes* infection (25). The granulocytes secrete all three cytokines in response to the infection, whereas the epithelial cells secrete IL-6 and low amounts of TNF-α but no IL-1ß. The transcription of the respective genes is also induced in these host cells. Mouse peritoneal macrophages also secrete IL-6 after L.monocytogenes infection (142). Hemolytic *L. monocytogenes* strains are less efficient in IL-6 induction than nonhemolytic mutants, probably due to cell damage caused by the high level of secreted listeriolysin. Inhibition of maturation of IL-1ß by *L. monocytogenes* in mouse peritoneal macrophages was recently proposed as a novel mechanism of how *L. monocytogenes* may escape the host cell response since infection of the macrophages by *L. monocytogenes* results in the intracellular accumulation of unprocessed IL-1ß precursor (143).

Using the mouse macrophage-like cell line $P388D_1$ and different mutants of *L. monocytogenes* to analyze cytokine induction after infection we could show (144) that viable *L. monocytogenes* rapidly induce IL-1α, IL-1ß, IL-6 and TNF-α mRNAs in these host cells, whereas killed *L. monocytogenes* induce IL-1ß mRNA only. Semiquantitative determination of these mRNAs by competitive PCR shows an increase of IL-6 transcription of more than 100-fold and about 25-fold for transcription of IL-1α and TNF-α. Nonhemolytic mutants which cannot escape into the cytoplasm and which do not multiply are unable to induce IL-1α, IL-6 and TNF-α but still induce IL-1ß mRNA. The amount of cytokines in the culture supernatant of the infected macrophages correlates well with the level of induced mRNAs, with the exception of IL-1α of which only low levels are found in the supernatant despite a significant induction of the IL-1α/ß mRNA. Mouse bone marrow--derived macrophages infected with *L. monocytogenes* also induce the proinflammatory cytokines (144). However, in these cells a nonhemolytic *L. monocytogenes* mutant induces the same types and amount of cytokines as the wild-type strain indicating that intracellular growth is not necessary for the transcriptional induction of these cytokines in bone marrow-derived macrophages (145). IL-6 is produced in the bone marrow-derived macrophages independently of IL-1 and TNF (145). The immunomodulating cytokines interleukin-10, interleukin-12 and the IL-1 receptor antagonist are also induced in bone marrow-derived macrophages after *L. monocytogenes* infection (145). Large amounts of the chemokine interleukin-8 (IL-8) are secreted by polarized human epithelial cell lines in response to *L. monocytogenes* infection (24). The induction of IL-8 secretion is caused by an increase of IL-8-specific mRNA synthesis and is not mediated by secreted TNF-α. Interestingly secretion of this cytokine occurs preferentially at the basolateral side whereas *L. monocytogenes* is taken up by the apical side. IL-8 may be an initial signal for the acute inflammatory response after bacterial invasion of mucosal surfaces. The transcription of cytokine receptor mRNAs in bone marrow-derived macrophages infected with *L. monocytogenes* seems to be regulated inversibly to the respective cytokines. TNF-receptor type I, IL-6-receptor and IFN-γ-receptor are transcriptionally downregulated shortly after infection of the macrophages by *L. monocytogenes*, whereas IL-1 receptor type II mRNA was unaffected (146).

Transcription of certain stress genes in mammalian host cells has been also analyzed in response to *L. monocytogenes* infection (147). Immediately after infection, the macrophage-like cell line J774 responds with enhanced synthesis of heat shock protein 70 (HSP70) mRNA and HSP70 protein. Phagocytosis of the bacteria appears to be necessary for this induction since cytochalasin D treatment abolishes the induction. The amount of HSP60 and HSP90 mRNAs remains relatively unchanged during infection. Another stress response protein, the protein tyrosine phosphatase (PTP), which seems to participate in signal transduction pathways, is significantly induced 2 to 4 hours post-infection. Transcription of both HSP70 and PTP mRNAs is much less induced when the macrophages are infected with nonhemolytic mutants of *L. monocytogenes* suggesting that escape of the bacteria into the cytoplasm is required for the induced transcription of the stress genes.

With the use of a modification of the previously described procedure of differential PCR (148), induction of host genes after infection by *L. monocytogenes* was determined in a more general way. This method allows the isolation of cDNAs representing fragments of genes which are differently expressed in infected and uninfected macrophages (149,150). By this procedure we obtained several cDNA clones derived from macrophage genes that are either transcriptionally activated or down-regulated after infection by *L. monocytogenes*. One of the cloned cDNAs was shown to be part of the gene for the above-mentioned protein tyrosine phosphatase (PTP). This result shows that differential PCR is a suitable method for the identification of differentially regulated genes in host cells infected with *L. monocytogenes*.

The enhanced expression of PTP in macrophages at a later time of infection is also an indication that signal transduction pathways may be modulated by a *L. monocytogenes* infection (147). Phosphorylation of MAP-kinase is observed after adherence of *L. monocytogenes* to epithelial Caco-2 cells and inhibition of phosphorylation also inhibits invasion (42). MAP-kinase is part of signal transduction pathways which link extracellular signals to gene expression. Differential gene expression is mediated by transcription factors like NF-κB (151). This transcription factor in particular is involved in transcription of many genes including most cytokine genes.

We have studied NF-κB DNA binding activity in response to *L. monocytogenes* infection in the macrophage-like cell line P388D$_1$ (152). A rapid enhancement of the NF-κB DNA binding activity is observed in these host cells within 10 to 20 min after addition of the bacteria. The NF-κB activation is not dependent on invasion since treatment of the host cells by cytochalasin B does not inhibit this activation. NF-κB activation is also not mediated by autocrine stimulation of the infected cells by IL-1 or TNF. Activation of NF-κB is also observed in Caco-2 cells after infection with *L. monocytogenes* but with somewhat slower kinetics than those observed in the macrophages (153). The DNA binding activity of two other transcription factors, AP-1 and NF-IL6, is not changed after infection of the two cell lines by *L. monocytogenes*, indicating that the observed activation of NF-κB by *L. monocytogenes* is a specific event (152).

Transfection of mammalian cells with virulence genes was recently used to study host cell response to individual virulence factors of *L. monocytogenes*. This approach allows the investigation of the host cell response to a single intracellularly expressed virulence protein without the background of all other bacterial virulence factors (154). As already described above, expression of the product of the *actA* gene results in morphological changes of the transfected cell and targeting of actin filaments to the ActA-coated mitochondria (see above) (91). Listeriolysin was also expressed in mammalian cells (155). This expression results in a dramatic change of cell morphology and the formation of foci consisting of tightly connected cells. The cytoskeleton of the *hly*-transfected mammalian cells is also altered and the cells exhibit a significantly enhanced proliferation rate. Due to the presence of its transport signal sequence, we assume that listeriolysin is transported *via* the golgi system to the cytoplasmic membrane of the host cell where it may trigger signal transduction pathway(s) which may ultimately lead to the observed enhanced cell proliferation.

CONCLUSIONS AND OPEN QUESTIONS

Considerable progress has been achieved in recent years concerning our knowledge of the genes determining virulence of *L. monocytogenes* and of the role which the virulence gene products play in the intracellular multiplication and survival. However, many problems concerning the virulence of *L. monocytogenes* still remain unsolved.

For instance, *L. ivanovii*, a species apathogenic for humans (1), behaves similarly to *L. monocytogenes* with respect to its intracellular life cycle in cell cultures (98). In addition, genes homologous to the known virulence genes of *L. monocytogenes* were also detected in this species (47,99,156) and the complete PrfA-regulated gene cluster identified in *L. monocytogenes* is apparently also present in *L. ivanovii* (47). However, *L. ivanovii* is only virulent for animals but not for humans and infection of the experimental mouse model by *L. ivanovii* appears to have a different outcome than that by *L. monocytogenes* (157). The LD_{50} of *L. ivanovii* is also higher than that of *L. monocytogenes* in the mouse model. What is the molecular explanation for this obvious difference in the pathogenic potential of these two *Listeria* species? Is it a different mechanism in regulation of the known virulence genes inside the infected cells or differences in the specific activity of the known virulence gene products? Are there yet unknown virulence factors in *L. monocytogenes* which are absent in *L. ivanovii*? Additional PrfA-dependent proteins are indeed observed in *L. monocytogenes*, some of which are positively regulated by PrfA while others are apparently negatively controlled by this transcriptional regulator (116).

Other putative virulence factors previously described but not yet studied in molecular terms are the monocytosis-producing activity (MPA) and the immunosuppression factor both of which were detected in *L. monocytogenes* isolates (158). There is a definite need for further characterizing these obviously interesting products and their genes.

Another unsolved but very important problem concerns the expression of the virulence genes inside the infected mammalian host cells. The expression pattern of the PrfA-dependent genes is already complex under *in vitro* growth conditions and very little is known of how PrfA controls these genes in the host cells. There is preliminary evidence for the preferential synthesis of listeriolysin inside the phagosome and of ActA inside the cytoplasm (112,159). But the precise timing in the expression of the virulence genes and the cellular signals which may control their intracellular expression are entirely unknown.

The growing number of publications on host cell response to a *L. monocytogenes* infection shows that this field, long ignored, is now becoming a topic of major interest as a suitable model system for studying the molecular basis of pathogen-host cell interactions. Besides the genes for cytokines and host stress proteins, the search now concentrates on new host genes which are differentially expressed during the various steps of a *L. monocytogenes* infection. The characterization of such host genes may help in understanding the strategies which the two partners are using in their intimate and sometimes very severe cross talks.

These studies may also show how signals are transmitted from pathogens to their host cells and how the host cells modulate gene expression of the pathogens. These central questions of pathogenesis by facultative intracellular bacteria are presently asked not only for *L. monocytogenes* but also for several gram-negative bacteria, such as *Shigella* and *Salmonella* (160,161) and may lead to exciting answers in the near future.

ACKNOWLEDGMENTS

We thank B. Blaha, A. Demuth and R. Gross for carefully and critically reading this manuscript and all members of our laboratory for allowing us to quote their unpublished results. We apologize to all those who contributed to our current knowledge on the infection biology of *L. monocytogenes* but were not mentioned in this article.

This work was supported by the Deutsche Forschungsgemeinschaft through the grant SFB 165-B4.

REFERENCES

1 Seeliger, H. P. R. and Jones, D. (1986) in Bergey's Manual of Systematic Bacteriology, Vol. 2, (Sneath, P.H.A., Mair, N.S., Sharpe, M.E. and Holt, J.G. eds.) pp. 1235-1245, The Williams and Wilkins Co., Baltimore, MD.

2 Jones, D. (1992) Abstr. 11th Symp. Probl. Listeriosis, pp. 7-8.

3 Gray, M.L. and Killinger, A.H. (1966) Bacteriol. Rev. 30, 309-382.

4 Gellin, B.G. and Broome, C.V. (1989) J. Amer. Med. Assoc. 261, 1311-1320.

5 Schuchat, A., Swaminathan, B. and Broome, C.V. (1991) Clin. Microbiol. Rev. 4, 169-183.

6 Flemming, D.W., Cochi, S.L., MacDonald, K.L., Brondum, J., Hayes, P.S., Plikaytis, B.D., Holmes, M.B., Audurier, A., Broome, C.V. and Reingold, A.L. (1985) N. Engl. J. Med. 312, 404-407.

7 Schlech, W.F., Lavigne, P.M., Bortolussi, R.A., Allen, A.C., Haldane, E.V., Wort, A.J., Hightower, A.W., Johnson, S.E., King, S.H., Nicholls, E.S. and Broome, C.V. (1983) N. Engl. J. Med. 308, 203-206.
8 Farber, J.M. and Peterkin, P.I. (1991) Microbiol. Rev. 55, 476-511.
9 Mackaness, G.B. (1962) J. Exp. Med. 116, 381-406.
10 Hahn, H. and Kaufmann, S.H.E. (1981) Rev. Infect. Dis. 3, 1221-1250.
11 Mitsuyama, M., Takeya, K., Nomoto, K. and Shimotori, S. (1978) J. Chem. Microbiol. 106, 165-171.
12 Audurier, A., Pardon, P., Marly, J. and Lautier, F. (1980) Ann. Microbiol. (Inst. Pasteur) 131B, 47-57.
13 Schlech, W.F. and Chase, D.P. (1985) Abstr. 9th Symp. Probl. Listeriosis p. 21.
14 Beuscher, H.U., Kuhn, M. and Goebel, W. (unpublished data).
15 Gaillard, J.L., Berche, P. and Sansonetti, P.J. (1986) Infect. Immun. 52, 50-55.
16 Kathariou, S., Metz, P., Hof, H. and Goebel, W. (1987) J. Bacteriol. 169, 1291-1297.
17 Portnoy, D.A., Jacks, P.S. and Hinrichs, D.J. (1988) J. Exp. Med. 167, 1459-1471.
18 Havell, E.A. (1986) Infect. Immun. 54, 787-792.
19 Gaillard, J.L., Berche, P., Mounier, J., Richard, S. and Sansonetti, P.J. (1987) Infect. Immun. 55, 2822-2829.
20 Kuhn, M., Kathariou, S. and Goebel, W. (1988) Infect. Immun. 56, 79-82.
21 Kuhn, M. and Goebel, W. (1989) Infect. Immun. 57, 55-61.
22 Tilney, L.G. and Portnoy, D.A. (1989) J. Cell. Biol. 109, 1597-1608.
23 Mounier, J., Ryter, A., Coquis-Rondon, M. and Sansonetti, P.J. (1990) Infect. Immun. 58, 1048-1058.
24 Eckmann, L., Kagnoff, M.F. and Fierer, J. (1993) Infect. Immun. 61, 4569-4574.
25 Arnold, R., Scheffer, J., König, B. and König, W. (1993) Infect. Immun. 61, 2545-2552.
26 Wood, S., Maroushek, N. and Czuprynski, C.J. (1993) Infect. Immun. 61, 3068-3072.
27 Ebel, F., Sibelius, U., Grimminger, F., Wehland, J. and Chakraborty, T. (1994) BioEngineering 10(3), 69.
28 Sokolovic, Z. and Goebel, W. (unpublished data).
29 Ly, T. M. C. and Müller, H.E. (1990) J. Med. Microbiol. 33, 51-54.
30 Bubert, A., Kuhn, M., Goebel, W. and Köhler, S. (1992) J. Bacteriol. 174, 8166-8171.
31 Hess, J., Gentschev, I., Szalay, G., Ladel, C., Goebel, W. and Kaufmann, S.H.E. (unpublished data).
32 Kuhn, M. and W. Goebel. (unpublished data).
33 Ruhland, G.J., Hellwig, M., Wanner, G. and Fiedler, F. (1993) J. Gen. Microbiol. 139, 609-616.
34 Wuenscher, M.D., Köhler, S., Bubert, A., Gerike, U. and Goebel, W. (1993) J. Bacteriol. 175, 3491-3501.
35 Köhler, S., Leimeister-Wächter, M., Chakraborty, T., Lottspeich, F. and Goebel, W. (1990) Infect. Immun. 58, 1943-1950.
36 Köhler, S., Bubert, A., Vogel, M. and Goebel, W. (1991) J. Bacteriol. 173, 4668-4674.
37 Gaillard, J.L., Berche, P., Frehel, C., Gouin, E. and Cossart, P. (1991) Cell 65, 1127-1141.
38 Dramsi, S., Dehoux, P. and Cossart, P. (1993) Mol. Microbiol. 9, 1119-1122.
39 Falkow, S. (1991) Cell 65, 1099-1102.

40 Karunasagar, I., Senghaas, B., Krohne, G. and Goebel, W. (1994) Infect. Immun. 62, 3554-3558.
41 Goebel, W., Leimeister-Wächter, M., Kuhn, M., Domann, E., Chakraborty, T., Köhler, S., Bubert, A., Wuenscher, M. and Sokolovic, Z. (1993) Zbl. Bakt. 278, 334-347.
42 Tang, P., Rosenshine, I. and Finlay, B.B. (1994) Mol. Biol. Cell 5, 455-464.
43 Davies, W.A. (1983) J. Reticuloendothelial Soc. 34, 131-141.
44 De Chastellier, C. and Berche, P. (1994) Infect. Immun. 62, 543-553.
45 Raybourne, R.B. and Bunning, V.K. (1994) Infect. Immun. 62, 665-672.
46 Lampidis, R., Gross, R., Sokolovic, Z., Goebel, W. and Kreft, J. (1994) Mol. Microbiol. 13, 141-151.
47 Gouin, E., Mengaud, J. and Cossart, P. (1994) Infect. Immun. 62, 3550-3553.
48 Vazquez- Boland, J.A., Kocks, C., Dramsi, S., Ohayon, H., Geoffroy, C., Mengaud, J. and Cossart, P. (1992) Infect. Immun. 60, 219-230.
49 Cossart, P., Vicente, M.F., Mengaud, J., Baquero, F., Perez-Diaz, J.C. and Berche, P. (1989) Infect. Immun. 57, 3629-3636.
50 Kathariou, S., Rocourt, J., Hof, H. and Goebel, W. (1988) Infect. Immun. 56, 534-536.
51 Smyth, C.J. and Duncan, J.L. (1978) in Bacterial Toxins and Cell Membranes (Jelaszewics, J. and Wadström, T., eds.) pp. 129-183, Academic Press Inc., New York, NY.
52 Parrisius, J., Bhakdi, S., Roth, M., Tranum-Jensen, J., Goebel, W. and Seeliger, H. P.R. (1986) Infect. Immun. 51, 314-319.
53 Geoffroy, C., Gaillard, J.L., Alouf, J.E. and Berche, P. (1987) Infect. Immun. 55, 1641-1646.
54 Bielecki, J., Youngman, P., Connelly, P. and Portnoy, D.A. (1990) Nature 345, 175-176.
55 Mengaud, J., Chenevert, J., Geoffroy, C., Gaillard, J.L. and Cossart, P. (1987) Infect. Immun. 55, 3225-3227.
56 Mengaud, J., Vicente, M.F., Chenevert, J., Pereira, J.M., Geoffroy, C., Gicquel--Sanzey, B., Baquero, F., Perez-Diaz, J.C. and Cossart, P. (1988) Infect. Immun. 56, 766-772.
57 Domann, E. and Chakraborty, T. (1989) Nucl. Acids Res. 17, 6406.
58 Michel, E., Reich, K.A., Favier, R., Berche, P. and Cossart, P. (1990) Mol. Microbiol. 4, 2167-2178.
59 Fuzi, M. and Pillis, I. (1962) Nature 196, 195.
60 Sun, A. N., Camilli, A. and Portnoy, D.A. (1990) Infect. Immun. 58, 3770-3778.
61 Kathariou, S., Pine L., George, V., Carlone, G.M. and Holloway, B.P. (1990) Infect. Immun. 58, 3988-3995.
62 Mengaud, J., Vicente, M.F. and Cossart, P. (1989) Infect. Immun. 57, 3695-3701.
63 Camilli, A., Goldfine, H. and Portnoy, D.A. (1991) J. Exp. Med. 173, 751-754.
64 Leimeister-Wächter, M., Domann, E. and Chakraborty, T. (1991) Mol. Microbiol. 5, 361-366.
65 Mengaud, J., Braun-Breton, C. and Cossart, P. (1991) Mol. Microbiol. 5, 367-372.
66 Goldfine, H. and Knob, C. (1992) Infect. Immun. 60, 4059-4067.
67 Gandhi, A.J., Perussia, B. and Goldfine, H. (1993) J. Bacteriol. 175, 8014-8017.
68 Camilli, A., Tilney, L.G. and Portnoy, D.A. (1993) Mol. Microbiol. 8, 143-157.
69 Schwan, W.R., Demuth, A., Kuhn, M. and Goebel, W. (1994) Infect. Immun. 62, 4795-4803.
70 Leighton, I., Threlfall, D.R. and Oakley, C.L. (1975) in Problems of Listeriosis, Vol. 6. (Woodbine, M. ed.) pp. 239-241, Leicester University Press, Leicester.

71 Geoffroy, C., Raveneau, J., Beretti, J.L., Lecroisey, A., Vazquez-Boland, J. A., Alouf, J.E. and Berche, P. (1991) Infect. Immun. 59, 2382-2388.
72 Goldfine, H., Johnston, N.C. and Knob, C. (1993) J. Bacteriol. 175, 4298-4306.
73 Raveneau, J., Geoffroy, C., Beretti, J.L., Gaillard, J.L., Alouf, J.E. and Berche, P. (1992) Infect. Immun. 60, 916-921.
74 Poyart, C., Abachin, E., Razafimanantsoa, I. and Berche, P. (1993). Infect. Immun. 61, 1576-1580.
75 Domann, E., Leimeister-Wächter, M., Goebel, W. and Chakraborty, T. (1991) Infect. Immun. 59, 65-72.
76 Mengaud, J., Geoffroy, C. and Cossart, P. (1991) Infect. Immun. 59, 1043-1049.
77 Sokolovic, Z., Riedel, J., Wuenscher, M. and Goebel, W. (1993) Mol. Microbiol. 8, 219-227.
78 Böckmann, R. and Goebel, W. (unpublished data).
79 Dabiri, G.A., Sanger, J.M., Portnoy, D.A. and Southwick, F.S. (1990) Proc. Nat. Acad. Sci. U.S.A. 87, 6068-6072.
80 Theriot, J.A., Mitchison, T.J., Tilney, L.G. and Portnoy, D.A. (1992) Nature 357, 257-260.
81 Sanger, J.M., Sanger, J.W. and Southwick, F.S. (1992) Infect. Immun. 60, 3609-3619.
82 Tilney, L.G., Connelly, P.S. and Portnoy, D.A. (1990) J. Cell Biol. 111, 2979-2988.
83 Tilney, L.G., DeRosier, D.J. and Tilney, M.S. (1992) J. Cell Biol. 118, 71-81.
84 Tilney, L.G., DeRosier, D.J., Weber, A. and Tilney, M. S. (1992) J. Cell Biol. 118, 83-93.
85 Tilney, L.G. and Tilney, M.S. (1993) Trends Microbiol. 1, 25-31.
86 Kuhn, M., Prévost, M.C., Mounier, J. and Sansonetti, P. J. (1990) Infect. Immun. 58, 3477-3486.
87 Domann, E., Wehland, J., Rohde, M., Pistor, S., Hartl, M., Goebel, W., Leimeister-Wächter, M., Wuenscher, M. and Chakraborty, T. (1992) EMBO J. 11, 1981-1990.
88 Kocks, C., Gouin, E., Tabouret, M., Berche, P., Ohayon, H. and Cossart, P. (1992) Cell 68, 521-531.
89 Kocks, C., Hellio, R., Gounon, P., Ohayon, H. and Cossart, P. (1993) J. Cell Sci. 3, 699-710.
90 Niebuhr, K., Chakraborty, T., Rohde, M., Gazlig, T., Jansen, B., Köllner, P. and Wehland, J. (1993) Infect. Immun. 61, 2793-2802.
91 Pistor, S., Chakraborty, T., Niebuhr, K., Domann, E. and Wehland, J. (1994) EMBO J. 13, 758-763.
92 Southwick, F.S. and Purich, D.L. (1994) Proc. Nat. Acad. Sci. U.S.A. 91, 5168-5172.
93 Theriot, J. A., Rosenblatt, J., Portnoy, D. A., Goldschmidt-Clermont, P. J. and Mitchison, T.J. (1994) Cell 76, 505-517.
94 Kocks, C. (1994) Current Biology 4, 465-468.
95 Cossart, P. and Kocks, C. (1994) Mol. Microbiol. 13, 395-402.
96 Dold, F.G., Sanger, J.M. and Sanger, J.W. (1994) Cell Motil. Cytoskel. 28, 97-107.
97 Brundage, R.A., Smith, G.A., Camilli, A., Theriot, J.A. and Portnoy, D.A. (1993) Proc. Nat. Acad. Sci. U.S.A. 90, 11890-11894.
98 Karunasagar, I., Krohne, G. and Goebel, W. (1993) Infect. Immun. 61, 162-169.
99 Kreft, J., Dumbsky, M. and Theiss, S., (unpublished data).
100 Leimeister-Wächter, M., Goebel, W. and Chakraborty, T. (1989) FEMS Microbiol. Lett. 65, 23-30.

101 Gormley, E., Mengaud, J. and Cossart, P. (1989) Res. Microbiol. 140, 631-643.
102 Leimeister-Wächter, M., Haffner, C., Domann, E., Goebel, W. and Chakraborty, T. (1990) Proc. Nat. Acad. Sci. U.S.A. 87, 8336-8340.
103 Mengaud, J., Dramsi, S., Gouin, E., Vazquez-Boland, J.A., Milon, G. and Cossart, P. (1991) Mol. Microbiol. 5, 2273-2283.
104 Chakraborty, T., Leimeister-Wächter, M., Domann, E., Hartl, M., Goebel, W., Nichterlein, T. and Notermans, S. (1992) J. Bacteriol. 174, 568-574.
105 Freitag, N.E., Youngman, P. and Portnoy, D.A. (1992) J. Bacteriol. 174, 1293-1298.
106 Kolb, A., Busby, S., Buc, H., Garges, S. and Adhya, S. (1993) Annu. Rev. Biochem. 62, 749-795.
107 Landschulz, W.H., Johnson, P.F. and McKnight, S.L. (1988) Science 240, 1759-1764.
108 Freitag, N.E., Rong, L. and Portnoy, D.A. (1993) Infect. Immun. 61, 2537-2544.
109 Freitag, N.E. and Portnoy, D.A. (1994) Mol. Microbiol. 12, 845-853.
110 Goebel, W., Kreft, J., Bohne, J., Demuth, A., Kestler, H. and Sokolovic, Z. (1994) in Bacterial Protein Toxins (Freer, J., et al., eds.) pp. 138-145, Gustav Fischer Verlag, New York, NY.
111 Domann, E., Wehland, J., Niebuhr, K., Haffner, C., Leimeister-Wächter, M. and Chakraborty, T. (1993) Infect. Immun. 61, 3073-3075.
112 Bohne, J., Sokolovic, Z. and Goebel, W. (1994) Mol. Microbiol. 11, 1141-1150.
113 Sokolovic, Z. and Goebel, W. (1989) Infect. Immun. 57, 295-298.
114 Sokolovic, Z., Fuchs, A. and Goebel, W. (1990) Infect. Immun. 58, 3582-3587.
115 Leimeister-Wächter, M., Domann, E. and Chakraborty, T. (1992) J. Bacteriol. 174, 947-952.
116 Sokolovic, Z., Riedel, J., Wuenscher, M. and Goebel, W. (1993) Mol. Microbiol. 8, 219-227.
117 Datta, A. R. and Kothary, M.H. (1993) Appl. Env. Microbiol. 59, 3495-3497.
118 Kuhn, M. and Goebel, W. (unpublished data).
119 Park, S. F. and Kroll, R. G. (1993) Mol. Microbiol. 8, 653-661.
120 Dramsi, S., Kocks, C., Forestier, C. and Cossart, P. (1993) Mol. Microbiol. 9, 931-941.
121 Haas, A. and Brehm, K. (1993) Gen. Eng. Biotechnol. 13, 243-269.
122 Haas, A. and Goebel, W. (1992) Free Rad. Res. Commun. 16, 137-157.
123 Welch, D.F., Sword, C.P., Brehm, S. and Dusanic, D. (1979) Infect. Immun. 23, 863-872.
124 Vasconcelos, J.A.P. and Deneer, H.G. (1994) Appl. Env. Microbiol. 60, 2360-2366.
125 Welch, D.F. (1987) Ann. Inst. Pasteur/Microbiol. 138, 265-268.
126 Haas, A. and Goebel, W. (1992) Mol. Gen. Genet. 231, 313-322.
127 Brehm, K., Haas, A., Goebel, W. and Kreft, J. (1993) Gene 118, 121-125.
128 Brehm K., Nichterlein, T. and Goebel, W. (unpublished data).
129 Godfrey, R.W. and Wilder, M.S. (1985) Infect. Immun. 47, 837-839.
130 Dallmier, A.W. and Martin, S.E. (1988) Appl. Env. Microbiol. 54, 581-582.
131 Dallmier, A.W. and Martin, S.E. (1990) Appl. Env. Microbiol. 56, 2807-2810.
132 Leblond-Francillard, M., Gaillard, J.L. and Berche, P. (1989) Infect. Immun. 57, 2569-2573.
133 Swartz, M.A., Welch, D.F., Narayanan, R.P. and Greenfield, R.A. (1991) Clin. Microbiol. 96, 130-133.
134 Van Dissel, J.T., Stikkelbroeck, J.J.M. and van Furth, R. (1993) Scand. J. Immunol. 37, 443-446.
135 Haas, A., Brehm, K., Kreft, J. and Goebel, W. (1991) J. Bacteriol. 173, 5159-5167.

136 Göhmann, S., Leimeister-Wächter, M., Schiltz, E., Goebel, W. and Chakraborty, T. (1990) Mol. Microbiol. 4, 1091-1099.
137 Kaufmann, S.H.E. (1993) Annu. Rev. Immunol. 11, 129-163.
138 Campbell, P.A. (1993) in Macrophage-Pathogen Interactions (Zwilling, B. S. and Eisenstein, T.K., eds.) pp. 313-328, Marcel Dekker. Inc., New York, NY.
139 Mielke, M.E., Ehlers, S. and Hahn, H. (1993) Immunobiol. 189, 285-315.
140 Mitsuyama, M., Igarashi, K.I., Kawamura, I., Ohmori, T. and Nomoto, K. (1990) Infect. Immun. 58, 1254-1260.
141 Havell, E.A. and Sehgal, P.B. (1991) J. Immunol. 146, 756-761.
142 Kretschmar, M., Nichterlein, T., Chakraborty, T., Aufenanger, J. and Hof, H. (1993) Med. Microbiol. Lett. 2, 95-101.
143 Beuscher, H.U., Roosen, S., Jordan, M., Otterness, I.G., Kuhn, M. and Röllinghoff, M. (unpublished data).
144 Kuhn, M. and Goebel, W. (1994) Infect. Immun. 62, 348-356.
145 Kuhn, M. and Goebel, (unpublished data).
146 Demuth, A., Kuhn, M. and Goebel, W. (unpublished data).
147 Schwan, W.R. and Goebel, W. (1994) Proc. Nat. Acad. Sci. U.S.A. 91, 6428-6432.
148 Liang, P. and Pardee, A.B. (1992) Science 257, 967-971.
149 Schwan, W.R. and Goebel, W. (1994) Abst. Annu. Meet. Amer. Soc. Microbiol. 93, 29.
150 Schwan, W.R. and Goebel, W. (unpublished data).
151 Baeuerle, P.A. (1991) Biochim. Biophys. Acta 1072, 63-80.
152 Hauf, N., Goebel, W., Serfling, E. and Kuhn, M. (1994) Infect. Immun. 62, 2740-2747.
153 Hauf, N., Goebel, W. and Kuhn, M. (unpublished data).
154 Wels, W., Baldrich, M., Chakraborty, T., Gross, R. and Goebel, W. (1992) Mol. Microbiol. 6, 2651-2659.
155 Demuth, A., Chakraborty, T., Krohne, G. and Goebel, W. (1994) Infect. Immun. 62, 5102-5111.
156 Haas, A., Dumbsky, M. and Kreft, J. (1992) Biochim. Biophys. Acta 1130, 81-84.
157 Hof, H. and Hefner, P. (1988) Infection 16 (Suppl. 2), 141-144.
158 Galsworthy, S.B., Gurofsky, S.M. and Murray, R.G.E. (1977) Infect. Immun. 10, 1105-1109.
159 Sokolovic, Z., Kestler, H. and Goebel, W. (unpublished data).
160 Maurelli, A.T. (1992) ASM News 58, 603-608.
161 Galán, J. (1994) Trends Cell Biol. 4, 196-199.

INDIRECT USE OF IMMOBILIZED METAL AFFINITY CHROMATOGRAPHY FOR ISOLATION AND CHARACTERIZATION OF PROTEIN PARTNERS

Michèle Sawadogo[1] and Michael W. Van Dyke[2]

[1]Department of Molecular Genetics
[2]Department of Tumor Biology
The University of Texas, M.D. Anderson Cancer Center
1515 Holcombe Boulevard
Houston, TX 77030

INTRODUCTION

Immobilized metal affinity chromatography (IMAC), originally called metal chelate chromatography, is an effective way to purify proteins based on their interaction with a matrix-bound chelated transition metal ion (1). Retention of natural proteins on IMAC columns is determined by the number and proximity of particular amino acid residues on the protein surface (2-4). Commonly the immobilized metal binds to the imidazole group of histidine residues, but interactions with accessible cysteine and tryptophan residues are also possible. A variety of desorption methods, whether steps or gradients, have been utilized, based on two different principles. Changes in pH will elute proteins from the column by affecting the protonation of the interacting residues. Alternatively, elution can be induced by agents such as imidazole that compete with the protein for the metal ion binding sites.

Use of IMAC has flourished in recent years due to the development of molecular biology. The idea, first applied in the cases of the cloned mouse dihydrofolate reductase (5) and human immunodeficiency virus *tat* gene products (6), was to express recombinant proteins as fusions with a short stretch of polyhistidines. In most cases, the oligohistidine tag is positioned either at the N- or the C-terminus, where it is most likely to be exposed on the protein surface. Since few natural proteins display very strong affinities for immobilized metal ions, this small tag allows facile and highly specific purification of recombinant fusion proteins by IMAC. With the use of this method, large amounts of purified recombinant protein can often be prepared in a single chromatographic step (see Refs. 7-10 for practical examples and cookbook recipes).

A most fascinating application of the polyhistidine tag stems from the strength and selectivity of the interaction between the tagged protein and the immobilized metal ion on

Genetic Engineering, Vol. 17, Edited by J.K. Setlow
Plenum Press, New York, 1995

the column. This unique interaction is insensitive to most of the components usually present during the purification of biomolecules. As a result, IMAC can also be used as a convenient means to affix a protein to a support for affinity chromatography and protein-protein interaction studies. Here we will review the various methods by which IMAC can be applied to discover new interacting protein partners or characterize the interactions between known partners.

THE CHEMISTRY OF IMAC

The governing principle underlying IMAC is the interaction between an electron donating group on a protein surface and a chelated transition metal possessing one or more free coordination sites. Under physiological conditions, the imidazole moiety of histidine residues provides the primary electron donating group (11). Chelators commonly used include the tetradentate ligand nitrilotriacetic acid (NTA) and the tridentate ligand iminodiacetic acid (IDA). Transition metal cations employed typically include the 3*d*-block elements Ni^{2+}, Co^{2+}, Cu^{2+} and Zn^{2+}. Ni^{2+} is the most commonly used transition metal cation, providing a coordination number of six, electrochemical stability under chromatographic conditions and borderline polarizability. Interactions between this ion and intermediate ligands such as imidazole are strong and exhibit relatively rapid and reversible kinetics, all important considerations in protein chromatography.

The octahedral coordination spheres of the most commonly used adsorbants, Ni^{2+}-IDA and Ni^{2+}-NTA, are shown schematically in Figure 1. Note that with Ni^{2+}-IDA the nickel ion is attached to the matrix through three coordinate bonds while with Ni^{2+}-NTA the ion is coordinated through four bonds. Thus removal of the nickel from an NTA matrix is more difficult than from IDA. In both cases the protein is coordinated to the chelated metal ion through two residues, typically histidines separated by three amino acids (His-X_3-His) as part of an α-helical domain (2). However, with Ni^{2+}-IDA the protein can potentially interact with any two of the three available coordination sites. The end result of this flexibility in choice of coordination site is a requirement for higher imidazole concentrations to elute proteins from a Ni^{2+}-IDA column (250 mM to 1 M) than from a Ni^{2+}-NTA column (approximately 150 mM).

The protein-Ni^{2+} interaction is primarily affected by competition with other electron-donating groups on borderline and soft bases present in the aqueous environment. Other than imidazole, these include such anions as azide (N_3^-) and mercaptans (RS^-). Most reagents used in the chromatography of proteins do not appreciably affect the metal-protein interaction and thus are compatible with IMAC. Examples include Cl^-, acetate, PO_4^{-3}, SO_4^{-2} and RNH_2, and also components containing these substituents, e.g., sulfate-containing buffers such as HEPES and detergents such as SDS. A listing of chemicals compatible for use with IMAC on Ni^{2+}-IDA is provided in Table 1. Note that similar maximum concentrations have been found for several of these compounds on Ni^{2+}-NTA (8).

There exist several practical considerations in the use of IMAC for the purification of protein complexes. IMAC is compatible with most buffer constituents that may be required to maintain protein complex integrity, e.g., moderate ionic strengths, physiological pHs, polyhydroxylated hydrocarbons (glycerol) and many protease inhibitors. Monomercaptan reducing agents, like 2-mercaptoethanol, may be used at concentrations as high as 10 mM if minimal amounts of imidazole are also present (20 mM imidazole for Ni^{2+}-IDA). Dimercaptans such at dithiothreitol or dithioerythreitol should be avoided, given their propensity to chelate 3*d* transition metals. Similarly, the use of metal chelators like EDTA and EGTA should be minimized. In our experience, however, submillimolar concentrations of these chelators do not appreciably affect chromatography on Ni^{2+}-IDA under standard conditions. In contrast, high concentrations of EDTA (i.e. greater than 50 mM) will strip

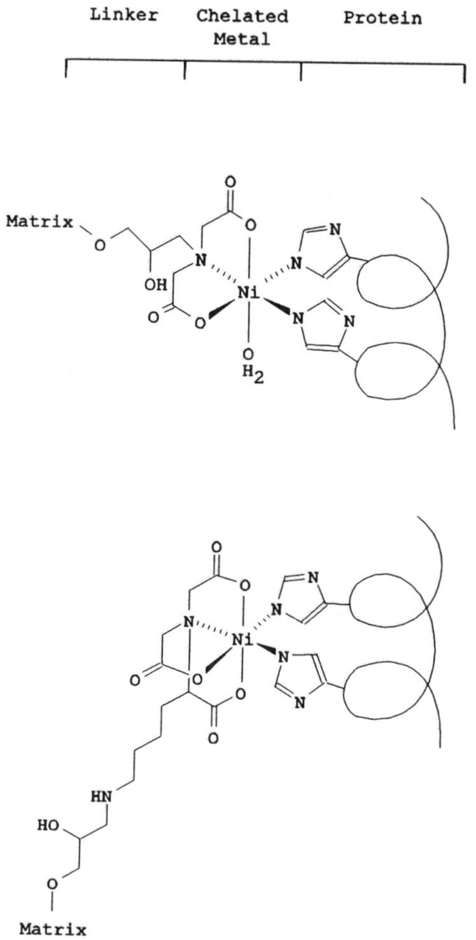

Ni^{2+}-IDA

Ni^{2+}-NTA

Figure 1. Schematic representation of Ni^{2+}-IDA and Ni^{2+}-NTA adsorption centers interacting with a histidine-tagged protein. Shown from left to right are the chemical linkages between the column matrix and the chelator, the coordination sphere surrounding the Ni^{2+} ion, and the imidazole moieties of histidine residues present in an exposed α-helical domain of a bound protein.

Table 1. Reagent Concentrations that Do Not Dissociate the His_6-Ni^{2+}-IDA Interaction

Cations:			
	Na^+		1 M
	K^+		1 M
	NH_4+	pH 7.9	2 M
	$Tris^+$	pH 7.3	0.5 M
	Mg^{2+}		1 M
Anions:			
	Cl^-		1 M
	$acetate^-$	pH 7	1 M
	CO_3^{2-}	pH 9.5	0.5 M
	$S0_4^{2-}$	pH 7.9	1 M
	$P0_4^{3-}$	pH 6.6	0.5 M
Zwitterions:			
	glycine	pH 8.5	0.1 M
	HEPES	pH 7.9	1 M
Alcohols and mercaptans:			
	ethanol		30 %
	glycerol		50 %
	2-mercaptoethanol		10 mM
Denaturants:			
	urea		3 M
Detergents:			
	(cationic)		
	cetyltrimethylammonium bromide		1 %
	(anionic)		
	sodium *N*-lauroylsarcosine		1 %
	sodium dodecyl sulfate		1 %
	sodium deoxycholate		1 %
	(nonionic)		
	Triton X-100		1 %
	Tween 20		1 %
	NP-40		1 %

the Ni^{2+} from the resin. Elution of His_6-tagged protein complexes is usually achieved by increasing concentrations of imidazole, from 100 to 200 mM for Ni^{2+}-NTA and 250 to 1000 mM for Ni^{2+}-IDA. These imidazole concentrations typically have little effect on the hydrogen bonding and hydrophobic interactions present in protein complexes.

PROPERTIES OF POLYHISTIDINE-TAGGED PROTEINS

Polyhistidine-tagged proteins expressed in various vectors and hosts can often be purified in a single chromatographic step with IMAC. Binding of the protein to the column can be done either in denaturing (e.g., 6 M guanidinium hydrochloride) or native conditions (7-10), yielding in both cases highly purified recombinant proteins suitable for many studies. Importantly, the exact same procedure can be used almost independently of the nature of the cloned protein. Note, however, that IMAC-purified recombinant proteins from *E. coli* expression systems are not entirely devoid of contaminating proteins of bacterial origin. In particular, superoxide dismutase, which is known to bind divalent metals, has been found to contaminate materials prepared under denatured conditions by NTA-IMAC (12). In addition, a metal-binding protease has been identified in extracts from *E. coli* (13). In cases where a greater degree of purity is desirable, purification by IMAC should therefore be combined with an additional conventional chromatography step. In our hands, excellent results are usually obtained by rapidly desalting the protein after IMAC on a P6DG column (BioRad, Hercules, CA) and performing a final purification on an FPLC Mono Q or Mono S column (Pharmacia, Uppsala, Sweden) (see for example Ref. 14).

One major advantage of the oligohistidine tagging method is that the foreign moiety is very small and usually does not alter the normal properties or enzymatic function of the recombinant protein. As a result, protease cleavage of the tag is in most cases unnecessary for functional assays to be carried out with the recombinant protein. Numerous examples of oligohistidine fusion proteins that are fully active in a variety of assays can be found in the literature. Among these, we noted several transcription factors active *in vivo* [e.g., the σ^{32} factor of *E. coli* (15) and the human immunodeficiency virus Tat protein (6)] or in *in vitro* reconstituted systems [e.g., the class II transcription initiation factors TFIID (16,17) and TFIIB (18), the RNA polymerase II elongation factor SII (14), the GATA factor (19), the glucocorticoid receptor (20), and the serum response factor and Elk-1 protein (21,22)]. Proteins involved in RNA binding and pre-mRNA splicing were also successfully expressed in bacterial polyhistidine fusion vectors (23,24). Copeland and Wang investigated the enzymatic mechanism of primase using subunits of the mouse primase individually expressed in *E. coli* as polyhistidine fusions (25). Other illustrative examples include the human androgen receptor produced in vaccinia virus that was found to be active *in vivo* (26), the Type I adenylylcyclase purified as a polyhistidine fusion protein from baculovirus (27), the extracellular portion of betaglycan found active in binding to transforming growth factor beta (28), or the yeast prepro-α-factor purified from *E. coli* that was shown to be active in translocation across yeast mitochondrial membranes, signal sequence cleavage and *N*-linked glycosylation (29).

PURIFICATION OF PROTEIN COMPLEXES

In vivo Formed Complexes

The purification of *in vivo* assembled protein complexes in which one subunit contains an oligohistidine tag can be thought of as a simple extension of the purification of individual oligohistidine-containing proteins (see Figure 2). The major differences stem from the need to maintain integrity of the protein complex. This requirement obviously precludes the use of denaturating conditions. Also in this case imidazole becomes the method of choice to elute protein complexes from the column, since it is much less likely to affect protein-protein interactions than other possible eluents, for example low pH buffers.

There are already several examples in the literature that illustrate the use of IMAC for purification of protein complexes. To study the regulation of heat-shock gene expression, Gamer et al. (15) introduced a histidine tag in the *E. coli* heat-shock transcription factor σ^{32}. Activity of the fusion protein was confirmed by complementation of a temperature-sensitive mutant strain. Both the C-terminally and N-terminally tagged σ^{32} were highly active *in vivo* as specific transcriptional activators of heat shock genes. Each of these proteins could be efficiently purified from crude lysates in a single IMAC step. More important, three proteins (DnaK, DnaJ and GrpE) were also found in the Ni^{2+} column eluate as a direct result of their physical interaction with σ^{32}. Use of IMAC in this system, combined with extracts from different mutant strains, allowed further characterization of the interaction of these proteins with the σ factor. For instance, it was shown that the interaction of DnaK and GrpE (but not DnaJ) with σ^{32} was sensitive to ATP, and that the DnaJ-σ^{32} and DnaK-σ^{32} associations occurred independently of each other (15). Another case where IMAC purification allowed preparation of material suitable for structural and enzymological analyses is a study done with the reverse transcriptase (RT) from human immunodeficiency virus type 1 (HIV-1). Chattopadhyay et al. expressed a recombinant version of the 66-kD precursor protein (p66) in *E. coli* with an N-terminal polyhistidine extension (30). Earlier studies had shown that purification of p66 from recombinant expression systems was hampered by the sensitivity of this protein to host proteases. In contrast, rapid one-step purification by IMAC allowed preparation of the unprocessed p66. Pure heterodimeric (p66/p51) RT was then generated by the action of recombinant HIV-1 protease. Noticeably, addition of amino acid residues to the N-terminus of RT had no effect on the polymerase function or RNase H activity of the protein.

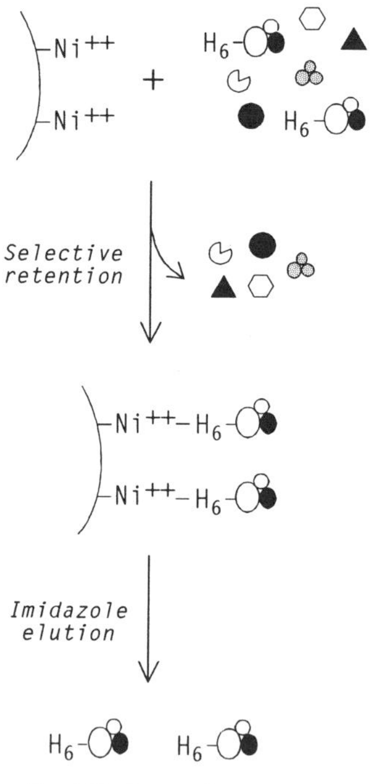

Figure 2. Purification of protein complexes by IMAC. The purification by IMAC of a multimeric protein complex in which one of the subunits contains an oligohistidine tag (H_6), is fundamentally identical to the purification of a tagged monomeric protein under conditions that maintain protein complex integrity (native conditions).

Two other examples demonstrate how tagging a single polypeptide with histidines can allow purification of protein complexes from yeast cells. One of the RNA polymerase II general transcription initiation factors, designated factor b in the yeast nomenclature, is a multisubunit protein composed of five polypeptides with two catalytic activities: a protein kinase and a DNA-dependent ATPase. This protein, which also plays an important role in nucleotide-excision repair of DNA, has proven extremely difficult to purify by conventional means. Feaver et al. took advantage of the unique properties of IMAC for the purification and characterization of factor b (31). Since one of the subunits had been cloned, an oligohistidine tag was added to its C-terminus and the recombinant gene was introduced into yeast cells. This tag greatly facilitated purification of the entire protein complex, yielding nearly homogeneous preparations in five chromatographic steps including IMAC. Peterson et al. used a similar system to provide physical evidence for the existence of a large multisubunit complex of the SWI/SNF gene products (32). The corresponding genes were known to be required for transcriptional induction of a large number of yeast genes; however, direct biochemical proof for the existence of a protein complex containing all five polypeptides was lacking. After the SWI2 protein was tagged at its C-terminus with a polyhistidine stretch, affinity purification on a Ni^{+2}-NTA column resulted in the isolation of the entire SWI/SNF complex.

In vitro Formed Complexes

Under some circumstances it may not be possible to assemble *in vivo* protein complexes containing a histidine-tagged subunit; for instance there may be no expression system available for the organism in which the protein complex normally exists. In these cases *in vitro* assembly of protein complexes could still permit their purification by IMAC. Two examples of this approach were demonstrated with the gene-specific transcription factor USF (33) and the RNA polymerase II general transcription initiation factor TFIIF (34).

Human upstream stimulatory factor (USF) is a dimeric protein composed of various combinations of either 43- or 44-kD subunits. Wild type 43-kD USF and a fusion protein containing an N-terminal oligohistidine domain and the C-terminal two thirds of USF (His_6-delUSF) were separately expressed in *E. coli* (33,35). Bacterial extracts containing these two proteins were mixed and submitted to a brief heat treatment to affect complex formation. Substantial quantities of denatured bacterial proteins were removed from the extracts by this procedure without appreciable loss of USF. Protein complexes containing His_6-delUSF homodimers and His_6-delUSF·wtUSF heterodimers were efficiently retained on a Ni^{2+}-IDA column and purified to homogeneity following a single imidazole step elution. Because a molar excess of the tagged protein was present in the original mixture, recovery of wtUSF was nearly quantitative (33). Following this procedure, both forms of USF were found to be fully functional with regard to sequence-specific DNA binding as determined by electrophoretic mobility shift assay.

TFIIF is a heterotetrameric protein consisting of two copies each of the RAP74 and RAP30 proteins. The cDNAs for both of these polypeptides have been cloned and shown to be efficiently expressed in *E. coli*. A vector, pVRAP30, was constructed to express a RAP30 fusion protein containing an N-terminal oligohistidine domain. This tag allowed the facile purification of His_6-RAP30 by IMAC and yielded a functionally active protein as determined by electrophoretic mobility shift assay. TFIIF complexes were assembled from partially purified His_6-RAP30 and RAP74 following sequential chemical denaturation and renaturation steps. Functional TFIIF complexes were then purified to near homogeneity with high yield in a single IMAC step (34).

Problems associated with the purification of *in vitro* assembled protein complexes usually stem from the relative efficiencies at which the individual protein subunits can be renatured prior to complex formation. Thus the easily renatured USF heterodimers were quantitatively recovered in a fully functional form (33) while not all RAP74 was sequestered into His_6-RAP30/RAP74 complexes (34). Ideally, purification of *in vitro* assembled protein complexes by IMAC would be best achieved either with completely renaturation-competent subunits or, preferably, with subunits that do not require extreme structural perturbations prior to complex formation.

IDENTIFICATION AND PURIFICATION OF PROTEIN PARTNERS

Preparation of affinity columns

An important consideration in designing an affinity chromatography column for purification of protein partners is the method used to affix the first protein to the matrix, which must involve either a covalent or a highly stable bond. Chemical crosslinking is convenient and relatively inexpensive and has the additional advantage of providing a nearly permanent means of attachment. However, most proteins contain many functional moieties that can interact with the typical activated matrix. Therefore a major drawback of chemical crosslinking is the inability to orient the protein molecules on the column so that they all present the desired face to the solution for interaction with the specific partners. The same shortcoming holds for attachment methods with polyclonal antibodies. In contrast, tethering of proteins to a matrix through immobilized monoclonal antibodies does allow control over the orientation of the protein on the column, but preparation and characterization of the necessary reagents is difficult and time consuming. Also, because the antibody itself adds a large proteinaceous surface on the column, the potential for nonspecific interactions is increased. Furthermore, since interaction of the protein with the antibody cannot be easily reversed, this method usually precludes the possibility of eluting the protein complex from the column in a native form.

With these considerations in mind, it seems that indirect use of IMAC offers a nearly ideal method for affinity chromatography based on protein-protein interactions. Because the oligohistidine tag is very small, nonspecific interactions are minimized. Also, with the ease and flexibility of recombinant protein construction, this tag can be introduced at the most appropriate location to orient the protein properly on the column. We already mentioned the numerous advantages afforded by the unique chemistry involved in IMAC. The high capacity of IMAC columns, up to several milligrams of protein per milliliter of resin, is another distinct advantage.

Possible Purification Schemes

One of the several possible ways by which IMAC can be used for affinity purification of protein partners is schematically represented in Figure 3. Here the cloned protein expressed as a polyhistidine fusion is indicated as H_6-A, while B refers to the interacting partner. An IMAC column is first charged to near saturation with the A protein. Next, the crude mixture containing protein B is passed through the column, with the use of a slow flow rate to allow the two proteins to interact. At this stage, it is necessary to determine the most appropriate wash buffers to remove weakly interacting contaminants from the column. Ideally, only the specific H_6-A·B complex should remain on the column. Two strategies are then possible for the elution conditions. High imidazole-containing buffers can be used to release the whole complex in a native form. Alternatively, the interacting partner can be selectively eluted by an increase in ionic strength (if the interaction between the two proteins is mostly electrostatic in nature) or by addition of detergents (in case of a predominantly hydrophobic interaction). Note that if the interacting partner is scarce or must be purified from a very complex mixture of proteins (e.g., a crude nuclear extract from mammalian cells), it would often be advisable to add an initial step in which the crude B fraction is first chromatographed over a guard IMAC column.

This would remove from the protein mixture those few polypeptides that would otherwise interact strongly with available nickel ions on the affinity column and therefore contaminate the final imidazole eluate.

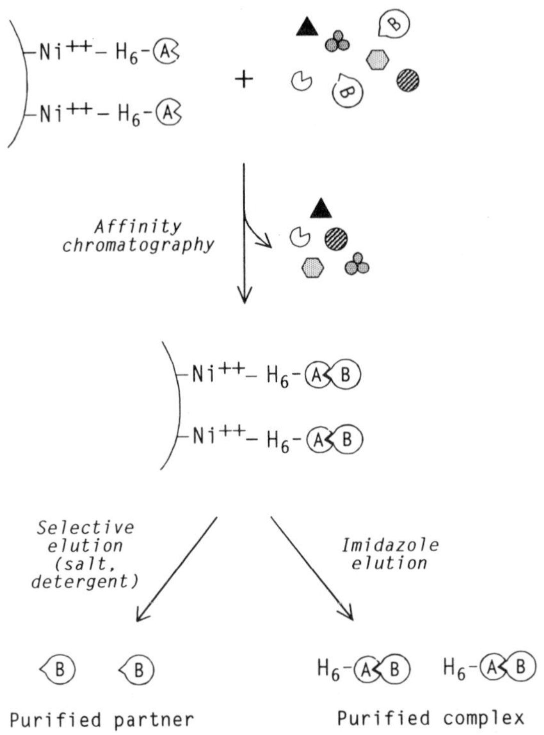

Figure 3. Possible strategy for the affinity purification of a protein partner (B) through its interaction with a cloned polypeptide expressed as an oligohistidine fusion (symbolized as H_6-A) and attached to an IMAC column.

Possible variations of the basic protocol are depicted in Figure 4. Instead of passing the extract through a column of immobilized A protein, the A·B complex can be preformed in solution by mixing the recombinant A protein with the B-containing fraction. This method would be preferable if the rate of formation of the A·B complex is slow or if the reaction only takes place with reasonable rates at elevated temperatures. Note that even if the B component is already complexed to natural A in the crude fraction, use of a large molar excess of recombinant A should drive the equilibrium towards B being primarily bound to the tagged A protein.

Practical Examples

Because the method and concepts are still relatively new, there are at this time only a few published examples of the indirect use of IMAC for the purification of protein partners. However, it is safe to predict that such successful illustrations will multiply in the near future. IMAC has proven very useful in studies of the mitochondrial chaperone system, as an alternative to more traditional approaches with co-immunoprecipitations, cofractionation, or coelectrophoresis. To investigate the function of mitochondrial hsp70 and identify possible partners, Bolliger et al. established a yeast strain with a mitochondrial hsp70 protein (mhsp70) containing a histidine tag at its C-terminus (36). IMAC performed

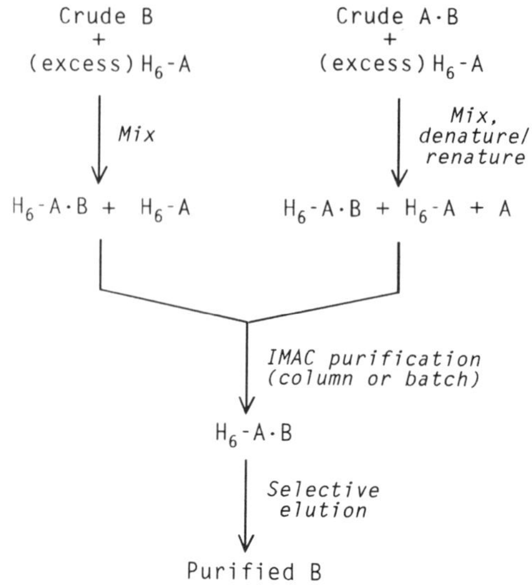

Figure 4. Alternative strategies to the method shown in Figure 3 for the isolation of protein partners. Note that the same schemes can be applied to small-scale batch procedures appropriate for protein-protein interaction studies.

with ATP-depleted mitochondrial extracts revealed binding of a 23-kD protein to mhsp70. Interaction of this protein with mhsp70 could be dissociated by addition of ATP. This strategy resulted in identification of the 23-kD protein as the mitochondrial homolog of the bacterial GrpE protein, identification which eventually led to the cloning of the corresponding yeast gene (36).

Boehmer and Lehman used IMAC to prove the existence of a tight interaction between two proteins from herpes simplex virus 1, the single-stranded DNA-binding protein (ICP8) and the origin-binding protein (UL9) (37). It was known from previous studies that ICP8 could stimulate the helicase activity of UL9. Use of affinity chromatography demonstrated the existence of a direct physical interaction between these two proteins. In particular, it was shown that a 37-kD C-terminal portion of UL9 tagged with histidines and immobilized on Ni^{2+}-NTA resin could selectively retain ICP8, indicating that the specific interaction between the two proteins was confined to the C-terminal DNA-binding domain of UL9 (37).

Beers and Gallis tagged ubiquitin (Ub) with histidines (38). Addition of HisUb to a wheat germ *in vitro* conjugation system resulted in covalent ligation of HisUb to wheat germ proteins, thereby allowing the resulting conjugates to be purified by IMAC. These authors also demonstrated the feasibility of using HisUb-IMAC columns as an affinity matrix to purify ubiquitin-specific hydrolases. A known yeast ubiquitin extension protein hydrolase was first expressed in *E. coli*. When the soluble fraction of these induced bacteria was passed over a column of immobilized HisUb, the hydrolase was specifically retained on the column and could be eluted by increasing the pH or by specific elution with wild-type ubiquitin. This purification was ultimately successful because the recognition site for the hydrolase is located in the C-terminus of ubiquitin and HisUb was attached through its N-terminus.

Note that the spectrum of partners that can be isolated through the use of oligohistidine fusion proteins is not limited to proteins. The same concept can be extended to the purification of any interacting molecule, ranging from nucleic acids to small cofactors. An elegant example can be found in the studies performed by Cross et al.

working with the rat MeCP2 protein (39). This protein is known to bind specifically to DNA methylated at CpG dinucleotides. These authors isolated the 85-amino acid methyl-CpG binding domain, cloned it with a polyhistidine tag at its N-terminus, and attached it to a Ni^{2+}-agarose matrix. When DNA with various degrees of methylation was passed over the column and subsequently eluted with a salt gradient, three distinct peaks could be separated containing unmethylated, partially methylated and completely methylated DNA. Binding affinities of DNA on the column were clearly dependent on methyl-CpG density. Following these observations, the same column was successfully used to prepare a representative library of genomic DNA fragments containing CpG islands (39).

Of all the conceivable uses of IMAC for affinity chromatography, there should be only a few cases where the polyhistidine-Ni^{2+} linkage would be too labile to allow selective elution of the interacting partner. Affinity purification of antibodies would fall in this category, since very harsh conditions are usually required to disrupt antibody-antigen interactions. In such cases a possible alternative is to affix the recombinant protein on the support by forming a kinetically inert ternary complex with immobilized iminodiacetato-Co(III) (40). The only reagents capable of dissociating a protein immobilized on such a resin are reducing agents that convert Co(III) to its labile +2 oxidation state. This method has been used successfully with the immobilized complex of recombinant papilloma virus type 16 E7 protein tagged with a chelating peptide, which served as an affinity matrix for purification of anti-E7 antibodies from rabbit serum (40).

PROTEIN-PROTEIN INTERACTION STUDIES

IMAC can also be used to characterize the interaction between two known proteins. Employing only small amounts of proteins, adsorption and elution steps can be performed batchwise in microcentrifuge tubes. In our laboratories, we commonly use this method to study interactions between a bacterially produced polyhistidine fusion protein and a radiolabeled polypeptide produced by *in vitro* transcription/translation reactions. This particular use of IMAC is in many ways similar to the traditional immunoprecipitation procedures, with the same technical difficulties in physically manipulating small amounts of affinity resin with reproducible recoveries. However, the IMAC batch method is directly applicable to any cloned protein and does not require the prior development of suitable specific antibodies.

Sporeno et al. used such a method in their analysis of the interaction between oncostatin M (OM) and the transmembrane glycoprotein gp130 (41). A soluble form of human gp130 containing a polyhistidine stretch was expressed in CHO cells and purified by IMAC. The dissociation constant of OM for this soluble form of gp130 was determined by displacement binding experiments with iodinated OM and increasing concentrations of unlabeled cytokine. The OM-gp130 complexes were collected on Ni^{2+}-NTA-agarose resin and centrifuged through a sucrose cushion. Li et al. (42) also successfully used batch adsorption to IMAC resin to determine the degree of interaction between the TATA box binding transcription factor TBP (expressed as a histidine-tagged fusion) and each subunit of a transactivation-mediating protein purified from silkworms.

We have used small-scale batch IMAC to characterize subunit interactions within the dimers of human transcription factor USF (33). Heterodimers were formed between two recombinant USF proteins: a full-length unfused species (wtUSF) and a histidine-tagged N-terminal truncated version of the same protein (His_6-delUSF). The resulting heterodimers were tethered to IMAC resin through the polyhistidine stretch, and conditions that would allow subunit dissociation (and therefore release of the wtUSF into solution) were investigated. This analysis was not only useful in ascertaining the conditions necessary for maintaining the integrity of USF dimers but also provided a confirmation of the variety of agents that could be used during IMAC chromatography without releasing the tagged protein-chelated metal interaction (33).

CONCLUDING REMARKS

IMAC, in conjunction with modern recombinant DNA technology, allows efficient immobilization of a tagged protein to a matrix with a consistent and defined spatial orientation. The unique nature of this immobilization method makes it immune to most biochemical reagents. The small size of the oligopeptide tag, typically six contiguous histidine residues, while sufficient to allow strong and specific interaction with a chelated metal, offers minimum interference with most protein functions, both enzymatic and physical. Together, these characteristics define a system that allows the assembly of tagged protein complexes, their facile purification in a native and functionally active state, and a means to dissect the interactions therein. In the future, biochemical studies involving interactions between polypeptides and other macromolecules will most likely see IMAC playing an increasingly important role.

ACKNOWLEDGMENTS

We thank Marilyn Szentirmay and Walter Pagel for reviewing the manuscript. Work from our laboratories was supported by Grants GM38212 from the National Institutes of Health, G1195 from the Robert A. Welch Foundation and 3110 from the Council for Tobacco Research.

REFERENCES

1 Porath, J., Carlsson, J., Olsson, I. and Belfrage, G. (1975) Nature 258, 598-599.
2 Sulkowski, E. (1989) BioEssays 10, 169-175.
3 Arnold, F.H. (1991) Bio/Technology 9, 151-156.
4 Porath, J. (1992) Protein Expression Purif. 3, 263-281.
5 Hochuli, E., Bannwarth, W., Döbeli, H., Gentz, R. and Stüber, D. (1988) Bio/Technology 6, 1321-1325.
6 Gentz, R., Chen, C.H. and Rosen, C.G. (1989) Proc. Nat. Acad. Sci. U.S.A. 86, 821-824.
7 Hochuli, E. (1990). in Genetic Engineering, Principles and Methods,Vol. 12 (J. Setlow, ed.) pp. 87-98. Plenum, New York, NY.
8 Crowe, J. and Henco, K. (1992) in The QIAexpressionist, 2nd ed., Qiagen Inc., Chatsworth, CA.
9 Winzerling, J.J., Berna, P. and Porath, J. (1992) Methods 4 (Companion to Meth. Enzymol.), 4-13.
10 Petty, K.J. (1993), in Current Protocols in Molecular Biology, (Ausubel, F.M., Brent, R.E., Moore, D.D., Seidman, J.G., Smith, J.A., and Struhl, K. eds.), section 10.11, pp. 8-22, Wiley-Interscience, New York, NY.
11 Yip, T.T., Nakagawa, Y. and Porath, J. (1989) Anal. Biochem. 183, 159-171.
12 Strüber, D., Matile, H. and Garotta, G. (1990) Immunol. Methods 4, 121-152.
13 Linder, P., Guth, B. Wülfing, C., Krebber, C., Steipe, B., Müller, F. and Plückthun (1992) Methods 4 (Companion to Meth. Enzymol.), 41-56.

14 Szentirmay, M.N. and Sawadogo, M. (1993) EMBO J. 12, 4677-4684.
15 Gamer, J., Bujard, H. and Bukau, B. (1992) Cell 69, 833-942.
16 Hoffmann, A. and Roeder, R.G. (1991) Nucl. Acids Res. 19, 21-22.
17 Parvin, J.D., Timmers, H.Th.M. and Sharp, P.A. (1992) Cell 68, 1135-1144.
18 Yamashita, S., Kokubo, T., Doi, K., Roeder, R.G., Horikoshi, M. and Nakatani, Y. (1993) Science 261, 463-466.
19 Aird, W.C., Parvin, J.D., Sharp, P.A. and Rosenberg, R.D. (1994) J. Biol. Chem. 269, 883-889.
20 Schmitt, J. and Stunnenberg, H.G. (1993) Nucl. Acids Res. 21, 2673-2681.
21 Janknecht, R., de Martynoff, G., Lou, J., Hipskind, R.A., Nordheim, A. and Stunnenberg, H.G. (1991) Proc. Nat. Acad. Sci. U.S.A. 88, 8972-8976.
22 Janknecht, R. and Nordheim, A. (1992) Gene 121, 321-324.
23 Kanaar, R., Roche, S.E., Beall, E.L., Green, M.G. and Rio, D.C.(1993) Science 262, 569-573.
24 Tsuchihashi, Z., Khosla, M. and Herschlag, D. (1993) Science 262, 99-101.
25 Copeland, W.C. and Wang, T.S.F. (1993) J. Biol. Chem. 268, 26179-26189.
26 De Vos, P., Schmitt, J., Verhoeven, G. and Stunnenberg, H. (1994) Nucl. Acids Res. 22, 1161-1166.
27 Taussig, R., Quarmby, L.M. and Gilman, A.G. (1993) J. Biol. Chem.268, 9-12.
28 Fukushiman, D., Bützow, R., Hildebrand, A. and Ruoslahti, E. (1993) J. Biol. Chem. 268, 22710-22715.
29 Bush, G.L., Tassin, A.M., Fridén, H. and Meyer, D.I. (1991) J. Biol. Chem. 266, 13811-13814.
30 Chattopadhyay, D., Evans, D.B., Deibel, M.R.Jr., Vosters, A.F., Eckenrode, F.M., Einspahr, H.M., Hui, J.O., Tomasselli, A.G., Zurcher-Neely, H.A., Heinrikson, R.L. and Sharma, S.K. (1992) J. Biol. Chem. 267, 14227-14232.
31 Feaver, W.J., Svejstrup, J.Q., Bardwell, L., Bardwell, A.J., Buratowski, S., Gulyas, K., Donahue, T.F., Frieberg, E.C. and Kornberg, R.D. (1993) Cell 75, 1379-1387.
32 Peterson, C.L., Dingwall, A. and Scott, M.P. (1994) Proc. Nat. Acad. Sci. U.S.A. 91, 2905-2908.
33 Lu, T., Van Dyke, M. and Sawadogo, M. (1993) Anal. Biochem. 213, 318-322.
34 Bowne S. and Van Dyke, M.W. (unpublished results).
35 Van Dyke, M.W., Sirito, M. and Sawadogo, M. (1992) Gene 111, 99-104.
36 Bolliger, L., Deloche, O., Glick, B.S., Georgopoulos, C., Jenö, P., Kronidou, N., Horst, M., Morishima, N. and Schatz, G. (1994) EMBO J. 13, 1998-2006.
37 Boehmer, P.E. and Lehman, I.R. (1993) Proc. Nat. Acad. Sci. U.S.A. 90, 8444-8448.
38 Beers, E.P. and Callis, J. (1993) J. Biol. Chem. 268, 21645-21649.
39 Cross, S.H., Charlton, J.A., Nan X. and Bird, A. (1994) Nature Genet. 6, 236-244.
40 Smith, M.C., Cook, J.A., Smanik, P.A., Wakulchik, M. and Kasher, M.S. (1992) Methods 4 (Companion to Meth. Enzymol.), 73-78.
41 Sporeno, E., Paonessa, G., Salvati, A.L., Graziani, R., Delmastro, P., Ciliberto, G and Toniatti, C. (1994) J. Biol. Chem. 269, 10001-10005.
42 Li, F.Q., Ueda, H. and Hirose, S. (1994) Mol. Cell. Biol. 14, 3013-3021.

STRUCTURE AND FUNCTION OF RNA PSEUDOKNOTS

C.W.A. Pleij

Leiden Institute of Chemistry
Gorlaeus Laboratories
Leiden University
The Netherlands

INTRODUCTION

Understanding the biological properties of RNA requires a detailed knowledge of the spatial folding of this macromolecule. Especially the discovery that natural RNA can function as a catalyst has not only radically changed our view on RNA, but has also made clear our relatively poor knowledge concerning its folding. By far the most important principle in the folding of single-stranded RNA is the formation of intramolecular double helical stem regions involving the Watson-Crick A-U and G-C base pairs. This gives rise to a collection of stem regions which is called the orthodox secondary structure of RNA.

As we have learned from the crystal structure of tRNA, the secondary structure at best forms the scaffold for the final and active shape which is in fact determined by tertiary interactions, like base intercalation, triple base interactions, hydrogen bond formation, etc. Another important feature of the spatial folding of RNA has been shown to be the formation of RNA pseudoknots. In this type of interaction, classical Watson-Crick base pairing takes place between a loop region in the secondary structure and a complementary sequence outside that loop. Although this principle of pseudoknot formation is based on the formation of the same kind of double helical stems as in the conventional secondary structure, the consequences for the folding of the RNA chain can be important and in some cases lead to interesting functional properties.

It has become clear during the last decade that RNA pseudoknots play a key role in a number of important processes in the cell, like translation, replication or self-splicing. This review will evaluate our current understanding of the structure and function of pseudoknots. For other recent reviews on this subject the reader is referred to Ref. (1-3).

Genetic Engineering, Vol. 17, Edited by J.K. Setlow
Plenum Press, New York, 1995

HISTORY AND DEFINITION

The first time that the term pseudoknot was used was in a paper by Studnicka et al. in 1978 (4). These authors, describing a program for the prediction of RNA secondary structure, proposed the concept of a "pseudoknot" to discern this type of folding from the related, but topologically different, "real knot" formation. In fact the idea of pseudoknotting of an RNA chain can already be found in the papers by Richards (5) and Ninio (6). At that time pseudoknot formation was only a theoretical possibility and examples of such structures in RNA were not known, let alone experimentally supported.

It was in 1982 that Rietveld et al. (7) proposed a model for the tRNA-like structure at the 3′ end of turnip yellow mosaic virus (TYMV) RNA, in which a pseudoknot structure played a crucial role in understanding its resemblance with a standard tRNA (Figure 1). Soon after this first "classic" pseudoknot, other plant virus RNAs appeared to harbour pseudoknot structures in their 3′ untranslated regions (UTR) (8-10), which in its turn led to the more general description of this folding principle (11).

An RNA pseudoknot is defined as a structure formed in a single-stranded RNA molecule upon Watson-Crick base pairing of a loop region with a complementary sequence outside this loop. This can be an interaction between a loop region and an unconstrained single-stranded region elsewhere, but it may also involve base pairing between two loop regions. The latter means that having hairpin, bulge, interior and multiple loops in an orthodox RNA secondary structure, one can discern in principle fourteen different types of pseudoknots (12). The simplest way to create a loop in an RNA chain is by folding a hairpin. Consequently, base pairing of the hairpin loop with a single-stranded region close to the hairpin gives rise to the most simple type of pseudoknot, also called classic (1) or H-type pseudoknot (13).

STRUCTURE

The structure of the classic pseudoknot is characterized by two stems S1 and S2 and two loop regions L1 and L2 (Figure 2). One has to note, however, that the pseudoknot

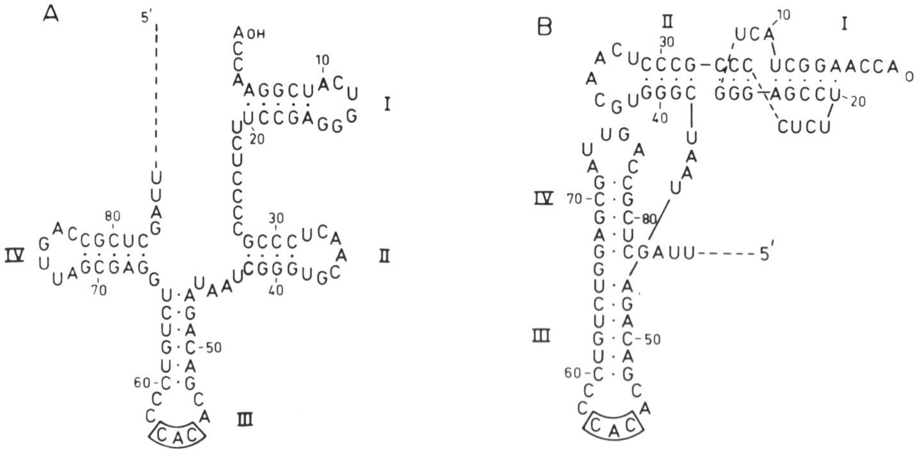

Figure 1. tRNA-like structure at the 3′ end of turnip yellow mosaic virus (TYMV) RNA. (A) secondary structure. (B) L-shape arrangement. Nucleotides and hairpins are numbered from the 3′ end. The pseudoknot is formed by the base pairing of the triple G sequence in hairpin loop I with the triple C sequence adjacent to stem II.

depicted in Figure 2 is a special case in that the two stems are contiguous. If unpaired nucleotides are present between S1 and S2, a more general type of pseudoknot is formed which in fact contains three loop regions (1,12). When the two stems are adjacent, however, coaxial stacking of the two segments can be envisioned as illustrated in Figure 2B. The first evidence for such a coaxial stacking was obtained from a study of the tRNA-like structure at the 3' end of the plant viral TYMV RNA (7). In order to explain its resemblance to $tRNA^{Val}$, a pseudoknot structure was proposed on the basis of structure mapping techniques and sequence comparisons (Figure 1).

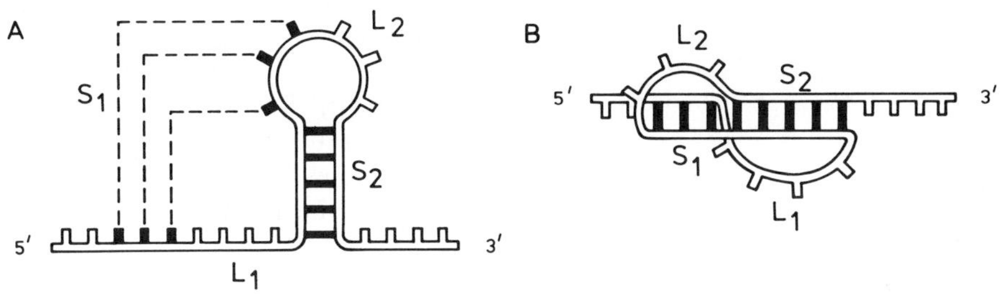

Figure 2. The formation of a classic pseudoknot involving a hairpin loop. (A) Conventional presentation. The dashed lines indicate the base pairing of nucleotides in the loop with a complementary sequence outside the loop. (B) Formation of the quasi-continuous helix upon coaxial stacking of the two stem segments. S1 and S2 represent the two stem regions, L1 and L2 the two connecting loops.

Shortly thereafter, pseudoknots of a similar type were discovered in the 3′ UTRs of other plant viral RNAs (8-10) and formed the starting point for a more general discussion of its structural features (11). In the latter paper we explored the minimal size requirements of the two connecting loops crossing the grooves of the quasi-continuous double helix which is formed upon the putative coaxial stacking of the two helical segments S1 and S2. It is obvious that this length will be dependent on the number of base pairs to be crossed, but the surprising outcome was that, due to the geometry of an RNA-A type double helix, loop L1 could be as short as two or even one nucleotide when spanning 5 to 7 base pairs (Figure 3). Examples of such phylogenetically and experimentally proven short loops have been found indeed (*see below*). The distance to be spanned over the deep groove is in this case only about 12 Å thanks to the close proximity of the two opposing strands (Figure 3). The loop L2 spanning the shallow groove on the other hand should minimally consist of three nucleotide residues when spanning 3 base pairs. A rule of thumb is that the first loop region encountered from the 5′ end is always the one which crosses the deep groove.

This all means that the most simple classic pseudoknot can be formed of stems of 5 to 6 and 3 base pairs respectively. Many examples of these have been found in all kinds of RNAs. The pseudoknot in the tRNA-like structure of TYMV RNA at the very 3′ end is a typical example (Figure 1). The 4 nucleotides in loop L1 seem to be more than actually needed for crossing the deep groove, but a comparison with related tymoviral RNA sequences has shown that this loop length varies from 5 to as short as 2 nucleotides (14). Site-directed mutagenesis of the tRNA-like structure has demonstrated that this loop can be shortened to even one single residue without affecting the aminoacylation efficiency (15). These loop size requirements were confirmed by model building of the TYMV RNA pseudoknot with molecular graphics (16).

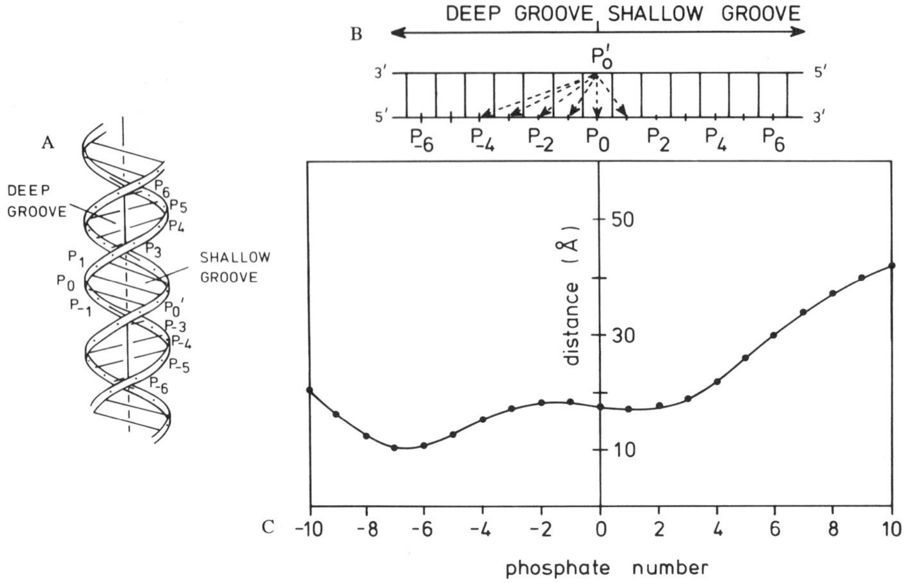

Figure 3. Distances between phosphate atoms in a regular RNA double helix of the A-type. The distance between a fixed phosphate on one strand (P_0') and the phosphates on the other strand is given. (A) Three-dimensional representation of the double helix showing the deep and shallow groove of the double helix. (B) Two-dimensional scheme. The arrows represent the distance between the fixed phosphate on one strand and the phosphate atoms on the other. (C) Graphical representation.

Another type of pseudoknot is obtained if the base pairing region in the hairpin loop is shifted to the other side of the loop. In that case the two stem regions also become adjacent and may coaxially stack. However, evidence for the existence of such pseudoknots has not yet been presented.

NMR AND THERMODYNAMIC STABILITY

A most important result about pseudoknot structure was obtained by Tinoco and coworkers (17,18). This group, using high resolution 2D NMR techniques, was able to get evidence for the coaxial stacking of the two stem regions in a model oligonucleotide (Figure 4). A nuclear Overhauser effect (NOE) between the iminoprotons of base pairs G3-C13 and C12-G22 confirmed this coaxial stacking, although some irregularities became apparent at the other side of the quasi-continuous helix where the two connecting loops emerge from the stem junction. This is not surprising, because of the unfavourable close proximity of the two loops and the possible sharp turns in the chain at this position.

Unfolding of the pseudoknot structure by raising the temperature or decreasing the Mg^{2+} concentration showed that the 5′ hairpin was the more resistant structure. This was unexpected because stem S1 in the pseudoknot structure consists of only 3 base pairs (see Figure 4). However, the resulting loop structure appeared to have peculiar conformational properties, including a C-A mismatch, explaining its higher stability than the 3′ hairpin. It will be evident that complete elucidation of one or more pseudoknotted structures with NMR or Röntgen diffraction techniques would be very helpful in evaluating structural details of pseudoknot folding and their dynamics. Unfortunately, the progress in this part of the analysis of pseudoknots is as slow as elsewhere in the study of RNA structures.

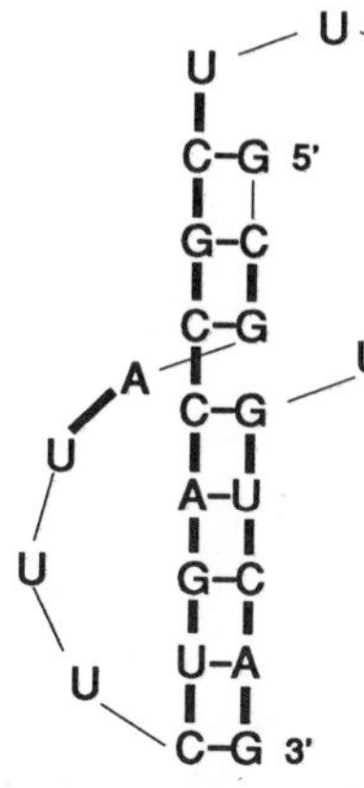

Figure 4. Schematic presentation of the structure of the RNA oligonucleotide PK5 as derived from a proton NMR study (adapted from (17)). The thick black bars represent nuclear Overhauser effects (NOEs) found between two base pairs. Thin lines indicate that no NOE interaction was observed. Note that also no NOE was found between the two G residues at the junction of the two stems S1 and S2.

Until now there has not been much insight into the thermodynamic consequences of pseudoknot formation, even in the case of the simple classic type. It will be obvious that the folding into pseudoknots is mainly driven by the decrease in enthalpy thanks to the formation of the base pairs in the second stem created. This contribution can be easily calculated from known nearest neighbour parameters (19) and could even take into account the coaxial stacking of the two stem regions. However, the stability of the pseudoknot is also greatly dependent on the (positive) contribution to the free energy by the formation of the single-stranded connecting loops. The values for each separate loop are not easy to determine experimentally and are presently unknown, despite some estimates used in programs for the computer prediction of RNA pseudoknots (see below). Nevertheless, we assume that especially the conformation of the loop spanning the shallow groove (loop L2, Figure 2) will be responsible for the relatively low stability of the classic pseudoknot. If this loop is short as in the case of TYMV RNA (Figure 1), its base residues are probably partially unstacked and pointing into the solvent, just because this groove is simply too shallow to harbor the loop nucleotides. This conclusion is in agreement with results obtained with chemical probing and model building (16). The situation for the other loop seems to be more favourable, because its residues can be buried in the deep groove quite easily, as model building has clearly shown (16,20). It is even possible that residues of loop L1 may contact the base pairs of stem S2, thus forming base triplets or other interactions.

Wyatt et al.,(18) have found that the change in free energy upon folding of their model oligonucleotide into a pseudoknot was only about 1.5-2 kcal/mol as compared to one of the two hairpins in the unfolded state. These values are in agreement with those estimated by Mans et al. on the basis of site-directed mutagenesis of the pseudoknot in TYMV RNA (15). The reason for this relatively low stability might be that S1 in both cases consists of only three base pairs , which can just compensate for the possibly unfavourable conformation of loop L2.

Wyatt et al. have also studied the influence of loop size and base composition on stability using a model oligonucleotide (18). A pseudoknot was found as the prevailing structure only when loop L1 contained three or more nucleotides and loop L2 four or more, although the equilibrium between pseudoknot and the 5′ or 3′ hairpin appeared to be dependent on base composition in a subtle way.

Gluick and Draper recently studied the folding of the complex pseudoknot structure from the leader of *Escherichia coli* α-mRNA (21). An unfolding pathway with seven different states was proposed and it was concluded that a simple model of RNA folding in which an optimum secondary structure forms first, followed by tertiary interactions that further stabilize the secondary structure, does not hold in this RNA pseudoknot. It remains to be seen if this finding also holds for more simple pseudoknot structures.

PREDICTION OF PSEUDOKNOTS

The formation of pseudoknots is just the addition of another stem in an otherwise regular secondary structure. Therefore most methods used for the determination of the secondary structure can be applied to the study of pseudoknots. Among them are structure probing with chemical modification and enzymatic digestion, sequence comparisons and covariation search and site-directed mutagenesis (13).

Some fundamental problems arise, however, when one wants to predict RNA pseudoknot structures by using computer programs. As mentioned above, Studnicka et al. (4) just came across pseudoknot formation in developing an algorithm for the prediction of RNA secondary structure. The most popular program so far, developed by Zuker and coworkers (22) and based on the calculation of the global energy minimum, does not allow for pseudoknot formation. This dynamic programming method would otherwise become very complex, because the final result would no longer be the sum of the energies of separate structural elements.

Two programs have been developed which are able to predict pseudoknots in an otherwise orthodox RNA secondary structure (23,24). These programs differ from the exhaustive search for an energy minimum, in that they simulate RNA folding by the stepwise addition of stems to a growing secondary structure. At each step the addition of a stem leading to a pseudoknot is possible.

A major problem for any program predicting pseudoknot structures at the moment is the lack of energy parameters for the connecting loops formed. Abrahams et al. (23) introduced a single value of 4.2 kcal/mol for all connecting loops, independent of size and base composition. This value produced the best results in predicting pseudoknot structures as proven by phylogenetic comparisons. It will be clear that such a single value is an over-simplification, especially in the case of somewhat more complicated pseudoknot structures with long connecting loops.

FUNCTIONAL ASPECTS

Pseudoknotted structures have now been found in almost every class of RNA. They are present in viral RNAs and messenger RNAs in both coding and non-coding regions, although they are very common at 5′ and especially 3′ termini. Several pseudoknots are present in ribosomal RNAs, where they seem to play a crucial role in protein biosynthesis. Perhaps the small tRNAs are an exception unless one considers the G-C pair formed upon interaction of the T- and D-loop to be a pseudoknot as well. In a few cases the function of the RNA appears to be directly dependent on the special stereochemical properties of the pseudoknot. It is conceivable, however, that many RNA pseudoknots make the RNA just more compact and more rigid in a way which is reminiscent of the S-S bridge in protein structures.

The functional significance of pseudoknots will be reviewed below. For earlier reviews about functional aspects see (1,2).

Viral RNAs

After the first description of the classic pseudoknot in the tRNA-like structure of TYMV RNA (Figure 1), plant viruses have provided many examples of this structural element, especially in the 3′ non-coding region. All three aminoacylatable groups of plant virus RNAs (tymo-, bromo- and tobamoviruses) have tRNA-like structures at their 3' end in which a pseudoknot plays an essential role in folding this domain. Aminoacylation was shown to be fully dependent on the presence of the pseudoknot (15,25), although it is still not clear what the function of the tRNA-like structure is in the life cycle of these plant viruses (for a review see Ref. (26)). In this respect it is interesting to note that Haenni and coworkers have shown that a 3′ terminal fragment of only 38 nucleotides of the plus sense TYMV RNA can be specifically recognized and replicated *in vitro* by the RNA dependent RNA replicase complex isolated from infected Chinese cabbage (27). It is therefore very interesting to know which part or feature of the pseudoknot structure is responsible for the binding of the replicase.

The histidine-accepting tRNA-like structure of tobacco mosaic virus RNA has an aminoacyl acceptor arm which strongly resembles that of TYMV RNA (Figure 5). It is remarkable, however, that this domain consists of only 11 stacked base pairs, whereas the standard elongator $tRNA^{His}$ contains 13 base pairs due to an extra base pair in the aminoacyl acceptor arm.

The pseudoknot-containing tRNA-like structure of BMV RNA has recently been submitted to a careful structure probing and modeling study which has led to an improved model for its three-dimensional structure, although its resemblance with $tRNA^{Tyr}$ remains largely elusive (28).

A recurrent characteristic of nonpolyadenylated plant viral RNAs is the presence of one or more pseudoknot structures in the 3′ UTR, just downstream of the stop codon. For instance, tobacco mosaic virus RNA contains three consecutive pseudoknots just between stop codon and the 3′ terminal tRNA-like structure (Figure 5) (29). These three pseudoknots are not only experimentally supported, but also strongly confirmed by sequence comparisons with related tobamoviral RNAs, including the satellite tobacco mosaic virus (30,31). In the tobamoviral odontoglossum ringspot virus (ORSV) RNA even 11 pseudoknots in a row are present due to duplications of some of the original pseudoknots (31).

Interestingly these pseudoknots in the 3′ UTR of TMV RNA seem to play a role in the initiation of protein synthesis like the polyA tail in cellular messenger RNAs (32). Gallie and coworkers have carried out an extensive mutational analysis of this so-called upstream pseudoknot domain (UPD) and have found that the higher order structure of the two 3′ proximal pseudoknots is essential for the regulation of the translation (33). A somewhat different conclusion was reached by Hall and coworkers for the pseudoknots in the 3′ UTR of BMV RNA. Deletion studies on genomic RNA 3 pointed to a minor role in translational control (34). Although a deletion study on this UPD clearly showed that this RNA segment was essential for viral multiplication its precise function remains to be established.

Pseudoknot structures have been reported to occur in both the 5′ and 3′ non-coding regions of picornavirus RNAs. For the pseudoknots proposed in the 3′ UTR it is conceivable that they play a role in the initiation of minus strand synthesis. Two different models were recently proposed for the folding of the 3′ UTR (35,36). One of these involved an interaction between two hairpin loops and was claimed to have tRNA-like features (35).

Pseudoknots of the classic type were proposed in the 5′ UTR of foot and mouth disease virus (FMDV) (37), encephalomyocarditis virus (EMCV) (38) and hepatitis A virus

(HAV) (39). Their position in the sequence is interesting in that they occur just upstream of the ribosomal landing pad and are associated with a long single-stranded region, which in the cardioviruses also involves a long polyC stretch. However, no function has as yet been ascribed to these pseudoknots.

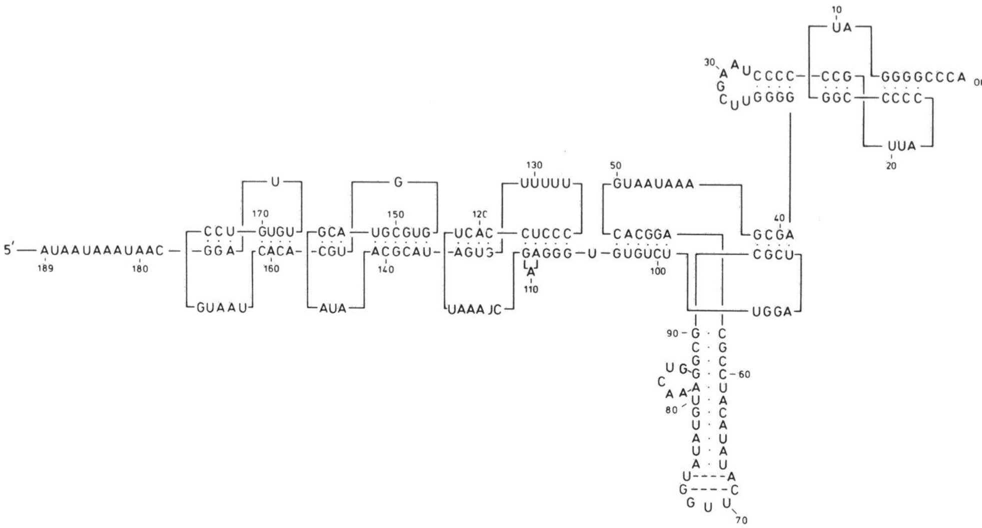

Figure 5. The folding of the 3′ UTR of tobacco mosaic virus RNA. The tRNA-like structure accepting histidine comprises nucleotides 1-105. The stalk region or upstream pseudoknot domain (UPD) with three consecutive classic pseudoknots runs from nucleotides 106-179. Numbering is from the 3′ end.

Pseudoknots and Translation

Prokaryotic messenger RNAs. The 5′ leaders of a few prokaryotic messenger RNAs have been shown to contain pseudoknotted structures, which are able to bind proteins encoded by these mRNAs, thereby autoregulating the expression of these proteins. This was first demonstrated for the gene 32 protein mRNA of bacteriophage T4 (40). Gene 32 protein binds to single-stranded nucleic acids in a cooperative manner. Although the leader of gene 32 mRNA is mainly single-stranded, it was shown that binding of gene 32 protein starts at low concentrations at the very 5′ end, just where a classic pseudoknot is located. Subsequently, cooperative binding of gene 32 protein to the adjacent single-stranded region takes place at higher concentrations. The protein finally also covers the ribosome binding site and thus represses the translation of its own mRNA. It is not yet clear which feature of the compact pseudoknot structure is responsible for the initial strong binding of the protein, but it may well be that it is related to the exposed and destacked character of the loop crossing the shallow groove of the pseudoknot, i.e., loop L2.

This type of autogenous regulation has also been demonstrated for two other messenger RNAs from *E.coli*. The translation of the messenger of the alpha-operon is regulated by the ribosomal protein S4, which binds to a complex, so-called double pseudoknot structure in the leader (41). From S4 binding studies and "toeprint" assays it was concluded that the repressor protein S4 does not displace the ribosome from the mRNA, but rather has an allosteric effect on the initiation complex of messenger RNA, 30S ribosomes and $tRNA_f^{Met}$, thereby preventing proper initiation (42). Similar results were obtained for the messenger RNA coding for S15, another protein of the small ribosomal subunit. By stabilizing the pseudoknot-containing conformation of the leader, it also blocks the ribosome in a preinitiation complex so that it shuts off its own synthesis (43).

Ribosomal Frameshifting and Readthrough. In general one can expect that coding regions of messenger RNAs will not harbor functionally important secondary or tertiary structures, because the ribosome needs to disrupt this structure first before codon-anticodon interaction can take place. In many cases, programmed ribosomal frameshifting and readthrough in viral RNAs appear to be dependent on pseudoknot structures downstream of the site where the translational "error" takes place (44). This mechanism is especially common in retroviruses, where the *pol* cistron is translated as a fusion protein with the upstream structural proteins by either a frameshift of the ribosome in the -1 direction or by suppression of a stop codon separating in-frame the two coding regions. Jacks and coworkers proposed in their "simultaneous slippage" model that both the bound aminoacyl tRNA and the peptidyl tRNA move one base into the -1 direction on a slippery heptanucleotide sequence having the following motif: XXXYYYN, where X is any base, Y is A or U and N is A,U, or C. This sequence makes it possible that upon shifting in the -1 direction and with the aminoacyl tRNA decoding the YYN codon, two out of three base pairs remain identical.

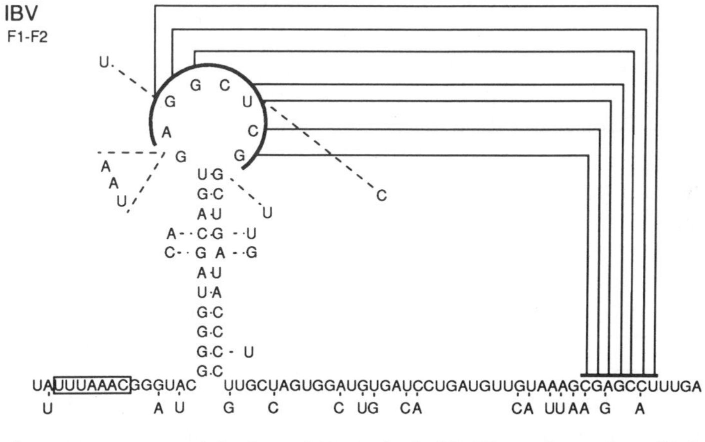

Figure 6. Secondary structure around the frameshift site in the F1-F2 overlap region of infectious bronchitis virus (IBV) RNA. The box indicates the shifty heptanucleotide sequence. The base substitutions and the UAA insertion correspond to the related sequence of mouse hepatitis virus (MHV) RNA. The solid lines indicate the pseudoknot-forming base pairing.

Brierley and coworkers demonstrated for the coronavirus IBV that a pseudoknot structure, located 6 nucleotides downstream of the slippery heptanucleotide in the orf 1A-orf 1B overlap, was essential for efficient -1 frameshifting (Figure 6) (45). In a later study it was shown that this pseudoknot could not be replaced by a simple hairpin of the same length and with the same base pairs as in the two stems of the pseudoknot, indicating that some specific structural feature of the pseudoknot itself is responsible for the frameshifting (46). ten Dam et al. found that many viral RNAs possessing overlapping reading frames have frameshift signals consisting of the heptanucleotide sequence motif and a pseudoknot at 4 to 8 nucleotides downstream of the slippery sequence (47). Pseudoknot-dependent -1 frameshifting has been proven in a number of cases, including plant viruses (48), animal viruses (49,50) and yeast viruses (51).

It is still not known how the pseudoknot structure influences the frameshifting, but a common explanation is that the pseudoknot is responsible for stalling the ribosome when decoding the slippery heptanucleotide sequence (52), giving the ribosome more time to respond to the signal. It may well be that the pseudoknot structure is more resistant to unwinding by a helicase activity in the ribosome.

In type C retroviruses the *pol* gene can only be expressed by readthrough of an amber stop codon, located in-frame between the *pol* and *gag* genes. ten Dam et al. (47) proposed that also in this case a downstream pseudoknot structure could function as a stimulator. This was confirmed by a mutational analysis of the sequence around the stop codon of Moloney murine leukemia virus RNA (53). It was found that a pseudoknot structure starting 8 nucleotides downstream of the amber codon was essential for readthrough. Moreover it turned out that the situation with readthrough is somewhat more complicated in that the identity of some sequences in the loop L2 and in the spacer between the stop codon and the pseudoknot were important as well.

Ribosomal RNAs

The small subunit ribosomal RNA contains three well-established pseudoknot structures which are strongly supported by phylogenetic evidence (54), while two of them have been experimentally tested by site-directed mutagenesis (55,56). The first of these involves the first hairpin at the 5′ end and connects the hairpin loop with a complementary stretch about 900 nucleotides downstream (Figure 7A) (11). This so-called central pseudoknot connects three major domains of the ribosomal RNA. There is strong direct evidence that this pseudoknot structure is not only important for the three-dimensional folding of this RNA, but that it also plays a key role in protein biosynthesis. Mutations weakening or disrupting this pseudoknot affect the formation of the 70S ribosomal complex (56). It is conceivable that this pseudoknot is involved in a major conformational switch during one or more steps in the elongation cycle.

The second well-supported pseudoknot in 16S rRNA is formed by base pairing between a hairpin loop in the 530 region and a bulge loop in the same hairpin (Figure 7B). This hairpin loop harbors a strongly conserved nucleotide sequence so that phylogenetic evidence for this base pairing could only be collected very recently. It is one of the few examples of a pseudoknot between a bulge loop and a hairpin loop (12). In fact three different stem regions can be discerned in the resulting structure and so far it is not clear how their mutual orientation or stacking is or might be. The functional role of this pseudoknot has become evident from a mutational study by Powers and Noller (55). Their results not only confirmed the existence of this pseudoknot but also showed that it is involved in the binding of the antibiotic streptomycin to the 16S rRNA, though in an indirect way. This antibiotic causes misreading upon binding to the ribosome, but mutations in this pseudoknot, leading to disruption or weakening of the base pairing, confer weak resistance to streptomycin.

A third more complicated pseudoknot structure is proposed to arise from base pairing of two nucleotides in a tetraloop with two complementary bases in a multiple loop some 300 nucleotides upstream (54). This pseudoknot structure has not yet been tested experimentally so that its existence is supported by phylogenetic comparisons only.

Many other pseudoknots have been proposed in the large subunit ribosomal RNA. In a recent review, Gutell et al. enumerated some 15 different tertiary interactions, which

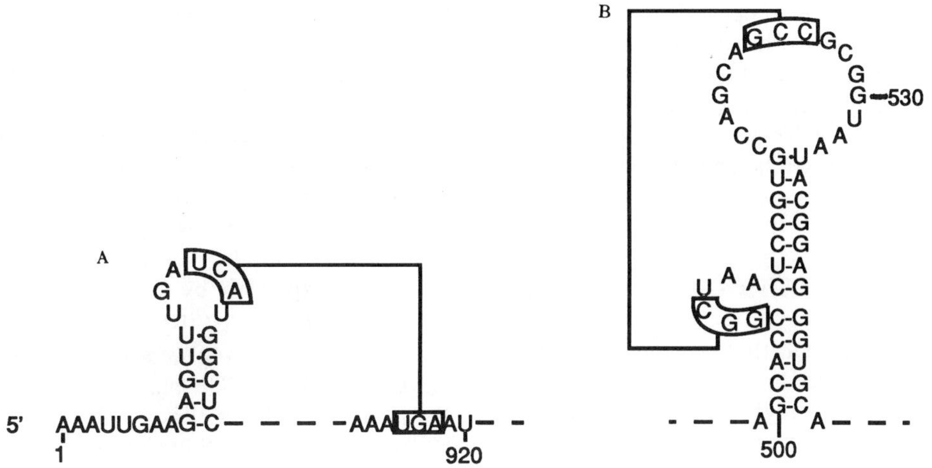

Figure 7. Two pseudoknot structures from the 16S ribosomal RNA from *E. coli*. (A) The central pseudoknot involving the 5′ terminal hairpin and a complementary sequence some 900 nucleotides downstream. (B) The pseudoknot from the 530 region. Note that the pseudoknot-forming tertiary interaction consists of base pairing between a hairpin loop and a bulge loop.

meet the definition of a pseudoknot, including loop-loop base pairing and single base pair interactions (54). All these pseudoknots were proposed on the basis of covariation search and convincing experimental evidence was obtained for only one of them. *In vitro* protein binding studies and mutations revealed the importance of the G1343:C1404 and U1344:A1403 pairings (*E. coli* numbering), bringing three helical domains together (57).

Catalytic RNAs

The folding of ribozymes and self-splicing RNAs in their active conformation depends in some cases on the formation of pseudoknotted structures. A well-known example is found in the core of the self-splicing group I introns. The interaction involves base pairing of one strand of an internal loop (between stems P7 and P8) and a single-stranded region upstream and was already proposed many years ago (11,58). The resulting stem region P3 was modeled as a coaxial stack with P7 and P8 and helps to fold the core of the intron into a compact, catalytically active structure. The guanosine cofactor hydrogen bonds to a G residue in the major groove of the P7 helix. For a detailed discussion of the three-dimensional structure of the group I introns, the reader is referred to Ref. (59). A second pseudoknot is formed in some group I intron sequences by base pairing of an internal loop with a hairpin loop and was supported by a site-directed mutagenesis study (60).

The catalytic RNA of RNase P from eubacteria contains two pseudoknots. A long range interaction between a multiple loop and an internal loop creates an eight base pair stem with one bulged U residue. This interaction brings a number of conserved base residues into proximity and is an important factor in creating the substrate binding pocket (61). The second pseudoknot is formed by base pairing of a bulge loop with a hairpin loop, though not in all eubacteria. In *B. subtilis* this pseudoknot is replaced by an additional hairpin structure, which is able to contribute to the final three-dimensional structure in a similar way as the pseudoknot (62). In this case it is suggested that the pseudoknot is simply an alternative to a standard secondary structure element and that it therefore has no special functional role.

Self-cleavage reactions occur in viral RNAs. For instance the genomic and antigenomic RNAs of hepatitis delta virus contain a self-cleavage site, which consists of some 90 nucleotides. Both cleavage sites can be folded in similar structures and harbor an RNA pseudoknot. The folding involves base pairing of a bulge loop with an upstream sequence in much the same way as the 5′ proximal pseudoknot in the tRNA-like structure of TMV RNA (compare Figure 5). Interestingly, the pseudoknot stabilizes the self-cleavage structure but is not essential for its formation (63). The proposed pairings in the self-cleaving structure have now been confirmed by structure mapping and mutational analysis. (64,65).

CONCLUDING REMARKS

It is clear that pseudoknot formation is an important structural motif in the three-dimensional folding of many RNA molecules. In general, pseudoknots may serve to bring together RNA regions which are far apart in the primary structure. They will make the RNA more compact and more rigid in much the same way as do S-S bridges in proteins. It prevents a large structure with only (branched) hairpins from having too much flexibility. Introduction of "crosslinks," especially when they are of long-range character, can impose the necessary rigidity to create a functionally active conformation. This is very likely to be the case for many of the pseudoknots in the ribosomal RNAs.

Sometimes the structural features of the pseudoknot itself are responsible for the particular function. This is evident in the case of the pseudoknot-dependent -1 frameshifting where replacement of the pseudoknot by a more or less equally stable hairpin structure gives a much less effective frameshifting.

Another attractive property of pseudoknots, especially in the case of the simple classic type, is the destacked, exposed nature of the loop spanning the shallow groove (loop L2, Figure 2). Pseudoknot formation may be one of the few ways to expose a series of hydrophobic base residues into the solvent, making them excellent candidates for protein binding sites. The autoregulation of some prokaryotic messenger RNAs may be a good example of such a function. More insight into the structural details of RNA pseudoknots is definitely needed, however, before we can evaluate and appreciate their important role in many different cellular processes.

REFERENCES

1 Westhof E., Jaeger L. (1992) Curr. Opin. Struct. Biol. 2, 327-333.
2 ten Dam, E., Pleij, K. and Draper, D. (1992) Biochemistry 31, 11665-11676.
3 Pleij, C.W.A. (1994) Curr. Opin. Struct. Biol. 4, 337-344.
4 Studnicka, G.M., Rahn, G.M., Cummings, I.W. and Salser, W.A. (1978) Nucl. Acids Res. 5, 3365-3387.
5 Richards, E.G. (1969) Eur. J. Biochem. 10, 36-42.
6 Ninio, J. (1971) Biochimie 53, 485-494.
7 Rietveld, K., Van Poelgeest, R., Pleij, C.W.A., Van Boom, J.H. and Bosch, L. (1982) Nucl. Acids Res. 10, 1929-1946.
8 Rietveld, K., Linschooten, K., Pleij, C.W.A. and Bosch, L. (1984) EMBO J. 3, 2613-2619,
9 Van Belkum, A., Abrahams, J.P., Pleij, C.W.A. and Bosch, 1. (1985) Nucl. Acids Res. 13, 7673-7686.
10 Pleij, C.W.A., Abrahams, J.P., Van Belkum, A., Rietveld, K. and Bosch, L. (1987) in Positive strand RNA viruses (Brinton M.A., Rueckert R., eds.) UCLA Symp. on Mol. Cell. Biol., Ser.54. pp. 299-316, Liss, New York, NY.

11 Pleij, C.W.A., Rietveld, K. and Bosch, L. (1985) Nucl. Acids Res. 13, 1717-1731.
12 Mans, R.M.W. and Pleij, C.W.A. (1993) in Nucleic Acids and Molecular Biology (Eckstein, F. and Lilley, D.M.J., eds.) Vol. 7, pp. 250-270, Springer-Verlag, Berlin, Heidelberg.
13 Pleij, C.W.A. and Bosch, L. (1989) Meth. Enzymol. 180a, 289-303.
14 Van Belkum, A., Bingkun, J., Rietveld, K., Pleij, C.W.A. and Bosch, L. (1987) Biochemistry 26, 1144-1151.
15 Mans, R., Van Steeg, M.H., Verlaan, P.W.G., Pleij, C.W.A. and Bosch, L. (1992) J. Mol. Biol. 223, 221-232.
16 Dumas, P., Moras, D., Florentz, C., Giegé, R., Verlaan, P., Van Belkum, A. and Pleij, C.W.A. (1987) J. Biomol. Struct. Dynam. 4, 707-728.
17 Puglisi, J.D., Wyatt, J.R and Tinoco, I. Jr. (1990) J. Mol. Biol. 214, 437-453.
18 Wyatt, J.R., Puglisi, J.D. and Tinoco, I. Jr. (1990) J. Mol. Biol. 214, 455-470.
19 Turner, D.H., Sugimoto, N. and Freier, S. (1988) Annu. Rev. Biophys. Biophys. Chem. 17, 167-192.
20 Major, F., Turcotte, M., Gautheret, D., Lapalme, G., Fillion, E. and Cedergren, R. (1991) Science 253, 1255-1260.
21 Gluick, T.C. and Draper, D.E. (1994) J. Mol. Biol. 241, 246-262.
22 Zuker, M. and Stiegler, P. (1981) Nucl. Acids Res. 9, 133-148.
23 Abrahams, J.P., Van der Berg, M., Van Batenburg, E. and Pleij, C.W.A. (1990) Nucl. Acids Res. 18, 3035-3044.
24 Gultyaev, A.P. (1991) Nucl. Acids Res. 19, 2489-2494.
25 Dreher, T.W. and Hall, T.C. (1988) J. Mol. Biol. 201, 41-55.
26 Mans, R.M.W., Pleij, C.W.A. and Bosch, L. (1991) Eur. J. Biochem. 201, 303-324.
27 Gargouri-Bouzid, R., David, C. and Haenni, A-L. (1991) FEBS Lett. 294, 56-58.
28 Felden, B., Florentz, C., Giegé, R. and Westhof, E. (1994) J. Mol. Biol. 235, 508-531.
29 Van Belkum, A., Abrahams, J.P., Pleij, C.W.A. and Bosch, L. (1985) Nucl. Acids Res. 13, 7673-7686.
30 Felden, B., Florentz, C., McPherson, A. and Giegé, R. (1994) Nucl. Acids Res. 22, 2882-2886.
31 Gultyaev, A.P., Van Batenburg, E. and Pleij, C.W.A. (1994) J. Gen. Virology 75, (2851-2856).
32 Gallie, D.R. and Walbot, V. (1990) Genes Dev. 4, 1149-1157.
33 Leathers, V., Tanguay, R., Kobayashi, M. and Gallie, D.R. (1993) Mol. Cell. Biol. 13, 5331-5347.
34 Lahser, F.C., Marsh, L.E. and Hall, T.C. (1993) J. Virol. 67, 3295-3303.
35 Pilipenko, E.V., Maslova, S.V., Sinyakov, A.N. and, Agol, V.I. (1992) Nucl. Acids Res. 20, 1739-1745.
36 Jacobson, S.J., Konings, D.A.M. and Sarnow, P. (1993) J. Virol. 67, 2961-2971.
37 Clarke, B.E., Brown, A.L., Currey, K.M., Newton, S.E., Rowlands, D.J. and Carroll, A.R. (1987) Nucl. Acids Res. 15, 7067-7078.
38 Duke, G.M., Hoffman, M., Palmenberg, A.C. (1992) J. Virol. 66, 1602-1609.
39 Brown, E.A., Day, S.P., Jansen, R.W. and Lemon, S.M. (1991) J. Virol. 65, 5828-5838.
40 McPheeters, D.S., Stormo, G.D. and Gold, L. (1988) J. Mol. Biol. 201, 517-535.
41 Deckman, I.C. and Draper, D.E. (1987) J. Mol. Biol. 196, 323-332.
42 Spedding, G. and Draper, D.E. (1993) Proc. Nat. Acad. Sci. U.S.A. 90, 4399-4403.
43 Philippe, C., Eyermann, F., Bénard, L., Portier, C., Ehresmann, B. and Ehresmann, C. (1993) Proc. Nat. Acad. Sci. U.S.A. 90, 4394-4398.
44 Farabaugh, P.J. (1993) Cell 74, 591-596.

45 Brierley, I., Digard, P. and Inglis, S.C. (1989) Cell 57, 537-547.
46 Brierley, I., Rolley, N.J., Jenner, A.J. and Inglis S.C. (1991) J. Mol. Biol. 229, 889-902.
47 ten Dam, E.B., Pleij, C.W.A. and Bosch, L. (1990) Virus Genes 4, 121-136.
48 Kujawa, A.B., Drugeon, G., Hulanicka, D. and Haenni, A-L. (1993) Nucl. Acids Res. 21, 2165-2171.
49 Chamorro, M., Parkin, N. and Varmus, H.E. (1992) Proc. Nat. Acad. Sci. U.S.A. 89, 713-717.
50 ten Dam, E., Brierley, I., Inglis, S. and Pleij, C. (1994) Nucl. Acids Res. 22, 2304-2310.
51 Dinman, J.D., Icho, T. and Wickner, R.B. (1991) Proc. Nat. Acad. Sci. U.S.A. 88, 174-178.
52 Somogyi, P., Jenner, A.J., Brierley, I. and Inglis, S.C. (1993) Mol. Cell. Biol. 13, 6931-6940.
53 Wills, N., Gesteland, R.F. and Atkins, J.F. (1991) Proc. Nat. Acad. Sci. U.S.A. 88, 6991-6995.
54 Gutell, R.R., Larsen, N. and Woese, C.R. (1994) Microbiol. Rev. 58, 10-26.
55 Powers, T. and Noller, H.F. (1991) EMBO J. 10, 2203-2214.
56 Brink, F.M., Verbeet, M.P. and De Boer, H.A. (1993) EMBO J. 12, 3987-3996.
57 Kooi, E.A., Rutgers, C.A., Mulder, A., Van 't Riet, J., Venema, J. and Raué, H.A. (1993) Proc. Nat. Acad. Sci. U.S.A. 90, 213-216.
58 Waring, R.B. and Davies, R.W. (1984) Gene 28, 277-291.
59 Michel, F. and Westhof, E. (1990) J. Mol. Biol. 216, 585-610.
60 Jaeger, L., Westhof, E. and Michel, F. (1991) J. Mol. Biol. 221, 1153-1164.
61 James, B.D., Olsen, G.J., Liu, J. and Pace, N.R. (1988) Cell 52, 19-26.
62 Haas, E.S., Morse, D.P., Brown, J.W., Schmidt, F.J. and Pace, N.R. (1991) Science 254, 853-856.
63 Perrotta, A.T. and Been, M.D. (1991) Nature 350, 434-436.
64 Perrotta, A.T. and, Been M.D. (1993) Nucl. Acids Res. 21, 3959-3965.
65 Kumar, P.K.R., Taira, K. and Nishikawa, S. (1994) Biochemistry 33, 583-592.

ROLE OF MOLECULAR CHAPERONES IN THE INITIATION OF PLASMID DNA REPLICATION

Dhruba K. Chattoraj

Laboratory of Biochemistry
National Cancer Institute
National Institutes of Health
Building 37, Room 4D-18
Bethesda, MD 20892

INTRODUCTION

The principles that govern protein folding, the process by which a polypeptide chain assumes its native form, remain largely unknown (1). It can be spontaneous or mediated by other proteins, collectively called molecular chaperones (2). By preventing nonproductive interactions, the chaperones can help in the assumption of secondary and tertiary structures of polypeptide chains and, in the case of multisubunit proteins, help their assembly into quaternary structures (3,4). Proteins that normally do not require chaperones for folding, may do so under conditions of stress which cause the proteins to denature and aggregate, since the chaperones can also refold partially denatured proteins into their native forms, prevent their aggregation, and can disaggregate protein aggregates (5-9). By controlling folding the chaperones can modulate virtually all subsequent fates of the polypeptide chains: phosphorylation (10), assembly into and disassembly from multimeric structures (11-13), transport across membranes (14), proteolysis of aberrant proteins (15,16) etc. Thus the knowledge of the mechanism of action of the chaperones will not only aid in our understanding of protein folding, but also of protein-protein interactions in general.

Several aspects of chaperone proteins are being investigated currently: induction and repression of their synthesis (17-19), recognition of substrate polypeptides (20,21), nucleotide binding, hydrolysis and release, and associated conformational changes that cause polypeptide binding and release (22-28), the requirement for multiple chaperones for a single function, the functioning of chaperone machines (29,30), structure-function analysis (31-32) and recycling (33). Overview of the recent progress on these topics can be obtained in refs. 34 and 35. In this article, the function of chaperones in DNA replication will be discussed.

Genetic Engineering, Vol. 17, Edited by J.K. Setlow
Plenum Press, New York, 1995

Bacteriophage λ Replication

Genetic Evidence for the Involvement of DnaK, DnaJ and GrpE. Involvement of chaperones in DNA replication was discovered in an attempt to identify host proteins necessary for the growth of *Escherichia coli* phage λ (36). Three of these proteins, DnaK, DnaJ and GrpE, were later shown to belong to a set of *E. coli* proteins, called heat shock proteins, that are universally conserved and whose rate of synthesis transiently increases after a shift to higher temperatures. The genes that encode about 20 or so identified heat shock proteins compose a regulon that is controlled by the gene product of *rpoH* (formerly called *htpR*) (37). The gene product is a sigma factor, σ^{32}, that recognizes specifically promoters of the heat shock genes (38). Mutations in *dnaK, dnaJ* and *grpE* block λ DNA replication at all temperatures, signifying that the level of expression in unshocked cells suffices for the replication of the phage (36). These studies also showed that chaperones can function without requiring stress. In fact, chaperones perform several housekeeping functions in addition to their role in minimizing protein damage during stress and restoring function of inactive proteins (34).

Replicon Structure and Function. Phage λ replicates bidirectionally from a fixed origin on negatively supercoiled templates. The basic replicon, i.e., the minimal region of the genome that can replicate with the characteristics of the intact genome, is about 2.4 kb long. Its important elements are a promoter, p_R, and three downstream genes, *cro, O* and *P* that together constitute an operon (Figure 1). The origin is embedded within the *O* gene. Transcription is necessary not only to synthesize the initiator proteins O and P, but also for activation of the origin (39,40). The Cro protein can bind to sites o_R that overlap p_R and thereby repress the activity of the promoter. Cro is, therefore, an autoregulatory protein and, by controlling the activity of p_R, the protein also controls the initiation frequency. Because of this control, the basic replicon of λ can be maintained as a circular plasmid in a growing culture with a copy number determined by the concentration of Cro protein (41). These mini-derivatives of the phage have also provided convenient substrates for replication studies *in vitro*.

Mutations in λ that could overcome the replication block in hosts mutated in any of the heat shock genes, *dnaK, dnaJ* and *grpE*, were mapped to gene *P*. Some of these *P* mutants could also overcome the replication defect due to mutations in the host gene *dnaB* (36). The DnaB protein is a helicase that is essential for initiation and elongation of DNA replication forks in *E. coli* and its phages and plasmids (42,43). These studies provided the initial genetic evidence that all three chaperones could be influencing the steps in initiation requiring both λP and host DnaB proteins. A deeper understanding of the functions of these chaperone proteins came from *in vitro* replication studies with purified proteins (13,44). These pioneering studies not only provided insight into their mechanism of action, but together with the work on *oriC* plasmids, helped define the early steps of initiation as strand-opening, which sets the stage for loading of DnaB, and finally, activation of the helicase function (43). These studies have been reviewed recently (43,45) and will be discussed here briefly.

Staging of the Initiation Reaction *in vitro*. Dissection of initiation of λ replication into discrete steps has been possible by isolation of stable intermediates in some cases (11-13). In step 1, λO protein dimers bind to the origin stably at four repeating sites (iterons) of hyphenated dyad symmetry forming a large nucleoprotein structure called an O-some (46; Figures 1,2). On negative supercoiled DNA the binding can make the adjoining A+T rich region reactive to single-strand specific nucleases (47) and chemical probes (R. McMacken, unpublished data). However, the fine structure of the reactive positions Figure

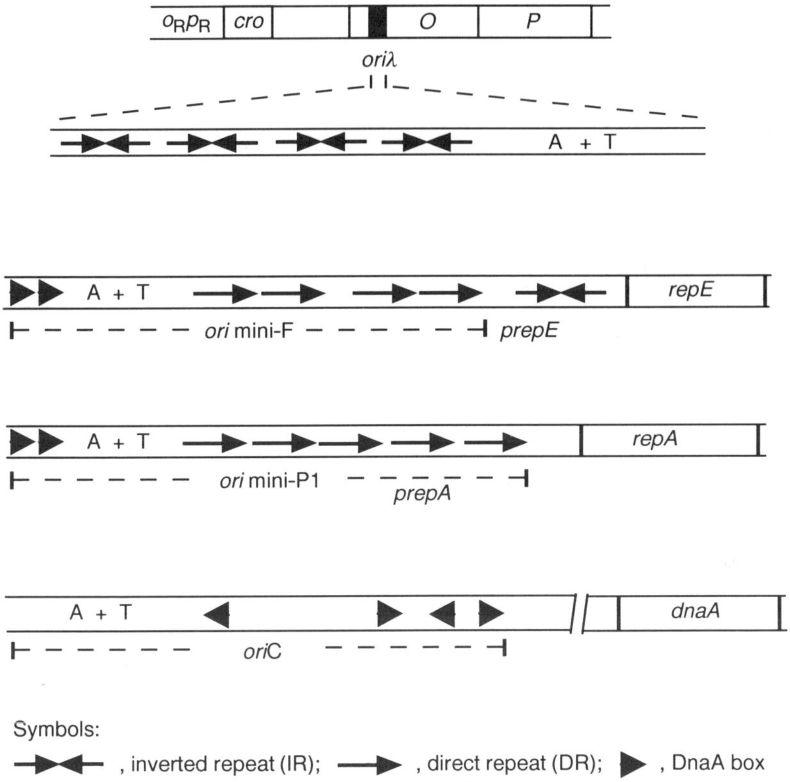

Figure 1. Maps of plasmid replicons showing features relevant to this review. The genetic map of λ replicon is shown at the top. Note that *ori*λ (shaded region) is embedded within the *O* gene. The origin is expanded to show the four inverted repeats (pairs of arrows) to which O protein dimers bind. Mini-F and mini-P1 plasmids are organized very similarly except for the promoter region. In mini-F the promoter, *prepE*, is outside the minimal origin and overlaps an inverted repeat to which RepE dimers bind for autorepression. RepE monomers bind to the direct repeats (long arrows) in the origin for initiation. In mini-P1, the promoter, *prepA*, is embedded within the direct repeats of the origin to which RepA monomers bind for initiation as well as autorepression of the promoter. In *oriC*, the four DnaA binding sites and their orientations are shown by arrowheads. The initiator gene, *dnaA*, is not contiguous with the origin. Note that DnaA binding sites are also present in mini-F and mini-P1 origins. A+T represents a stretch of about 50 bp that are rich in A and T, and the stretch is a common feature of replication origins. The maps are not drawn to scale but the minimal origins include no more than about 250 bp.

depends on the probe, indicating that the strands may not be fully separated by O binding alone. In step 2, the complex of λP-DnaB proteins joins the O-some to form the *ori*-O-P-DnaB complex. P is required for loading of DnaB for two reasons. DnaB normally complexes with DnaC for loading onto the *E. coli* origin. P can dissociate the DnaB-DnaC complex and the P-DnaB complexes are recruited at the λ origin through the O-P interaction. Both genetic and physical evidence for specific pairwise interactions between O-P and between P-DnaB have been obtained (43,48,49). The stability of the complex most likely arises from the fact that P can simultaneously bind to O and DnaB. Also the available data suggest that P alone binds weakly to O whereas the P-DnaB complex binds much more tightly to O (12). DnaB is nonfunctional in this complex because P quenches all known activities of DnaB: ATPase, general priming, single-stranded DNA-binding and helicase (43,48). The dissociation of P-DnaB complexes requires the chaperones. In step 3, DnaJ binds the complex stably. Interactions between DnaJ and DnaB have been reported

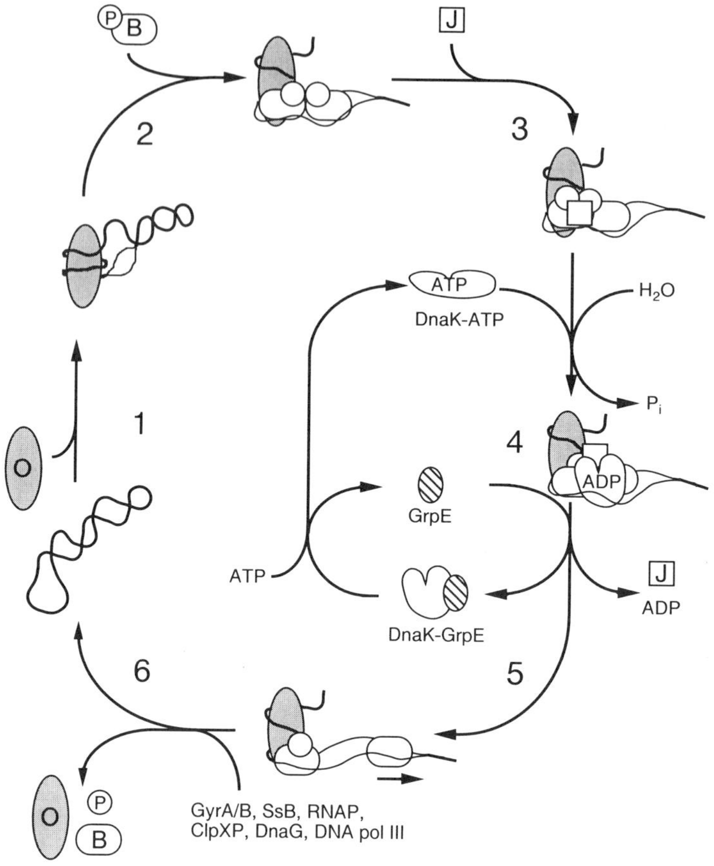

Figure 2. A model of initiation of λ replication based on *in vitro* studies with purified components. In step 1, λO binds to λ*ori* and causes a localized change in the topology of the strands on a supercoiled template. This allows DnaB to load at the origin as a complex with λP (step 2). In the P-DnaB complex, DnaB cannot function as a helicase. In order for initiation to proceed, the P inhibition of DnaB must be overcome by the combined action of the chaperones DnaK, DnaJ and GrpE in the presence of ATP. First, DnaJ can join the complex, most likely by interacting with the P and DnaB proteins (step 3). Next an ATP-DnaK complex is believed to recognize the P-DnaJ complex (step 4). Substrate binding and the presence of DnaJ activates the intrinsic ATPase activity of DnaK. The complex at step 4 is unstable. If sufficient DnaK is present, a partial disassembly reaction ensues that dissociates some P from DnaB so that the helicase can now be active in initiating strand unwinding. In the presence of GrpE, the levels of DnaK required drop substantially. GrpE helps by targeting DnaK to the P-DnaJ complex and also by releasing bound nucleotides from DnaK, and thereby recycling the protein. Unwinding and replication at this stage is unidirectional largely to the right (step 5). Most likely this is due to incomplete disassembly and the roadblock created especially by the O-*ori* complex. These strong interactions can be broken by transcription across the origin, and rebinding of O to *ori* can be prevented by selective proteolysis of free O proteins with ClpXP protease. SSB (single-stranded DNA-binding protein) is not essential for the unwinding reaction but helps maintain the DNA in the unwound form for the subsequent priming reaction. DNA gyrase maintains the negative superhelical tension of the template and thereby helps unwinding by DnaB. DNA polymerase III extends the primers formed by DnaG and switches the replication to the elongation mode. Stoichiometries of the components in different complexes are largely unknown. In step 2, two DnaB complexes are shown to indicate that bidirectional unwinding requires one complex at each fork. DnaK is cartooned in two different conformations to highlight the role of conformational changes in triggering the activation of DnaB. (The figure is courtesy of M. Zylicz).

(49,50). DnaJ most likely also contacts with the P protein of the complex. Refolded P, but not the native P, has been shown to form a binary complex with DnaJ in glycerol gradients (50). Weak complexes of DnaJ with native P can also be detected by crosslinking with glutaraldehyde (49). The freeing of DnaB from P can be accomplished by the addition of DnaK and ATP, but the reaction is more efficient when GrpE is also included in the reaction (steps 4 and 5). The helicase can then move away from the origin, unwinding DNA as it moves. Significant unwinding of supercoiled DNA requires the presence of DNA gyrase. SSB protein maintains the single-stranded state so that primase and polymerase III holoenzyme can add to the prepriming structure and initiate replication. The disassembly reaction even in the presence of GrpE appears to be partial. Complete and permanent disassembly requires the presence of RNA polymerase and protease ClpXP (step 6; 51; M. Zylicz, unpublished data).

Mechanism of Chaperone Action. Several aspects of chaperone function have been revealed from the studies on λ replication. The most striking fact was the requirement of three chaperones for a single event, release of DnaB from P protein. This is noteworthy, since in other replicons to be discussed later, the comparable reaction can proceed without the chaperones. In λ, all three chaperones contribute to the ordered assembly and disassembly reactions (11-13,49). An ordered assembly reaction at the origin may ensure that P-DnaB complexes are not disrupted while P is in the process of recruiting DnaB to the origin. Indeed, the P-DnaB complex cannot be broken by the chaperones without their being localized at the λ*ori* (M. Zylicz, unpublished data; R. McMacken, unpublished data). To understand the role of the individual components involved in initiation (*ori*, O, P, DnaB, DnaJ, DnaK, GrpE and ATP) several partial reactions have been studied. The reaction possibilities are enormous, since most of the proteins can associate with each other as well as to more than one partner simultaneously. Moreover, some of the proteins can be present in more than one conformation (22) and in different multimeric forms (52,53). For example, P can react with O, DnaB, DnaJ and DnaK possibly using different domains or by changing its conformation (49). DnaK can participate in the disassembly reaction either as a monomer or in higher oligomeric forms, depending upon whether GrpE is present in the system (12,49). The presence of GrpE allows DnaK to function at a lower concentration, at which DnaK is predominantly monomeric. The conformation, affinity and stoichiometry of the proteins may therefore depend upon the reaction conditions, and this can explain how the same combination of proteins can be used both in the assembly and disassembly pathways.

Although much remains to be understood, several points are rather clear.

1. There is an absolute requirement for ATP in the disassembly reaction (12,54). DnaK seems to be the crucial player. Its ATPase activity (55), believed to be the primary driving force for its chaperone activity, most likely can cause a conformational change in the protein which in turn changes the conformation of P, resulting in the breakage of the P-DnaB contact (22,23). Substrate binding stimulates the ATPase activity and the requirement of the coupling of the ATPase activity to substrate binding for the chaperone function of DnaK has been suggested by studies with mutant DnaK protein (27). The nature of the conformational change in P is not known, but an involvement of proteolysis or of the intrinsic kinase activity of DnaK seems unlikely (12).

2. DnaK is required in higher molar ratio (at least 10-fold) than the other proteins in the reaction. In itself this is not surprising, since DnaK is one of the most abundant proteins of *E. coli*. The requirement for high levels of DnaK could be for

binding to the P-DnaB complex and sequestering of released P so that it cannot rebind (13). Multimers of DnaK are favored at high concentrations and these show higher affinity for P than do the monomers of DnaK (49). There is also the possibility that the concentration of free DnaK is much lower than the total DnaK concentration because of its ability to bind many other replication proteins with low affinity. Finally, DnaK can exist in multiple conformations (free or associated with ATP or ADP), of which only some are proper for a reaction.

3. GrpE can reduce the requirement for DnaK by 10-fold (12,13). In a mixture of GrpE and DnaK, GrpE increases the release of ATP and ADP from DnaK without affecting the rate of hydrolysis and thereby may have a role in recycling DnaK (33). It can also play more direct roles in the assembly/disassembly pathways. It increases the affinity of monomeric DnaK for P but more importantly for the P-DnaJ complex, which is presumably a more relevant intermediate (49). Additionally, GrpE has been shown to dissociate the P-DnaJ-DnaK ternary complex (50). Thus both the affinity and fidelity of DnaK-catalyzed reactions could be improved in the presence of GrpE.

4. DnaJ increases the weak intrinsic ATPase activity of DnaK and, together with GrpE, helps recycling of DnaK (33). Unlike GrpE, DnaJ plays an essential role in the assembly/disassembly pathways *in vitro*. It can form binary complexes with P, DnaK and DnaB, stabilize the O-P complex, and increase the specificity of the DnaK interaction with P protein in the presence of GrpE as stated earlier. The most crucial role of DnaJ may be to modulate the conformational state of DnaK by controlling its ATPase activity.

RNA Polymerase, ClpXP Protease and the Directionality of Replication. One of the shortcomings of the *in vitro* system is that, until recently, the replication was found to be largely unidirectional rather than bidirectional, the norm *in vivo* (51). This discrepancy has been attributed primarily to incomplete disassembly of the preprimosomal complexes (12,45). In fact DnaB helicase activation is possible without P coming off from the complex when GrpE is present (45). The protein that resists dissociation from DNA the most is the λO protein (12). The presence of significant amounts of proteins at the origin could create a block for replication to proceed leftwards and may explain why replication is found to be largely unidirectional to the right *in vitro*. The rightward bias is almost certainly due to the positioning of the A+T region to the right of the O binding sites. It should be stressed, however, that unidirectional replication was also found to be leftwards in a significant fraction of the replicating molecules whether or not GrpE was present (51). In this study GrpE had no effect in increasing the frequency of bidirectional replication as was claimed in a separate study (56). The key for bidirectional replication was found to be transcription by RNA polymerase (51). Another relevant fact was that ClpXP protease could facilitate the disassembly process (M. Zylicz, unpublished data). It has been found that the molecules of O, which are not bound to DNA, are rapidly degraded by the ATP-dependent ClpXP protease (57, 58). It appears that transcription is required to dislodge O from DNA (51) and the proteins are prevented from rebinding to *ori* by proteolysis (M. Zylicz, unpublished data). [Interestingly, the ClpX subunit alone can act as a chaperone, in that it can bind to O and prevent it from aggregation (especially at high temperatures) in an ATP dependent fashion and thereby can stimulate the *ori*λ binding activity of O (M. Zylicz, unpublished data).] The roles of RNA polymerase and a protease in completing a disassembly reaction and making it irreversible and their consequence in bidirectional replication were not appreciated until recently.

Mini-F Plasmid Replication

Control of Plasmid Copy Number. Mini-F carries the basic replicon of the *E. coli* sex factor F and replicates with low copy number characteristic of the bacterial chromosome. Like λ, the replicon has iterons (Figure 1), where the plasmid encoded initiator protein, RepE, binds for initiation, but, unlike λ, iteron binding is also involved in the control of the initiation frequency. Initiation and control by iterons is common among bacterial plasmids and mini-F is the prototype of a family of plasmids where iterons play the dual roles (reviewed in ref. 59). An array of iterons similar to those of mini-F is also used in the episomal replication of Epstein-Barr virus (60). Initiator binding to iterons most likely forms higher-order structures analogous to the λ O-some, and are believed to be required for the initiation reaction (43,61). When RepE is limiting, its binding to one origin sequesters the protein from other origins and this is one mechanism by which the iterons control initiation frequency (62,63). The primary mechanism of limiting the initiator concentration is by transcriptional autoregulation as the iterons or parts of it can serve as operators (Figure 1; 64,65). Thus the same protein serves the roles of initiators and repressors, whereas the roles are carried out by different proteins in λ. When the protein concentration is not limiting, a different mechanism has been proposed for negative control of replication of the iteron-family of plasmids (66,67). This alternate mechanism is based on the fact that in addition to DNA binding, the initiator proteins can pair (synapse) iterons belonging to two origins *in trans* and this is believed to cause steric hindrance to the initiation reaction at both the origins ("handcuffing"; 59) by preventing the formation of higher-order structures.

Role of Chaperones DnaK, DnaJ and GrpE. Involvement of chaperones in plasmid replication was first evident in studies of mini-F plasmids. In *rpoH* mutant cells, mini-F was unstable (68). However, a pseudorevertant of an *rpoH* mutant that exhibited apparently normal induction of major heat shock proteins in the absence of functional σ^{32}, still failed to support mini-F replication (68). This suggested that the requirement of σ^{32} in mini-F replication, unlike the requirement in λ replication, is direct. Subsequently, it was shown that the transcription of the plasmid-encoded initiator gene, *repE*, is dependent upon σ^{32}, both *in vivo* and *in vitro* (65). In cells lacking σ^{32}, mini-F replication was possible, provided RepE could be supplied *in trans* from a foreign promoter, not dependent upon σ^{32}. However, the level of RepE needed was in excess of the level found in a wild-type cell. The requirement for excess RepE is apparently to compensate in Δ*rpoH* cells, for the lower level of the three chaperones, DnaK, DnaJ and GrpE, that normally participate in replication of mini-F (69-71). Thus, σ^{32} contributes by promoting synthesis of both RepE and the three chaperone proteins. As will be discussed later, the chaperones increase the formation of active RepE post-translationally and thereby ensure that a normal level of RepE synthesis suffices for replication. In the absence of the chaperones, the RepE needs to be synthesized at higher levels so that enough active RepE can form spontaneously. Although the same set of chaperones participate in λ and mini-F replication, their roles are at very different stages of the initiation reaction.

Chaperone-independent Initiator Mutants. Mini-F mutant plasmids able to replicate in Δ*rpoH* or in *dnaJ* mutant backgrounds have been isolated in order to understand the roles of these host factors in replication of the plasmid (72,73). In many mutant plasmids, RepE was found to be overproduced either due to changes in the promoter/operator region of the *repE* gene or due to changes in a small region of the *repE* coding region (codons 92-134 out of a total 250 codons). The majority of the mutant plasmids were of the latter type and the reason for RepE overproduction was because the proteins were defective in their autorepression activity. Thus, chaperone-independence can be obtained by overproduction of wild-type RepE as was found earlier (71-73) or of mutant RepE proteins with decreased repressor activity.

A second phenotype of the chaperone-independent mutants was their higher copy number. This phenotype was partly associated with RepE overproduction. In fact, an inverse correlation of the repressor and initiator activities was apparent in many cases. However, it was also clear that in some mutants the initiator activity could be higher without decreasing the repressor activity. Therefore, chaperone-independent initiation is possible without overproduction in hyperactive initiator mutants. For the convenience of discussion the high-copy mutants where RepE was overproduced will be called class I mutants and where the repressor activity was mostly unaffected will be called class II mutants. Some understanding of the basis of the phenotypes of mutants belonging to both the classes has been obtained from *in vitro* studies of the mutant proteins and similar studies with RepA protein of plasmid P1 as discussed below.

Mechanism of Chaperone Action. The iterons at the mini-F origin are arranged in a direct repeat fashion (Figure 1). The direct repeats (DRs) are each 19 bp long and lack any symmetry feature commonly found in DNA binding sites for many transcriptional repressors and activators and other replication initiators, including λO protein (62,74). In contrast, the operator site of *repE* does contain an inverted repeat (IR) of a 10 bp sequence, 8 positions of which are homologous to the DRs (75). This difference in sequence arrangement is significant since it appears that monomers of RepE can bind to DRs whereas RepE dimers bind to an IR (76). In other words, RepE monomers serve as initiators whereas RepE dimers serve as repressors. Class I RepE mutants that show high initiator activity show increased binding activity to DRs most likely because a greater fraction of them exists in the monomeric form. This has been shown most dramatically with a RepE mutant that had practically lost the repressor activity. Whereas the wild type RepE and most other mutants were found in the dimeric form (77), this particular mutant (RepE54) was almost exclusively monomeric and showed binding to DRs only (76). The picture is reasonably clear that supplying an adequate amount of monomeric RepE is a prime requirement for mini-F replication and the chaperones help in doing precisely that.

There are several ways the chaperones could help RepE binding to DRs (Figure 3).

I) The chaperones could help in the folding of nascent polypeptide chains of RepE (78). Such an activity of chaperones homologous to DnaK and DnaJ has been demonstrated with a mammalian translation system (79). In this model the fraction of properly-folded RepE (active fraction) would increase in the presence of chaperones and the distribution between the monomeric and dimeric forms would depend upon the affinity of the active monomers for each other. The chaperone-independent mutants in this model are folding mutants that have gained efficiency to fold spontaneously, but they are unaffected in dimerization. In other words, the mutants should bind to both DRs and IRs proficiently without requiring chaperones. None of the characterized mutants belonged to this category but this could be attributed to the selection strategy used to isolate the mutants.

II) The chaperones may function by dissociating dimers and thereby could increase the fraction of monomeric RepE (80). In this model, the dimerization domain is the target of the chaperones and the chaperone-independent mutants would be defective in dimerization. This model accounts for the properties of the Class I RepE mutants adequately.

Models I and II do not have to be mutually exclusive. It is possible that the chaperones both dissociate dimers and fold inactive monomers into active forms.

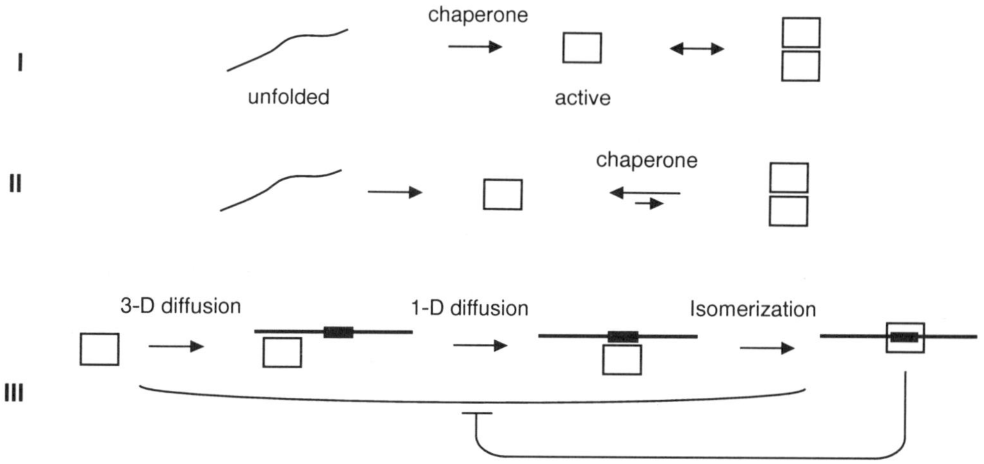

Figure 3. Modulation of DNA binding of plasmid initiator proteins of mini-F and mini-P1 by chaperones (I & II) or by the DNA-pairing activity of the initiators (III). In model I, the chaperones increase DNA binding by helping proper folding of monomeric polypeptides through prevention of nonproductive interactions. Only active (properly folded) monomers can serve as initiators. In model II, the chaperones increase monomer fraction by mediating dimer dissociation. In model III, an initiator-iteron complex can interfere with the formation of a second complex by protein-protein interactions. The possible steps of interference could be during 3-D diffusion of the free initiator, or during 1-D sliding of the nonspecifically bound initiator or during the isomerization of specifically bound initiators to their final forms. The predictions are that the chaperone-independent mutants would be reduced either in nonproductive interactions and thereby would fold unimpeded (model I), or reduced in dimerization (model II), or reduced in protein-protein interactions invoked in DNA-pairing (model III).

Role of DNA-pairing in DNA Binding. Models I and II are not adequate to explain the behavior of class II chaperone-independent mutants (RepE10, 18, 26, 40 and CopA1; 73,77). Compared to the wild type, these mutants confer higher copy number, increased binding to DRs but little or no reduction in binding to IR. According to models I and II, the mutants are expected to bind to both DRs and IRs efficiently (model I), or, to bind to DRs efficiently at the expense of binding to IRs (model II). Increased binding to DRs without decreased binding to IRs (model III) is possible by increasing the true affinity of the monomers for DRs. Studies on P1 initiator mutants indicate that the affinity for DRs can be increased by reducing the pairing activity of the initiators (81). Thus in the absence of chaperones, in addition to increased synthesis, initiator binding to DRs can possibly be improved by more efficient folding, by reducing the affinity of monomers for dimerization and by reducing the activity of the pairing domain.

Mini-P1 Plasmid Replication

Replicon Structure. The plasmid replicon is organized similarly but not identically to that of mini-F (Figure 1) and is dependent upon DnaK, DnaJ and GrpE for efficient replication both *in vivo* and *in vitro* (52,82,83). The plasmid has DRs only. They are used for both initiation of replication and repression of *prepA*, the promoter for the replication initiator gene, *repA*, since the promoter is completely contained within the DRs of the origin (84). The repressor and initiator forms of RepA are, therefore, one and the same (85). Unlike mini-F, *prepA* is transcribed from a σ^{70} promoter, and, there is no direct requirement of σ^{32} in mini-P1 replication.

Activation of DNA Binding. Studies with mini-P1 first showed that the three chaperones help plasmid replication by increasing RepA binding to the DRs and that this could be their only role in the replication of the plasmid. *In vivo*, the conditions that allowed adequate initiator binding in the absence of chaperones, overproduction of wild type RepA or the presence of appropriate mutant RepAs without overproduction (similar to class II mutants of mini-F), also allowed efficient replication (85,86). *In vitro*, DnaK and DnaJ (87), and in some conditions all three chaperones (88), dramatically increased RepA binding to DRs. Once RepA is bound to origin DNA, the RepA-origin complexes can be purified from the chaperone proteins and replication initiated in crude extracts lacking the chaperone proteins (89). RepA activation has also been achieved with urea and the urea-activated RepA supported mini-P1 replication as efficiently as the chaperone-activated RepA in crude extracts lacking the chaperones (80,89). Thus, both *in vivo* and *in vitro* results indicate that as long as RepA binding to the origin is adequate, the presence of chaperones is not obligatory for any subsequent steps of the initiation process. In this sense, the role of the chaperones in mini-F and in mini-P1 appear to be very similar.

Mechanisms of Activation of DNA Binding. The mechanism of activation has been proposed to be dimer dissociation (model II, Figure 3) since inactive RepA is found to be dimeric, whereas RepA activated by urea or by the chaperones or isolated from RepA-iteron complexes, is found to be monomeric when analyzed by sizing columns (80). Since urea and chaperones are equally effective in activation, the results indicate that activation does not involve covalent modification of the protein. *In vitro* work has also indicated that activation requires dilute solutions of RepA, indicating that monomerization could be a prerequisite for activation. However, a more quantitative analysis indicated that monomerization may not be sufficient for activation, since dilute solutions of RepA where the protein should be essentially monomeric (>99%) still required the chaperones for activation (78). The monomer fraction was deduced from the monomer-dimer equilibrium constant of the protein, as measured by analytical ultracentrifugation. The properties of chaperone-independent mutant initiators of P1 which show increased DRs binding and a high copy number phenotype in the absence of chaperones, like the class II mini-F mutants (73,77), also support the view that activation may require refolding of inactive monomers (81). Of the three mutants characterized *in vitro*, at least one was not enriched for (active) monomer fraction. This mutant was increased in true affinity for binding to DRs and the increase in affinity was due to an increase in the value of the association rate constant, k_{on}. In the other two mutants, the association rate was also higher due either to increased active fraction (models I & II) or to a combination of increased active fraction and increased k_{on}. All three mutants were defective in DNA pairing. This observation readily explained their high-copy-number initiator phenotype, and indicated that the pairing defect was directly or indirectly responsible for increasing DNA binding activity of RepA and, thereby, chaperone-independence.

The variety of the phenotypes of both mini-P1 and mini-F initiator mutants suggest that origin binding is regulated by multiple mechanisms. Three ways of modulating the binding activity post-translationally have been discussed (Figure 3). However, the mechanisms proposed in all three models need to be more firmly established. The limitation of the folding model (model I) is that it is based on equilibrium measurements and does not indicate the rates with which the monomers associate and the dimers dissociate. If the rates are unusually slow, dilution will not ensure monomerization and the role of the chaperones and the chemical denaturants could be simply in increasing the rate of the dissociation reaction (model II). If the chaperones simply dissociate RepA dimers, then the question remains of how the monomers are prevented from reassociation when the chaperone action is made ineffective (e.g., by lowering temperature to 0°C; 87). Once again if the monomers associate slowly, then there would be no need to invoke any refolding (model I). The requirement for dilute solutions to achieve RepA activation may simply be to slow down reassociation of monomers.

The mechanism proposed in model III is the least intuitive and also the most obscure in terms of details. Nonetheless, the idea is novel and stresses the significance of the pairing activity in DNA binding. The gratuitous finding of the high-copy phenotype in majority of the mutants of mini-F and mini-P1 underscores the fact that the binding affinity is more easily improved by reducing dimerization or pairing activity rather than increasing the activity of DNA-binding domain itself. The bias could be simply due to relative ease of loss *versus* gain of function. Finally, it should be noted that the pairing defect *per se* may not be the reason for increased value of k_{on}. The mutant proteins could be changed in conformation in a way that favored the association reaction, and fortuitously, initiation.

oriC Replication

Life without Chaperones. *E. coli* can be alive but not well when several of the genes related to chaperone functions are deleted: *rpoH* (90), *dnaK* (91,92), *dnaJ* (93) and *grpE* (94). The cells become temperature sensitive, sport suppressor mutations and in some cases can only survive in their presence (91-95). Cell survival at moderate temperatures may be possible due to the presence of homologous genes that provide for some functional redundancy (53,96-99). The extent to which the growth defects are due to defects in DNA replication is not clear, but chromosome segregation and cell division are definitely defective in Δ*dnaK* cells (91,92). A cell division defect is also seen in *dnaJ* null mutants (93). The conditional viability of the deletion mutants indicates that the deleted genes cannot be essential for *E. coli* replication. *In vitro* DNA replication of plasmids carrying the normally-used origin of *E. coli* replication, *oriC*, with purified components also does not require any chaperone protein (100,101). However, an indirect role of chaperones, and DnaK in particular, in chromosomal replication has been suggested by the following observations.

Involvement of Chaperones in Chromosomal Replication. The *dnaK111* mutation prevents *oriC* specific initiation at high temperature (102). When shifted back to permissive temperature, DNA replication initiates in a synchronous fashion from *oriC*, as is the case with the wild type cells. A similar phenotype is also found for the *dnaK756* mutant (103). In Δ*dnaK52* cells, the rate of DNA synthesis is three-fold lower than the wild type even at permissive temperature (91,92). Further, the DnaK, DnaJ and GrpE proteins are required for optimal expression of the *dnaQ* gene, which codes for an essential subunit of the DNA polymerase III enzyme (104). These results suggest that the chaperones can contribute positively to chromosomal replication. The *in vitro* results also support this idea as will be discussed. On the other hand, DnaA protein can negatively regulate the transcription of *rpoH* gene and thereby could regulate the expression of the chaperone genes (105).

Modulation of DnaA Activity by Multiple Chaperones. The DnaA initiator protein for *E. coli* chromosomal replication, when isolated from wild-type cells, is found as a mixture of monomeric and aggregated forms (106). The aggregates formed are most likely due to association with phospholipids, and these DnaA-phospholipid complexes are inactive in replication. In the presence of ATP, purified DnaK can dissociate the aggregates into monomers and thereby activate them for *in vitro* replication. Monomer formation has also been achieved with GroEL (J. Marszalek and J. Kaguni, unpublished data). It has been suggested that phospholipids may be involved in membrane association of DnaA *in vivo* and

DnaK may dissociate this complex (106,107). Phospholipids can also catalyze the release of bound ATP or ADP from DnaA in the absence of DnaK, but whether DnaK influences the reaction has not been determined (108). The nucleotide binding has been observed to regulate DnaA protein activity *in vitro* (108). DnaK (and GroEL) can also associate directly with wild type DnaA (109) and can protect the protein in dilute solutions (K. Skarstad and A. Kornberg, unpublished results cited in ref. 110). It thus appears likely that the chaperones could be involved in modulating the activity of DnaA either by altering its oligomeric state (106), nucleotide exchange rate (108,111), membrane association and thereby position in the cell (106,108), or its interactions with other initiation proteins, such as DnaR and DnaB (112,113). If the chaperones can up-regulate DnaA activity and DnaA can down-regulate the expression of *rpoH* and hence the *dnaKdnaJ* operon (105), the two controls can set up a control circuit that can result in a homeostatic regulation of the activities of both DnaA and DnaK-DnaJ.

Studies with Mutant DnaA Protein. Missense mutants of DnaA protein can also be the target of chaperones *in vivo* (114,115) and *in vitro* (116). Overexpression of both *groEL* and *groES* genes can restore the colony-forming ability of *dnaA*46 bacteria at high temperature (114,115,117). At low temperature GroELS oversupply causes overreplication of the *dnaA*46 mutant (but not of the wild-type) chromosome without increasing the level of the DnaA protein (117). The overreplication causes a cold-sensitive phenotype (114,115,117) similar to that seen for the *dnaAcos* mutant of *E. coli* (118). The *dnaAcos* mutant has an intragenic suppressor of the *dnaA*46 mutation that allows growth at 42°C. Again, the reason for overinitiation at low temperature is not due to an increase of DnaAcos concentration with respect to the wild type (119). These results indicate that the chaperones can induce conformational changes that mimic the effect of the suppressor mutations. Since oversupply of GroELS does not affect the initiation of wild-type cells, these chaperones may not recognize wild-type DnaA *in vivo.*

The replication activity of both DnaA46 and DnaA5 proteins has been revived by incubation with DnaK and GrpE at low temperature (116). The mutant proteins are temperature sensitive *in vitro* because of their thermolabile interactions with DnaK and GrpE. This result may also explain why the temperature sensitivity of these proteins is not corrected by the chaperones *in vivo*. From genetic studies described above (114, 115), it would be expected that GroELS should also correct the replication defect of DnaA46 *in vitro*, but this expectation has not been borne out (J. Marszalek and J. Kaguni, unpublished data). These studies highlight the subtle differences in conformation of DnaA that allow both DnaK and GroEL to be active in one kind of reaction (monomerization of $DnaA^+$ aggregates), but not in the replication activity of DnaA46 where DnaK and GrpE cannot be substituted with GroEL.

Chaperones and Replication Control

The role of chaperones in regulation of DNA replication (control of plasmid copy number during steady-state growth of a culture) has not been addressed seriously. Although activation of DnaB function is crucial for λ DNA replication, it is not considered to be the rate-limiting step in replication. The primary determinant of the copy number of λ-derived plasmid replicons is the concentration of λCro protein which modulates the activity of the λp_R as stated earlier. Thus, although λ does not replicate in the absence of the chaperones DnaK, DnaJ and GrpE, these proteins are not responsible for determining the average λ plasmid copy number in the culture. Maintenance of plasmid copy number requires functions that can adjust the rate of initiation in individual cells born with abnormally high or low copy numbers.

In mini-F, σ^{32} is required for the transcription of the initiator gene, *repE*. Since the chaperones DnaK, DnaJ and GrpE are involved in the negative regulation of σ^{32}, the chaperones could control mini-F copy number by controlling the concentration of RepE, considered to be rate limiting for mini-F replication (17,62,72,73). Since, the cellular concentration of σ^{32} is limiting, there may not be enough σ^{32} to replicate all plasmids in cells born with abnormally high copy numbers. This limitation of course will be less applicable to cells born with low copy numbers. Thus the basic requirements of a replication regulator seem to be satisfied by σ^{32}. One might expect that the absence of chaperones DnaK, DnaJ and GrpE should increase mini-F copy number, but in reality the copy number decreases. The positive role that the chaperones play in activation of RepE post-translationally appears to be more important for replication than their regulatory role through σ^{32}.

In mini-P1, the primary mechanism of replication control is believed to be through the initiator-mediated pairing of iterons (66). The pairing reaction has been demonstrated only in the presence of chaperones and, therefore, it is likely that the chaperones do not inhibit the reaction (81). Secondly, the stringency of replication control that increases with increase in iteron concentration was shown to be identical in wild type and in Δ*dnaJ* cells. Together, these results do not support a regulatory role of chaperones in P1 replication (81). Even if the chaperones are not involved in replication control, characterization of the basis of the high-copy phenotype of the chaperone-independent mutants has contributed and will continue to aid in our understanding of both DNA binding and replication control.

In *oriC* replication, unlike bacterial plasmid replication, the number of initiations is not controlled. Instead, the timing of initiation in the cell cycle is controlled and, irrespective of the number, all origins initiate synchronously once per cell cycle. In these respects, the control of *oriC* replication is similar to that of the origins of eukaryotic cells. Controlling the activity of DnaA seems to be one of the most likely means by which *oriC* replication is controlled. As argued earlier, DnaK can control the activity of DnaA, but the level of DnaK is invariant throughout the *E. coli* cell cycle (107). In contrast, the DnaK level in *Caulobacter crescentus* varies during the cell cycle, reaching a maximal level just prior to initiation of replication under normal growth conditions (120). In human cells also, the transcription of the gene for Hsp70, a DnaK homolog, is temporally controlled (121). It is induced at the onset of S phase, at which time the Hsp70 also migrates to the nucleus. Thus, although perhaps not in *E. coli*, the chaperones appear to play a role in the control of chromosomal replication in other organisms.

Concluding Remarks

The role of chaperones in plasmid replication can be essential, as in λ, or stimulatory, as in mini-F, mini-P1 and possibly *oriC*. Studies on chaperone involvement have been limited to a few replicons belonging primarily to the world of *E. coli*. Nonetheless, they have contributed to our understanding of the details of initiation reactions and have revealed many facets of chaperone functions as well. The studies that have led to the elaborate scheme of loading of DnaB at the λ origin and activation of its helicase function will surely serve as a paradigm in future studies with other replicons (122). These studies also helped delineate the stages of an initiation reaction and the basic scheme has already proven applicable to *oriC* and SV40 replication (43,123). In mini-F, mini-P1 and *oriC*, the chaperones do not participate in the initiation reaction directly, unlike the case in λ, but contribute to replication by controlling the DNA binding activity of the initiator proteins. DNA binding has been considered a crucial step in control of replication of mini-F and mini-P1. Recent demonstration of chaperone activity of the ATP-binding subunits of Clp proteases and involvement of proteases in the activities of λO and P1RepA initiators are certainly going to provide new insights in the field of DNA replication (35).

Knowledge of several of the presently known chaperone functions owe their origin to studies on replication. 1. The studies on λ demonstrated cooperativity among chaperone members, DnaK, DnaJ and GrpE: how they could be involved in both assembly and disassembly of nucleoprotein complexes and how ATP, by changing protein conformations, could regulate the assembly/disassembly reactions. The net effect is a dissociation reaction that liberates DnaB helicase from the inactive λP-DnaB complex at the origin. 2. In P1 and F, the same chaperones also convert inactive initiator proteins to active forms, but act on the substrate protein directly without their being part of a complex nucleoprotein structure. In these plasmids, inactivation by dimerization and activation by chaperone-mediated monomerization were unprecedented before these studies. Another highlight was the demonstration that the chaperones can influence a function as general as DNA binding. 3. Activation of DnaA by conversion of DnaA aggregates into monomers is a clear example of yet another function of chaperones, i.e., disaggregation of preexisting but naturally-occurring aggregates. 4. The recent findings of chaperone action of protease subunits ClpX and ClpA on λO and P1RepA, respectively, and degradation of the same substrates by the proteases ClpXP and ClpAP, respectively, are novel (35). These systems have the potential to reveal the signals that help the chaperones decide between refolding and degradation of misfolded proteins. 5. Although in many cases the chaperone trio, DnaK, DnaJ and GrpE, work as a team, there are examples where one or two of the proteins could be dispensed with *in vitro*. For example, λ replication without GrpE (11,13), activation of mini-F RepE without DnaK and GrpE (77), P1RepA activation without GrpE (78,87), disaggregation of DnaA without DnaJ and GrpE (106), activation of DnaA5 and DnaA46 without DnaJ (116), etc. However, such dispensability may be conditional. The GrpE requirement can be made obligatory by reducing DnaK concentration in the case of λ (12,13), or reducing Mg^{2+} in case of P1RepA (88). 6. There could also be overlap of chaperone functions in some cases. For example, disaggregation of DnaA with DnaK or GroEL (106; J. Marszalek and J. Kaguni, unpublished data), activation of RepA by DnaK and DnaJ, or by ClpA (35), activation of λO by DnaK, DnaJ, or ClpX (M. Zylicz, unpublished data), etc. To make sense out of the variety, specificity and redundancy of chaperone functions, we must have a better understanding of the signals for chaperone recognition and how the conformational changes are effected. The conformation may be right for one chaperone, but may require accessory factors for others to achieve the required affinity and specificity (45). The resolution of these questions will be the key to solving the remaining puzzle of the central dogma (124):

$$\text{DNA} \leftrightarrow \text{RNA} \rightarrow \text{Polypeptide} \leftrightarrow \text{Folded protein.}$$

ACKNOWLEDGMENTS

The author is grateful to B. Bukau, J. Kaguni, R. McMacken, G. Walker, T. Yura and M. Zylicz for providing unpublished results, preprints and reprints, and to E. Crooke, J. Dibbens, J. Kaguni, R. McMacken, S. Wickner, M. Yarmolinsky and M. Zylicz for comments on initial drafts of this manuscript.

REFERENCES

1 Lattman, E.E. and Rose, G.D. (1993) Proc. Nat. Acad. ScI. U.S.A. 90, 439-441.
2 Ellis, J. and van der Vies, S.M. (1991) Annu. Rev. Biochem. 60, 321-347.

3 Hendrick, J.P. and Hartl, F.-U. (1993) Annu. Rev. Biochem. 62,349-384.
4 Georgopoulos, C. and Welch, W.J. (1993) Annu. Rev. Cell. Biol.9, 601-635.
5 Gragerov, A.I., Martin, E., Krupenko, M.A., Kashlev, M.V. and Nikiforov, V.G. (1992) FEBS Lett. 291, 222-224.
6 Gaitanaris, G.A., Papavassiliou, A.G., Rubock, P., Silverstein,S.J. and Gottesman, M.E. (1990) Cell 61, 1013-1020.
7 Skowyra, D., Georgopoulos, C. and Zylicz, M. (1990) Cell 62, 939-944.
8 Schröder, H., Langer, T., Hartl, F.-U. and Bukau, B. (1993) EMBO J. 12, 4137-4144.
9 Ziemienowicz, A., Skowyra, D., Zeilstra-Ryalls, J., Fayet, O.,Georgopoulos, C. and Zylicz, M. (1993) J. Biol. Chem. 268, 25425-25431.
10 Sherman, M.Y. and Goldberg, A.L. (1993) Proc. Nat. Acad. Sci.U.S.A. 90, 8648-8652.
11 Alfano, C. and McMacken, R. (1989) J. Biol. Chem. 264, 10699-10708.
12 Alfano, C. and McMacken, R. (1989) J. Biol. Chem. 264, 10709-10718.
13 Zylicz, M., Ang, D., Liberek, K. and Georgopoulos, C. (1989) EMBO J. 8, 1601-1608.
14 Phillips, G.J. and Silhavy, T.J. (1990) Nature 344, 882-884.
15 Strauss, D.B., Walter, W.A. and Gross, C.A. (1988) Genes Dev.2, 1851-1858.
16 Sherman, M.Y. and Goldberg, A.L. (1992) EMBO J. 11, 71-77.
17 Bukau, B. (1993) Mol. Microbiol. 9, 671-680.
18 Liberek, K. and Georgopoulos, C. (1993) Proc. Nat. Acad. Sci.U.S.A. 90, 11019-11023.
19 Kanemori, M., Mori, H. and Yura, T. (1994) J. Bacteriol. 176,5648-5653.
20 Flynn, G.C., Pohl, J., Flocco, M.T. and Rothman, J.E. (1991) Nature 353, 726-730.
21 Landry, S.J., Jordan, R., McMacken, R. and Gierasch, L.M. (1992) Nature 355, 455-457.
22 Liberek, K., Skowyra, D., Zylicz, M., Johnson, J. and Georgopoulos, C. (1991) J. Biol. Chem. 266, 14491-14496.
23 Banecki, B., Zylicz, M., Bertoli, E. and Tanfani, F. (1992) J.Biol. Chem. 267, 25051-25058.
24 Richarme, G. and Kohiyama, M. (1993) J. Biol. Chem. 268, 24074-24077.
25 Park, K., Flynn, G.C., Rothman, J.E. and Fasman, G.D. (1993) Protein Sci. 2, 325-330.
26 Palleros, D.R., Reid, K.L., Shi, L., Welch, W.J. and Fink, A.L.(1993) Nature 365, 664-666.
27 Buchberger, A., Valencia, A., McMacken, R., Sander, C. and Bukau, B. (1994) EMBO J. 13, 1687-1695.
28 Schmid, D., Baici, A., Gehring, H. and Christen, P. (1994) Science, 263, 971-973.
29 Georgopoulos, C. (1992) Trends Biochem. Sci. 17, 295-299.
30 Langer, T., Lu, C., Echols, H., Flanagan, J., Hayer, M.K. and Hartl, F.U. (1992) Nature 356, 683-689.
31 Wang, T.-F., Chang, J.-H. and Wang, C. J. Biol. Chem. (1993) 268, 26049-26051.
32 Wall, D., Zylicz, M. and Georgopoulos, C. (1994) J. Biol. Chem. 269, 5446-5451.
33 Liberek, K., Marszalek, J., Ang, D., Georgopoulos, C. and Zylicz, M. (1991) Proc. Nat. Acad. Sci. U.S.A. 88, 2874-2878.
34 Morimoto, R.I., Tissières, A. and Georgopoulos, C. (1994) in The Biology of Heat Shock Proteins and Molecular Chaperones, Cold Spring Harbor Laboratory Press, New York, NY.
35 Craig, E.A., Weissman, J.S. and Horwich, A.L. (1994) Cell, 78,365-372.

36 Friedman, D.I., Olson, E.R., Georgopoulos, C., Tilly, K., Herskowitz, I. and Banuett, F. (1984) Microbiol. Rev. 48, 299-325.
37 Neidhardt, F.C., VanBogelen, R.A. and Lau, E.T. (1983) J. Bacteriol. 153, 597-603.
38 Grossman, A.D., Erickson, J.W. and Gross, C.A. (1984) Cell 38,383-390.
39 Yamamoto, T., McIntyre, J., Sell, S.M., Georgopoulos, C., Skowyra, D. and Zylicz, M. (1987) J. Biol. Chem. 262, 7996-7999.
40 Mensa-Wilmot, K., Carroll, K. and McMacken, R. (1989) EMBO J.8, 2393-2402.
41 Womble, D.D. and Rownd, R.H. (1986) J. Mol. Biol. 191, 367-382.
42 LeBowitz, J.H. and McMacken, R. (1986) J. Biol. Chem. 261, 4738-4748.
43 Baker, T.A. and Wickner, S.H. (1992) Annu. Rev. Genet. 26, 447-477.
44 Mensa-Wilmot, K., Seaby, R., Alfano, C., Wold, M.C., Gomes, B.and McMacken, R. (1989) J. Biol. Chem. 264, 2853-2861.
45 Zylicz, M. (1993) Philos. Trans. Roy. Soc. London 339(1289), 271-277.
46 Dodson, M., Roberts, J., McMacken, R. and Echols, H. (1985) Proc. Nat. Acad. Sci. U.S.A. 82, 4678-4682.
47 Schnös, M., Zahn, K.,Inman, R.B. and Blattner, F.R. (1988) Cell 52, 385-395.
48 Mallory, J.B., Alfano, C. and McMacken, R. (1990) J. Biol. Chem. 265, 13297-13307.
49 Osipiuk, J., Georgopoulos, C. and Zylicz, M. (1993) J. Biol. Chem. 268, 4821-4827.
50 Hoffmann, H., Lyman, S.K., Lu, C., Petit, M-A. and Echols, H.(1992) Proc. Nat. Acad. Sci. U.S.A. 89, 12108-12111.
51 Learn, B., Karzai, A.W. and McMacken, R. (1993) Cold Spring Harbor Symp. Quant. Biol. 58, 389-402.
52 Wickner, S.H. (1990) Proc. Nat. Acad. Sci. U.S.A. 87, 2690- 2694.
53 McCarty, J.S. and Walker, G.C. (1994) J. Bacteriol. 176, 764-780.
54 Liberek, K., Georgopoulos, C. and Zylicz, M. (1988) Proc. Nat.Acad. Sci. U.S.A. 85, 6632-6636.
55 Zylicz, M., LeBowitz, J.H., McMacken, R. and Georgopoulos, C.(1983) Proc. Nat. Acad. Sci. U.S.A. 85, 6632-6636.
56 Wyman, C., Vasilikiotis, C., Ang, D., Georgopoulos, C. and Echols, H. (1993) J. Biol. Chem. 268, 25192-25196.
57 Wojtkowiak, D., Georgopoulos, C. and Zylicz, M. (1993) J. Biol.Chem. 268, 22609-22617.
58 Gottesman, S., Clark, W.P., de Crecy-Lagard, V. and Maurizi, M.(1993) J. Biol. Chem. 268, 22618-22626.
59 Kittel, B.L. and Helinski, D.R. (1993) in Bacterial Conjugation(Clewell, B.D., ed.), pp. 223-242, Plenum Press, New York, NY.
60 Frappier, L., Goldsmith, K. and Bendell, L. (1994) J. Biol. Chem. 269, 1057-1062.
61 Mukhopadhyay, G. and Chattoraj, D.K. (1993) J. Mol. Biol. 231,19-28.
62 Tsutsui, H., Fujiyama, A., Murotsu, T. and Matsubara, K. (1983) J. Bacteriol. 155, 337-344.
63 Chattoraj, D., Cordes, K. and Abeles, A. (1984) Proc. Nat. Acad. Sci. U.S.A. 81, 6456-6460.
64 Kline, B.C. (1985) Plasmid 14, 1-16.
65 Wada, C., Imai, M. and Yura, T. (1987) Proc. Nat. Acad. Sci.U.S.A. 84, 8849-8853.
66 Pal, S.K. and Chattoraj, D.K. (1988) J. Bacteriol. 170, 3554-3560.
67 McEachern, M.J., Bott, M.A., Tooker, P.A. and Helinski, D.R. (1989) Proc. Nat. Acad. Sci. U.S.A. 86, 7942-7946.
68 Wada, C., Akiyama, Y., Ito, K. and Yura, T. (1986) Mol. Gen. Genet. 203, 208-213.

69 Zhou, Y.N., Kusukawa, N., Erickson, J.W., Gross, C.A. and Yura,T. (1988) J. Bacteriol. 170, 3640-3649.
70 Ezaki, B., Ogura, T., Mori, H., Niki, H. and Hiraga, S. (1989) Mol. Gen. Genet. 218, 183-189.
71 Kawasaki, Y., Wada, C. and Yura, T. (1990) Mol. Gen. Genet. 220, 277-282.
72 Kawasaki, Y., Wada, C. and Yura, T. (1991) J. Bacteriol. 173, 1064-1072.
73 Ishiai, M., Wada, C, Kawasaki, Y. and Yura, T. (1992) J. Bacteriol. 174, 5597-5603.
74 Papp, P.P., Chattoraj, D.K. and Schneider, T.D. (1993) J. Mol.Biol. 233, 219-230.
75 Tokino, T., Murotsu, T. and Matsubara, K. (1986) Proc. Nat. Acad. Sci. U.S.A. 83, 4109-4113.
76 Ishiai, M., Wada, C, Kawasaki, Y. and Yura, T. (1994) Proc. Nat. Acad. Sci. U.S.A. 91, 3839-3843.
77 Kawasaki, Y., Wada, C. and Yura, T. (1992) J. Biol. Chem. 267,11520-11524.
78 DasGupta, S., Mukhopadhyay, G., Papp, P.P., Lewis, M.S. and Chattoraj, D.K. (1993) J. Mol. Biol. 232, 23-34.
79 Frydman, J., Nimmesgren, E., Ohtsuka, K. and Hartl, F.U. (1994) Nature 370, 111-117.
80 Wickner, S., Hoskins, J. and McKenney, K. (1991) Proc. Nat. Acad. Sci. U.S.A. 88, 7903-7907.
81 Mukhopadhyay, G., Sozhamannan, S. and Chattoraj, D.K. (1994) EMBO J. 13, 2089-2096.
82 Tilly, K. and Yarmolinsky, M. (1989) J. Bacteriol. 171, 6025-6029.
83 Bukau, B. and Walker, G.C. (1989) J. Bacteriol. 171, 6030-6038.
84 Chattoraj, D.K., Snyder, K.M. and Abeles, A.L. (1985) Proc. Nat. Acad. Sci. U.S.A. 82, 2588-2592.
85 Sozhamannan, S. and Chattoraj, D.K. (1993) J. Bacteriol. 175,3546-3555.
86 Tilly, K., Sozhamannan, S. and Yarmolinsky, M. (1990) New Biol.2, 812-817.
87 Wickner, S., Hoskins, J. and McKenney, K. (1991) Nature 350, 165-167.
88 Skowyra, D. and Wickner, S. (1993) J. Biol. Chem. 268, 25296-25301.
89 Wickner, S., Skowyra, D., Hoskins, J. and McKenney, K. (1992) Proc. Nat. Acad. Sci. U.S.A. 89, 10345-10349.
90 Zhou, Y.-N., Kusukawa, N., Erickson, J.W., Gross, C.A. and Yura, T. (1988) J. Bacteriol. 170, 3640-3649.
91 Bukau, B. and Walker, G.C. (1989) J. Bacteriol. 171, 6030-6038.
92 Bukau, B. and Walker, G.C. (1991) in Heat Shock (Maresca, B. and Lindquist, S., eds.), pp. 55-66, Springer-Verlag, Berlin.
93 Sell, S.M., Eisen, C., Ang, D., Zylicz, M. and Georgopoulos, C.(1990) J. Bacteriol. 172, 4827-4835.
94 Ang, D. and Georgopoulos, C. (1989) J. Bacteriol. 171, 2748- 2755.
95 Tobe, T., Kusukawa, N. and Yura, T. (1987) J. Bacteriol. 169,4128-4134.
96 Kang, P.J. and Craig, E.A. (1990) J. Bacteriol. 172, 2055-2064.
97 Ueguchi, C., Kakeda, M., Yamada, H. and Mizuno, T. (1994) Proc.Nat. Acad. Sci. U.S.A. 91, 1054-1058.
98 Kawula, T.H. and Lelivelt, M.J. (1994) J. Bacteriol. 176, 610-619.
99 Seaton, B.L. and Vickery, L.E. (1994) Proc. Nat. Acad. Sci. U.S.A. 91, 2066-2070.
100 Kaguni, J.M. and Kornberg, A. (1984) Cell 38, 183-190.
101 Funnell, B.E., Baker, T.A. and Kornberg, A. (1986) J. Biol. Chem. 261, 5616-5624.
102 Sakakibara, Y. (1988) J. Bacteriol. 170, 972-979.
103 Ohki, M. and Smith, C.L. (1989) Nucl. Acids Res. 17, 3479- 3490.
104 Foster, P.L.and Marinus, M.G. (1992) J. Bacteriol. 174, 7509-7516.

105 Wang, Q. and Kaguni, J.M. (1989) J. Biol. Chem. 264, 7338-7344.
106 Hwang, D.S., Crooke, E. and Kornberg, A. (1990) J. Biol. Chem.265, 19244-19248.
107 Hupp, T.R., Keasling, J.D., Cooper, S. and Kaguni, J.M. (1994) Res. Microbiol. 145, 99-109.
108 Sekimizu, K. and Kornberg, A. (1988) J. Biol. Chem. 263, 7131-7135.
109 Malki, A., Hughes, P. and Kohiyama, M. (1991) Mol. Gen. Genet. 225, 420-426.
110 Skarstad, K. and Boye, E. (1994) Biochim. Biophys. Acta 1217,111-130.
111 Crooke, E., Castuma, C.E. and Kornberg, A. (1992) J. Biol. Chem. 267, 16779-16782.
112 Sakakibara, Y. (1993) J. Bacteriol. 175, 5559-5565.
113 Marszalek, J. and Kaguni, J.M. (1994) J. Biol. Chem. 269, 4883-4890.
114 Fayet, O., Louarn, J.-M. and Georgopoulos, C. (1986) Mol. Gen.Genet. 202, 435-445.
115 Jenkins, A.J., March, J.B., Oliver, I.R. and Masters, M. (1986) Mol. Gen. Genet. 202, 446-454.
116 Hupp, T.R. and Kaguni, J.M. (1993) J. Biol. Chem. 268, 13143-13150.
117 Katayama, T. and Nagata, T. (1991) Mol. Gen. Genet. 226, 491-502.
118 Kellenberger-Gujer, G., Podhajska, A.J. and Caro, L. (1978) Mol. Gen. Genet. 162, 9-16.
119 Katayama, T. and Kornberg, A. (1994) J. Biol. Chem. 269, 12698-12703.
120 Gomes, S.L., Gober, J.W. and Shapiro, L. (1990) J. Bacteriol.172, 3051-3059.
121 Milarski, K.L. and Morimoto, R.I. (1986) Proc. Nat. Acad. Sci.U.S.A. 83, 9517-9521.
122 Stillman, B. (1994) J. Biol. Chem. 269, 7047-7050.
123 Waga, S. and Stillman, B. (1994) Nature 369, 207-212.
124 Highwater, L.E., Sadis, S.E. and Takenaka, I.M. (1994) in The Biology of Heat Shock Proteins and Molecular Chaperones, pp. 179-207. Cold Spring Harbor Laboratory Press, New York, NY.

STRUCTURE, FUNCTION AND ENGINEERING OF *Bacillus thuringiensis* TOXINS

Mark A. Thompson, H. Ernest Schnepf and Jerald S. Feitelson

Department of Molecular Biology
Mycogen Corporation
4980 Carroll Canyon Road
San Diego, CA 92121

INTRODUCTION

Subspecies of *Bacillus thuringiensis* (*B.t.*) are an extensively studied group of Gram-positive, spore-forming bacteria. These organisms produce an enormous variety of intracellular proteinaceous crystalline inclusions known as δ-endotoxins after exponential growth is complete. The proteins comprising the crystal are in the vanguard of active ingredients for biological control of a growing number of commercially important agricultural and household pests, as well as vectors of animal and human disease (1). Extensive screening programs have discovered different *B.t.* strains with activity against species within seven different orders of insects, as well as against protozoa, flatworms, nematodes and mites (2).

In all cases examined to date, the parasporal inclusion crystal must be ingested by the target organism to be toxic. Therefore, these proteins are oral toxicants, and have no demonstrated contact activity. The best studied cases involve the Lepidoptera order of insects, where the protein is solubilized in the high pH of the midgut and activated by gut proteases. The activated, truncated toxins then bind to epithelial cells lining the midgut, causing disruption of the cellular membrane leading to osmotic imbalance, cell lysis and ultimately insect death. In some cases, specific toxin receptors have been identified.

The ultimate cytotoxic lesion of these toxins has not yet been characterized; however, the preponderance of evidence is leading to a "barrel-stave" multimer pore-forming mode of action. Toxins of this general type are found in a number of bacteria and eukaryotic organisms, and may contribute to their virulence or defense (3). Indeed, thuringiolysin, produced by *B.t.*, is from another "barrel-stave" pore-forming family (the sulfhydryl-activated cytolysins). These "barrel-stave" lytic toxins have reasonable cytotoxic activities, yet are not nearly as potent as the most active toxins of the "intracellular enzymatic" type, such as diphtheria, tetanus and botulism toxins (3-5).

Of great importance to the use of these toxins in commercial applications is their very high specificity. Many *B.t.* δ-endotoxin proteins kill only the larval stages of a small

Genetic Engineering, Vol. 17, Edited by J.K. Setlow
Plenum Press, New York, 1995

number of related insect species, usually within the same order, family, or even genus. Though ecologically advantageous compared to conventional chemical pesticides, *B.t.*'s narrow spectrum of activity also limits its effectiveness in many field situations, since farmers often have unrelated pest species attacking their crop.

Another disadvantage of the most commonly used *B.t.* products employing microbial spore-crystal preparations is their very brief field persistence. Their half-life on leaf surfaces is often less than one day, requiring frequent spray applications for acceptable efficacy. Many research groups are attempting to address this problem by expressing *B.t.* toxin genes in alternative hosts, such as plants (6) and other bacteria (7).

Unlike traditional chemical insecticides, to date there has been only one lepidopteran species, diamondback moth *(Plutella xylostella)*, that has exhibited field resistance to *B.t.* despite its extensive commercial use (8). This exceptional case involved frequent, heavy *B.t.* applications in tropical locations where many pest generations occurred during a single growing season. Indian meal moth (*Plodia interpunctella*) may have been preselected in the field for resistance, but this was never proven (8). Commercial products active against Coleoptera and Diptera have not yet been affected by bioinsecticide field resistance.

The attractiveness of these toxins for protein engineering results from the combination of these characteristics: (a) lack of inherent human toxicity combined with relatively high target specificity, (b) apparent requirement for several molecules per cell to achieve a lethal lesion, and (c) a variety of related structures and targets from which to draw.

PURPOSE OF CHAPTER

Because knowledge about *B.t.* toxin structure and function is rapidly growing, there are grounds for high optimism about the prospects for using genetic engineering techniques to overcome many of the significant limitations to the more widespread use of *B.t.*-based products. Four applications probably account for the greatest value from applying these methods: (a) higher specific activity against a given pest, (b) broadening target host range, (c) increasing protein stability, and (d) managing pest resistance. Much of the current thinking about *B.t.* toxins and their cognate genes is summarized below, focusing on both theoretical and practical prospects for future generation products.

The taxonomy of *B.t.* δ-endotoxins is the subject of current international interest, particularly with the explosive growth of novel *B.t.* toxin gene sequences deposited in computer databases and in the patent literature (see Figure 1). Höfte and Whiteley published a review in 1989 suggesting a classification scheme based primarily on activity spectrum, but with secondary consideration given to amino acid primary sequence (9). Extensive computer analyses of toxin primary structures have revealed important facets of their molecular anatomy, including internally conserved sequence blocks, delineation of specificity regions, and possible evolutionary relationships between toxins.

The high resolution 3-D structure solved for a truncated beetle-active toxin (CryIIIA) (10) has allowed many hypotheses about structure-function relationships to be tested. A number of segment-swapping experiments has also provided information on those toxin regions responsible for target specificity. Insect midgut membrane preparations, termed brush border membrane vesicles (BBMV), provide a source of membrane-bound receptors for assays of *in vitro* binding and membrane permeability. Physiological experiments, with such receptor-enriched membrane preparations, have increased our knowledge about mode of action and possible mechanisms of resistance to *B.t.* toxins.

Much of this information is compiled and some possibilities are outlined in this chapter for semi-empirical and targeted structural changes to *B.t.* toxin genes for improved performance: higher specific activity, expanded host range, increased protein stability, and management of resistance.

Several review articles have recently appeared summarizing *B.t.* toxin mode of action (11-15), and relevance to the producing bacterium (16).

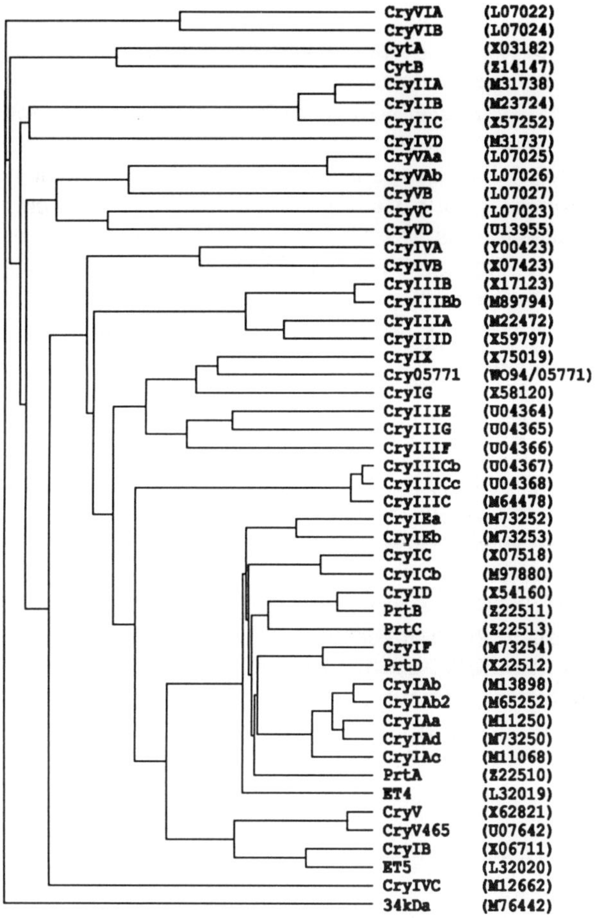

Figure 1. Possible evolutionary relationships between *B.t.* toxins. The PILEUP routine of the GCG computer package (83) was used to compare primary sequences of 50 full-length toxins. Amino acid sequences were translated from the nucleotide sequences indicated by the GENBANK or EMBL database accession numbers, or PCT patent application. This algorithm assumes a single rooted tree and a constant mutational clock. The CryVI, Cyt and 34 kD toxins probably derived from different evolutionary roots because their sequences appear to be unrelated to other Cry toxins.

STRUCTURE BY SEQUENCE HOMOLOGY

The first research showing an insight into structural differences between crystal protein sequences appeared in 1986 and 1987. These represented sequences of what are now known as the *cryIA* gene (9,17-20). It was apparent that CryIAb could be represented as a functional natural recombinant of the two other lepidopteran-active toxins in strains HD-1 (CryIAa) and HD-73 (CryIAc). Comparison of these genes and their respective proteins suggested the presence of 4 segments. From N-terminus to C-terminus, analysis showed: (a) residues 1 to ~280 were nearly identical between the three proteins, (b) residues ~280 to ~455 were variable, with the CryIAb and CryIAc sequences being nearly identical, (c) residues ~456 to 610 were variable, with the CryIAb and CryIAa sequences being nearly identical, and (d) residues ~610 to ~1170 were remarkably similar. In this final segment, CryIAc showed some variability between residues 610 and 722, and CryIAb contained a deletion at about residue 800 and an additional block of sequence differences near the C-terminus.

Additional comparison of these sequences with the next group of less similar sequences of non-lepidopteran toxins resulted in a rather different view of toxin relatedness, as summarized in (9). First, several of these naturally-occurring proteins corresponded quite well in sequence to only the genetically truncated toxin (see below) of the lepidopteran-active proteins, indicating that a 130 kD protoxin was not necessary for either crystal formation, processing, or toxicity for some members of this gene family. Second, the toxin portion of this family of proteins was quite diverse in sequence with the exception of 5 conserved blocks of homology [See (9)]. Interestingly, the two variable regions found in the lepidopteran toxins noted above corresponded rather well to the segments of the protein between conserved blocks 2 and 3, and from conserved block 3 through conserved block 5.

Partial function was ascribed to large toxin regions by genetic deletion studies (17-21) and biochemical processing analyses (22-25, summarized in 9). These studies suggested that with the 3′ deletions, the essential toxin-encoding segments of crystal protein genes extended through conserved block 5, with little sequence required past this point to obtain activity. At the 5′ end, truncations of only a few codons past the N-terminus of a minimal toxin protein (typically obtained by proteolysis) were also inactive, while those having endpoints preceding the cleavage site at about residue 29 were active. Significantly, the N-terminal half of the 130 kD protein was the toxin. This toxin could be envisioned as having three functional domains: a relatively similar "activity" domain, and two variable regions that could determine insecticidal spectrum (17-19). Thus, the large toxins are generally described as existing in two forms: the large proprotein (~130 kD) with a protease-sensitive C-terminal end called the "protoxin," and the smaller, approximately half-sized functional "toxin" core resulting from proteolysis.

PROTOXIN STRUCTURE AND FUNCTION

Crystal proteins of ~130 kD have substantially higher sequence similarity in the protoxin segment from the end of conserved block 5 through the C-terminus (9). The speculated function of the long C-terminal segments of these proteins is to aid in formation of an ordered crystalline array in *B.t.* Additionally, the presence of most of the cysteine residues in the C-terminal "protoxin" segment, perhaps involved in intermolecular disulfide bridges, suggests that they may explain the alkaline, reducing conditions often required for solubility of these proteins. Since these C-terminal segments, and often short portions of the N-terminus as well, are missing in the essential toxic fragments of these proteins, their role in solubilization and efficient activation could be substantial.

The role of the protoxin portion of these proteins in mediating toxicity was not extensively examined. Evidence suggests it may limit toxicity in some cases. Only proteolytically truncated proteins were cytotoxic to some insect cell lines (26). For certain *B.t.* strains and certain insects, dramatic improvements in activity could be observed by dissolving crystals, or proteolytically activating them, prior to bioassay (27). However, with several combinations of *B.t.* strains and insects, certain protoxins promoted the availability of other protoxins (28). For some Lepidoptera and for other insects having a gut pH near neutrality, the requirement for either the presence of spores in addition to crystals, or for solubilization and/or pre-activation, can be dramatic (29-31).

Structurally, the protoxin segment has been most thoroughly analyzed in lepidopteran-active gene products. The cysteine residues, near the N-terminus and in the C-terminal region of the HD-73 protein, apparently all form disulfide bonds. Bonds crosslink to the same positions of a second protoxin, forming the protoxin dimers making up the crystals (32). Monoclonal antibodies reactive with the native toxin of a *B.t.* subsp. *kurstaki* crystal often reacted with the native protoxin as well (33). This indicates that the toxin is at least somewhat exposed in the protoxin. Proteolysis of protoxins occurred sequentially from the C-terminus, generating a series of ~10 kD peptides (34). While the preceding study indicated that there was little conformational change of the toxin region during the proteolytic activation process, a subsequent examination of the HD-73 protein with the use of differential scanning calorimetry (35) suggests that at least one domain of the toxin becomes more thermally stable following activation. Thus far, however, no mechanism has been proposed for the lack of cytotoxicity of the unproteolysed protoxin (26,35).

DIVERSITY AND SEQUENCE DISTRIBUTION AMONG LEPIDOPTERAN TOXINS

A search of GENBANK Release 83 and recent patent literature revealed 23 different genes encoding proteins that either have been called CryI (i.e., Lepidoptera-active; ref. 9) or seem to fall within a CryI-related structural group. The differences between these sequences have been characterized by: (a) vertical bars to show individual similarities or differences between aligned sequences (9,17-19,21), (b) by block diagrams that attempt to summarize similarities to various related proteins (15), and (c) most recently by relatedness tree diagrams that show relative percent relatedness among a group of proteins (1,14). While each of these has its merits and disadvantages in conveying the complexity of sequence information present in a gene family of this size, the following discussion will use the relatedness tree method (36) along with the 3-D structure information (10) on a domain-by-domain basis to convey the segmental nature of relatedness among these proteins, before a discussion of the activities of various mutant and engineered derivatives.

Figure 2 shows a relatedness tree diagram for the full length "CryI-like" proteins. There are clearly clustered groups of sequences for each of the previously-noted CryIA through CryIG, as well as two new sequences having less than 85% sequence similarity. For CryIA, IB, IC, ID, IE and IF there are additional class-specific sequences having >85% sequence similarity. These include the well-known CryIAa through CryIAc (9), and new variants of CryIB, IC, ID, IE and IF. Additionally, there are three new sequences having ~80% sequence similarity to CryIA, IC, ID, IE and IF: ET4, PrtA and PrtC. CryV (37), CryV465, CryIB and ET5 seem to be less related to the group above, and are almost as unrelated to these other lepidopteran-active proteins as they are to the coleopteran-active proteins (1,14). Indeed, CryIB (38) and CryV reportedly have coleopteran activity. The two sequences clustering with CryIG are relatively unrelated to the rest of these proteins (see Figure 1), and they are clustered here primarily through their similar protoxin segments.

Figure 3 shows a similar analysis of only the presumed toxin portions of these proteins. Note that the overall topology of the figure is similar to Figure 2; however, the CryIA, IC, ID, IE, IF, PrtA, PrtC and ET4 are related by 60% to 70% sequence similarity and the sub-groupings have >80% sequence similarity. The CryIB and CryIG sequences are again quite different, and CryIB clusters better with the coleopteran-active (CryIII)

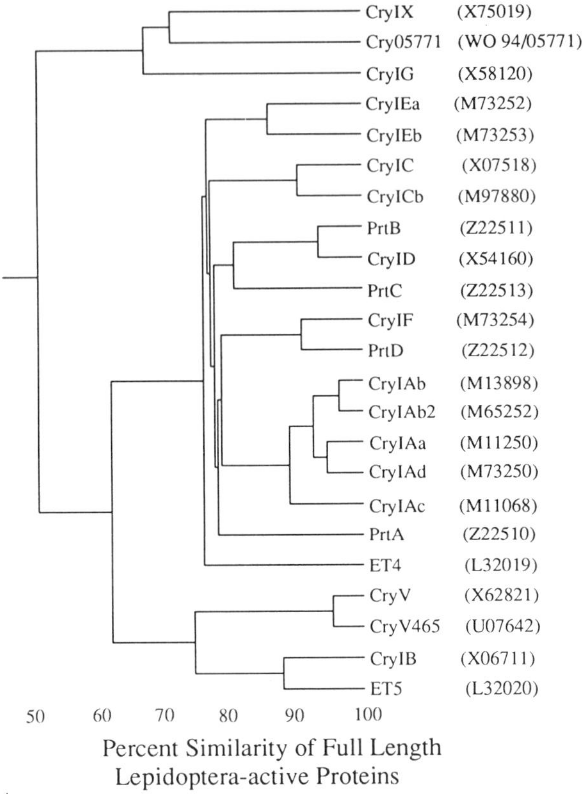

Figure 2. Full-length amino acid sequences translated from the nucleotide sequences from the indicated GENBANK or EMBL database accession numbers or PCT patent application were aligned with the use of routine PILEUP from the GCG computer package (83).

proteins, while CryIG does not cluster well with any other protein. The Cry05771 and CryIX proteins that clustered with CryIG as full-length proteins, now appear rather different.

Similar analyses using estimates of the individual domains of these proteins are shown in Figures 4 through 6. The domain I group is defined here as encompassing approximately residues 30 through 250. The domain II set includes approximately residues 250 through 455, and domain III extends from approximately residues 455 through 610. The relatedness tree of domain I (Figure 4) shows a similar topology to the tree of whole toxin domains (Figure 2), but with some clustering of the CryIA, ID and IE sequences. This grouping of sequences of 75 to 85% similarity may have led to the description of the CryID and CryIE proteins as having CryIA-like character (15). However, the CryIA group is >95% related in this domain, while the CryIC, IF, PrtA and PrtC sequences are 60% to 75% related to the CryIA proteins over a similar sequence. Another notable feature of this analysis is the clustering of the Cry05771 domain I with the CryIB-like group.

As shown in the domain II analysis of Figure 5, the CryIA, IC, ID, IE, IF, PrtA, PrtC and ET4 groupings are each related at 46% to 63% sequence similarity. The slightly greater similarity (63% *versus* 55%) of the CryIC-like and CryID-like proteins to each other may be what was implied by the CryIC-like portion of CryID (15). The CryIA group is divided at 78% sequence similarity between the nearly identical CryIAc/CryIAb domain

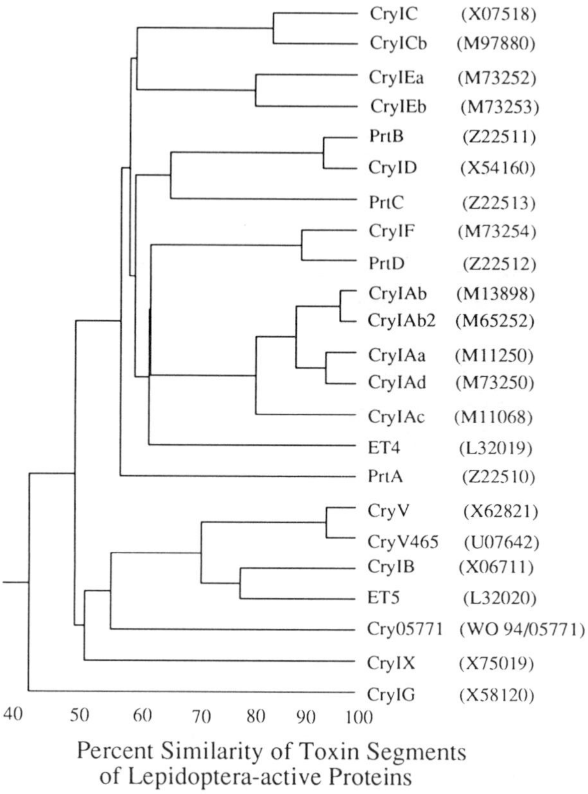

Figure 3. Amino acid sequences translated from the nucleotide sequences from the indicated GENBANK or EMBL database accession numbers or PCT patent application were truncated to the minimal toxin-encoding polypeptide (9) and aligned with the use of routine PILEUP from the GCG computer package (83).

II and the 95% similar CryIAa/CryIAd domain II. There is an additional group of 4 sequences related by ~70% or more to the CryIB domain II.

By contrast, a similar analysis of domain III (Figure 6) shows quite a different clustering topology. The CryID cluster keeps similar topology at 96% sequence similarity, but shows closer relationship to domain III of one of the new proteins, PrtC, at 87% sequence similarity. Surprisingly, several new CryI-like proteins, PrtA, ET4 and a new CryIF-like protein, PrtD, have domain III sequences with >78% sequence similarity to the domain III sequences of the CryIB-like proteins CryV, CryV465 and ET5. The CryIAa and CryIAb domains III are identical (17), with the CryIAd domain clustering very closely (97% similar).

The CryIC and CryIEa domains III are closely related, as reported earlier (Figure 6, ref. 39); however, the domains III of CryICb and CryIEb are also very similar. The latter observation suggests that CryIEa and CryICb could have arisen from ancestral crossovers between CryIEb and CryICa, as proposed for CryIAb (see above). Two other exceptional features of the tree in Figure 6 are: (a) the remarkable unrelatedness of the CryIAc domain III, and (b) the degree to which the domain III of CryIB clusters with CryIF and CryIA (~65% similarity).

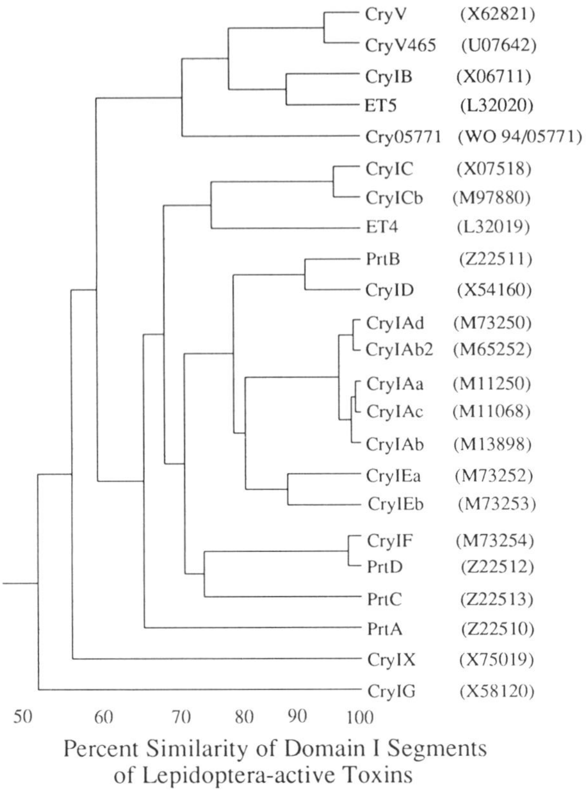

Figure 4. Amino acid sequences translated from the nucleotide sequences from the indicated GENBANK or EMBL database accession numbers or PCT patent application were truncated to correspond to domain I of CryIIIA (10) and aligned with the use of routine PILEUP from the GCG computer package (83).

FUNCTION INFERRED FROM THREE-DIMENSIONAL STRUCTURE

The three-dimensional structure of the truncated *B.t.* CryIIIA toxin, resolved to 2.5Å, provided a substantially improved foundation for evaluating genetically-based structural alterations to probe, and perhaps even improve toxin function (10). CryIIIA consists of three domains. Domain I contains a 7-helix bundle, likely responsible for channel formation in the membrane. Domains II and III are predominantly beta sheet. The "Greek key" motif of domain II is proposed to bind the insect gut receptor, while the suggested role of the "jelly roll" structure of domain III is toxin stabilization. Structure-function relationships of each domain are examined more closely below.

DOMAIN I - THE PUTATIVE CHANNEL-FORMING REGION

The structure of domain I begins with an N-terminal ~230 amino acid alpha-helical barrel, having a central helix corresponding to conserved block 1, surrounded by six largely amphipathic helices. The final two helices correspond to a portion of conserved block 2. Nearest neighbor helices adopt an antiparallel arrangement. By analogy with 4-

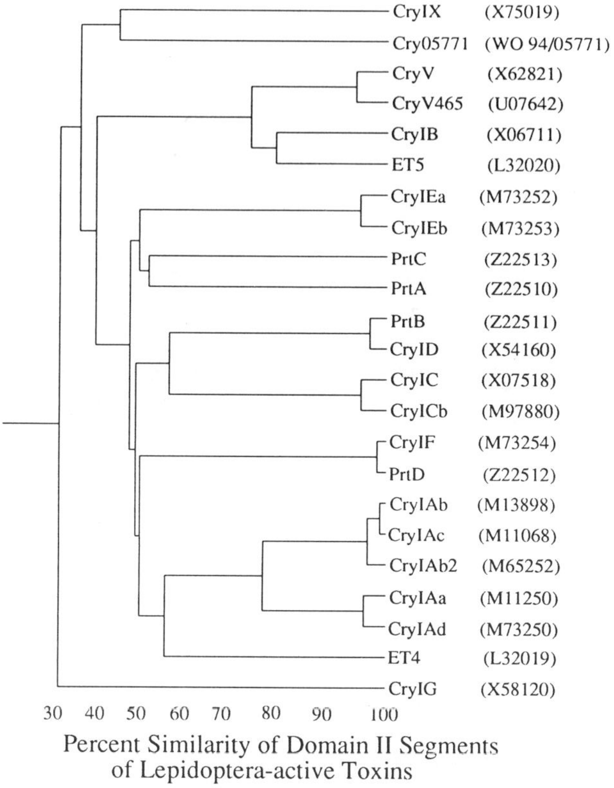

Figure 5. Amino acid sequences translated from the nucleotide sequences from the indicated GENBANK or EMBL database accession numbers or PCT patent application were truncated to correspond to domain II of CryIIIA (10) and aligned with the use of routine PILEUP from the GCG computer package (83).helix bundle model peptides, this may confer the most energetically favorable arrangement (40), contributing to the stability of *B.t.* toxins.

helix bundle model peptides, this may confer the most energetically favorable arrangement (40), contributing to the stability of *B.t.* toxins.

A role for these amphipathic helices in membrane insertion is consistent with other models of peptide insertion into membranes (3,10). An examination of the individual helices of CryIIIA was made for structural features consistent with pore- or channel-forming properties. Generally, amphipathic helices may form channels provided the helix length is sufficient and polar amino acid distribution falls within defined parameters.

A length of about 20 amino acids per helix seems to be generally accepted as the minimum required for transmembrane channel formation (41,42). A minimum value was redefined with synthetic helices of 11, 15 and 18 residues optimized for only alpha-helical structure (43). Both the 15- and 18-residue helices formed cation channels. The authors additionally suggest that a reorientation of the hydrophobic helical faces, i.e., from central to peripheral, occurs to generate pores. All CryIIIA helices, except helix 2, are of sufficient length.

Helix 2 was excluded from subsequent analyses since it consists of two 14 amino acid helices disrupted by a stretch of 5 amino acids containing a helix-disrupting proline residue. It is unclear if the lengths of these helices are sufficient for channel formation. By analogy, however, segment S3 of some cation channels contains a conserved proline

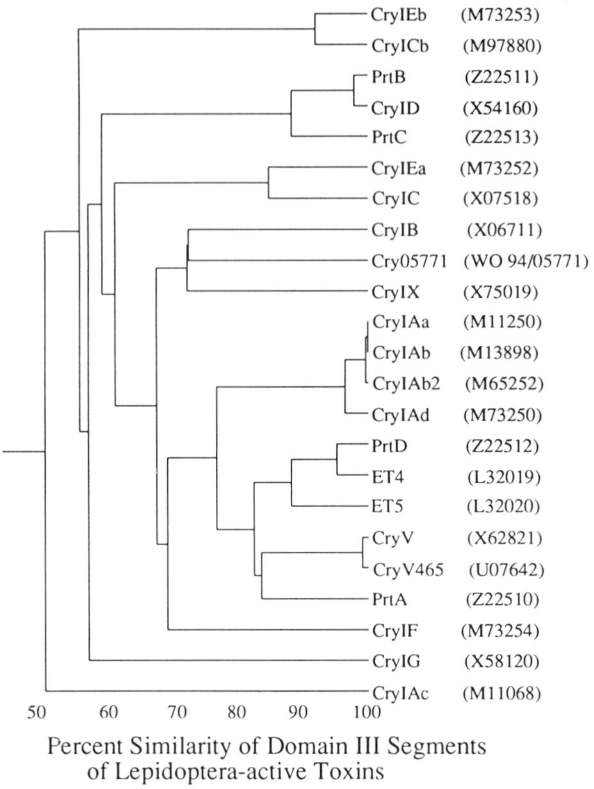

Figure 6. Amino acid sequences translated from the nucleotide sequences from the indicated GENBANK or EMBL database accession numbers or PCT patent application were truncated to correspond to domain III of CryIIIA (10) and aligned with the use of routine PILEUP from the GCG computer package (83).

residue (42), inferring that helix 2 of CryIIIA could be part of a channel.

A comparison of CryIIIA helices with a particular class of amphipathic helices may promote better understanding of mechanism of action. Amphipathic helices may be separated into seven general classes (44): A (apolipoproteins), H (polypeptide hormones), L (lytic polypeptides), G (globular proteins), K (calmodulin-regulated protein kinases), C (coiled-coiled proteins) and M (transmembrane proteins). Some key criteria distinguishing class L (lytic, e.g., Hymenoptera venoms) from class M (transmembrane, e.g., bacteriorhodopsin) proteins are: mean charged residue density per 11-mer of helix, and mean positive to negative charge ratio. Values were calculated for helices 1, 3, 4, 5, 6 and 7 for CryIIIA. Normalized to an 11-mer helix, positively-charged residues range from 0.5 to 1.5, negatively charged residues from 0.5 to 2.2, total charged residues vary from 1.0 to 3.7, and the positive/negative ratio extended from 0.68 to 1. These values align more closely with lytic peptides than with transmembrane proteins. An additional parameter to consider is the lys/arg ratio, which is high for L class peptides. Helices 1, 3 and 5 contain lysine only, helix 6 has a 5:1 ratio, and helices 4 and 7 contain only arginine. There is an indication that substitution of arginine for lysine may reduce lytic properties (45), so replacement of arginine residues with lysine in helices 4, 6 and 7 may improve toxin potency.

Finally, polar residue distribution is another characteristic which correlates with channel-forming function. In contrast to the A and G amphipathic helix classes, the residues of the L class form a narrower polar face, while the M (transmembrane) class has the narrowest face (~60°). Helical wheel analysis was done with a residue angle of 100° and a residue pitch of 1.5Å. Helices 1 and 5 show polar residue distribution over about a 140° angle, similar to that described for a model lytic peptide (45). However, helix 5 contains polar amino acids on the non-polar face, and vice-versa. These may participate in stabilization of domain I. Polar residues in Helix 6 are distributed over a face of 140° to 160°. Again, polar residues disrupt the non-polar face. By contrast, polar amino acids of helices 3 and 4 subtend an angle greater than 200°, similar to class G and class A proteins (45). Finally, the polar residues of helix 7 tend to cluster at the N-terminus, while a predominantly hydrophobic patch is found at the C-terminal end. Non-polar amino acid distributions of helices 5, 6 and 7 may play a key role in membrane insertion.

Helix function may be deduced by mutagenesis and synthetic *in vitro* systems. Properties of helix 5, a highly conserved region (block 1) among toxins, was more extensively analyzed by site-directed mutagenesis and with synthetic peptides. Substitution mutations introduced into alpha helix 5 maintain membrane or receptor binding, but gave reduced toxicity (46). Additionally, these mutations resulted in reduced K^+-dependent transport into membrane vesicles (47). Synthetic helix 5 was able to incorporate into membranes, then self-aggregate and form ion channels (48). Further, it effected red blood cell lysis. A peptide containing an alanine to proline mutation was shown to have reduced channel-forming activity, as expected. By contrast, synthetic helix 7 was unable to form *de novo* pores. However, when tested in combination with synthetic helix 5, it appeared to be able to coassemble into channels (49). Together, these data support the participation of helix 5 in channel formation.

In summary, at least five helices of domain I of CryIIIA appear to have gross structural characteristics favorable for pore or channel formation. Definition of roles of individual amphipathic helices will benefit from the types of synthetic approaches and vesicle reconstitution studies already described.

DOMAIN II - RECEPTOR-BINDING REGION

The second domain of the CryIIIA toxin is predominantly beta sheet, characterized as the "Greek key" motif. In the linear sequence, it runs from the middle of conserved block 2 into the beginning of conserved block 3. This domain, and especially the final 120 amino acids, is homologous to the first of the two variable regions observed between the HD-1 toxins, as noted above. Li et al. (10) pointed out that receptor binding is best correlated with this domain, as determined by analysis of naturally-occurring toxins. Subsequent analysis of *in vitro*-generated chimeric proteins (50) is consistent with this hypothesis. Two protruding loops oriented in parallel with the helical bundle of domain I were suggested to be an attractive structure for receptor binding.

DOMAIN II - *IN VITRO* CHIMERIC GENES TO DEFINE TOXIN SPECIFICITY

The insecticidal spectrum of different cloned genes has been noted in several reports (9,11,51,52). The first report with chimeric genes to investigate a sizable differential activity on insects was that of Ge et al. (53). Specificity to silkworm (*Bombyx mori*) was localized to the C-terminal 2/3 of what became known as domain II of CryIAa. Although activity was reported as essentially plus or minus, the selectivity of CryIAa over CryIAc toward this insect is sufficient (200-fold) that the data were convincing. A follow-up report indicated that specificity correlated well with high-affinity receptor binding with the use of brush border membrane vesicles (50).

The same (54), or a similar (55) set of chimeric toxins was used to locate specificity-determining regions toward cabbage looper (*Trichoplusia ni*). Codons 332 to 450 (i.e., the C-terminal 2/3 of what became known as domain II) of CryIAc determined specificity to this insect. The same investigations (54,55) also located specificity-determining regions toward tobacco budworm (*Heliothis virescens*). Both studies concluded that sequences defining both domains II and III were required for maximal activity toward *H. virescens*. Experimentally, however, Ge et al. (54) used double reciprocal crossovers between both *cryIAc* and *cryIAa* between codons 332 and 612, while Schnepf et al. (55) exchanged mostly *cryIAa* segments into the codon 332 to 722 segment of *cryIAc*. Thus, with one exception, the 11 differences in *cryIAa* relative to *cryIAc* prior to codon 332 were not examined (55), while several other constructs contained the domain III of *cryIAa* fused to the processing region (612-722) of *cryIAc* having 12 amino acid differences from the corresponding *cryIAa* sequence (54). The most notable difference between the results of these two studies is that a construct having the *cryIAc* 450 to 612 segment replacing the corresponding *cryIAa* gene segment was 30 times as active against *H. virescens* as CryIAc (54). Interestingly, a similar construct differing only by the 12 changes in CryIAc from CryIAa between codons 612 and 722 was intermediate in activity between CryIAa and CryIAc (55). Additionally, constructs having the 11 changes in CryIAc away from CryIAa prior to codon 332 with (55) or without (54) the 12 differences from CryIAa between codons 612 and 722 were either as active as CryIAc (54), or were intermediate in activity between CryIAa and CryIAc (54,55).

With chimeric genes composed of *cryIC* and *cryIAb*, specificity was definitively transferred between Egyptian cotton leafworm (*Spodoptera littoralis*) and *H. virescens* (56). A chimeric gene having all but about 10 codons of domain I from CryIC fused to the remainder of domain I and all of domains II and III from CryIAb produced a protein having activity on *H. virescens* 55-fold below that of CryIAb but well above the undetected activity of CryIC. This chimeric protein, like CryIAb, had no activity on *S. littoralis*. A nearly reciprocal chimeric having domain I from CryIAb and domains II and III from CryIC had activity about 9-fold below that of CryIC, and well above the undetectable activity of CryIAb on *S. littoralis*. The latter chimeric, like CryIC, had no activity on *H. virescens*. Because of instability to proteases, only the latter chimeric could be evaluated for receptor binding; however, the binding had high affinity and was competitive with CryIC but not CryIAb. Therefore, insecticidal specificity and binding were shown to reside in regions of CryIC homologous to domains II and III of CryIIIA.

More limited substitutions and deletions in domain II have suggested the location of at least part of the binding determinant of CryIAa (57). A hexa-alanine substitution or deletion of residues 365-371 reduced binding of this toxin to *B. mori* by 10-fold and essentially rendered the toxin inactive to *B. mori*. By contrast, a penta-alanine substitution of residues 385-389 had effects on neither binding nor toxicity. However, 4 additional alanine block substitution mutants and 5 additional block deletion mutants between residues 357 and 404 were too unstable to analyze. While this analysis is rather sparse, it suggests that the portion of CryIAa corresponding to the β6-loop2-β7 segment of CryIIIA is involved in receptor binding.

Selective loss of activity toward tobacco hornworm (*Manduca sexta*) has been observed in several cases. The mutation A92D resulted in >500-fold loss of activity toward *M. sexta*, and a similar dramatic loss toward *T. ni*, when incorporated into either the CryIAa, CryIAb or CryIAc protoxins (47). By contrast, when incorporated into the same toxins, activity loss toward *H. virescens* was 6-fold for CryIAb, 28-fold for CryIAc and >16-fold for CryIAa. In vesicle binding assays, A92D CryIAc toxin competed with labeled CryIAc toxin for binding to *H. virescens* vesicles (at somewhat reduced affinity) but did not compete for binding to *M. sexta* vesicles. The integrity of the mutant protein during the binding assays was not reported. Since this mutation is in domain I, these data may suggest that domain II is not the exclusive determinant of receptor binding.

Another region of CryIAc, residues 429 to 447, differs in some 13 positions from CryIAa. When the CryIAa variant of this sequence was substituted into CryIAc, a dramatic decrease in activity was observed toward *M. sexta* (55,58), while an activity intermediate between CryIAc and CryIAa was observed toward *H. virescens* (54,55). The mutant protein was stable enough to proteolysis that binding studies with *B. mori* vesicles could be performed (50). Evidence was presented that binding of this mutant protein to *M. sexta* BBMV was relatively unaffected compared to CryIAc (59). Since CryIAa and CryIAc have equal activities on *M. sexta*, the reduction in activity of this and one or two other chimerics was suggested to result from a structural incompatibility in a protein domain (55). It remains to be determined whether the shift in species selectivity of the mutant toxins noted here reflects the proteolytic demands of the host, or a more fundamental species-specific interaction at the post-receptor-binding level.

Clearly progress is being made defining regions responsible for species specificity. However, the analyses to date have mostly involved large-scale changes that could distort the overall structure of the protein, often making it protease-sensitive. The field awaits high resolution studies detailing structures and sequences responsible for receptor binding.

DOMAIN III - THREE PUTATIVE FUNCTIONS

Domain III is a beta-sandwich structure proposed to stabilize the toxin (10), possibly by protection from proteolysis. It contains most of conserved block 3 and all of conserved blocks 4 and 5. In linear sequence, it appears to be the second hypervariable region identified by comparison of the HD-1 toxins, truncated at the end by conserved block 5. Additionally, domain III may contain information for cell lysis, and may also function as a ligand.

A recent report (60) suggests channel-forming properties may extend into this domain. Mutagenesis in a set of 11 residues in conserved block 4 altered channel formation characteristics. This region forms an amphipathic beta sheet. Since a sheet of 10 residues is able to span the membrane (3), channel-forming potential exists. It has been suggested that at least 8 amphiphilic beta-strands are required to form a barrel (3). This interpretation is controversial (42). If true, it implies the need for either 8 molecules of domain III or recruitment of amphipathic beta-strands from other regions, or possibly assembly into existing membrane proteins [conceptual models in (12)]. Dissection and reconstitution studies may resolve the role of this conserved motif.

Several lines of evidence infer domain III may be a ligand for gut receptors. Beta sandwich structures are observed as receptor-binding domains in the 3-dimensional structure of diphtheria toxin (61) or *Pseudomonas* exotoxin A (62). Additionally, two toxins, CryIAb and CryIAc differing primarily in domain III, show profound differences in receptor binding properties and activities toward certain insects. CryIAc toxin binds to gypsy moth (*Lymantra dispar*) vesicles with 10 times the affinity of CryIAb, yet has 1/400th the activity (11). CryIAc toxin had a 10-fold lower binding affinity for European corn borer (*Ostrinia nubilalis*) vesicles than CryIAb and a 10-fold lower activity on the insect (52). In both cases the receptor populations were similar for the two toxins, and in both cases the two toxins competed for the same receptor. In the case of these two insects, further exploration will define whether it is the few differences between the domain II regions, or the more varied domain III regions that account for differences in activity. As noted in the discussion of chimeric proteins above, domain III of CryIAc is also a partial specificity determinant for *H. virescens*.

The construction of chimeric proteins between CryIEa and CryIC has shown that domain III of CryIC is a significant determinant of activity toward beet armyworm (*Spodoptera exigua*) and cabbage moth (*Mamestra brassicae*) (63). In this case, a chimeric protein having the binding specificity of CryIE (conferred by domains I and II of CryIE) gained activity on the two insects only when domain III was spliced from CryIC.

PROGRESS IN IDENTIFICATION OF THE RECEPTOR

Several reports have dealt with the identification of the macromolecular nature of a receptor for *B.t.* toxins independently of experiments using competition with ^{125}I-labeled toxin for saturable receptors on BBMV (see above). Some of these studies have pointed to the importance of glycosidic residues in binding (64-66). Progress on the purification and identification of the protein portion of the receptor has included the identification of polypeptides by size (65-69), antigenic characterization (67) and identification of specific purified receptor molecules (69-71). Two recent reports (70,71) identified aminopeptidase N as comprising at least a portion of the receptor population for CryIAc. In contrast with the ^{125}I-labeled toxin competition experiments with BBMV implicating domain II in receptor binding (see above), one interpretation of the ligand blot experiments (67) suggests that differences in domain III target alternative receptor polypeptides. Subsequent characterization of the CryIAb (69) and CryIAc (70,71) receptors, indicating they are different sizes, is consistent with that interpretation.

MAKING THE TOXIN TOXIC - COUPLING RECEPTOR BINDING AND CHANNEL FORMATION

A general model for signal transduction was proposed to occur through an alpha helix 7/beta sheet 1 ($\alpha 7/\beta 1$) boundary deformation (12). Hairpins between helices 4/5 and 6/7 have been proposed as candidates to initiate pore formation, with helices 6/7 favored because of proximity to domain II. Additionally, proteolytic nicking is suggested to participate in insertion (72). Notably, helix 7 of CryIIIA has a hydrophobic patch at the C-terminus. Deformation caused by binding may expose this patch for direct membrane interactions.

Models for helix insertion from an aqueous environment into the hydrophobic membrane may be extrapolated from analogous bacterial systems. Colicins contain three domains: one forms a channel, another binds receptor, and the third is involved in translocation (for reviews, see refs. 4,73). Electron spin resonance indicates insertion is mediated through hydrophobic interactions (74), likely through hydrophobic helices 8 and 9. By analogy with the colicin structure, it would seem reasonable to propose that helix 5 has a similar function. Candidate amino acids for targeted mutagenesis would be homologues of CryIIIA valine 191 through leucine 194. Additional hydrophobic stretches near helix ends that may be appropriate for mutagenesis include: leucine 76 through valine 89 at the end of helix 1 into the beginning of helix 2, and leucine 279 through leucine 285 at the end of helix 7. Notably, these latter two hydrophobic regions are proximal in the crystal structure (10). It is possible to envision a large conformational change and toxin unfolding by alteration of this region, since these two helices bracket the other helices in the primary structure of Domain I.

Several testable hypotheses of this signal transduction model include: (a) assuming loops bind receptors, as the distance from the helical bundle increases, toxin effectiveness decreases, (b) engineering loops nearest the helical bundle (e.g., through phage-display libraries) for increased receptor affinity may increase toxicity, and (c) decreased strength of $\alpha 7/\beta 1$ interaction may increase toxicity up to a point, as long as ability to form a crystal and resist proteolysis is unimpaired. This may be designed to be dependent on gut pH, that is, insect-specific. Finally, (d) increased flexibility of the 'hinge' between $\alpha 7/\beta 1$ may increase toxicity by allowing the helix bundle to swing out more easily.

PROTEIN FUSIONS WITH *B.t.* TOXINS TO IMPROVE PERFORMANCE

Genetic fusions provide a means of analyzing expression, allowing genetic selection, establishing structure-function relationships, or generating proteins with novel properties. Translational fusions were made with truncated *B.t.* toxin and neomycin phosphotransferase for use as a selectable marker in plants (75,76). The fusion proteins retained insect toxicity. A partial survey of the literature revealed that several types of *B.t.* toxin or toxin subfragment fusions were proposed or implemented: fusions to other *B.t.* toxins or with various enzymatic toxins or insect virus proteins.

FUSIONS WITH OTHER *B.t.* TOXINS

Two minimal toxins of CryIAb and CryIC were translationally fused (77). In an *in vitro* assay, activity of the hybrid protein overlapped toxicities of each individual component.

FUSIONS WITH ENZYMATIC TOXINS

ADP-ribosylating toxins or ribosome-inactivating proteins are cytotoxic at levels extrapolated to be a few molecules per cell. In these toxins, the segment mediating binding to eukaryotic cells may be substituted with other ligands (78,79). Hybrids with receptor-binding subfragments of δ-endotoxins may generate insect-specific toxins of high potency (80).

FUSIONS WITH VIRAL PROTEINS

Polyhedrin is an abundant protein of the occluded form of nuclear polyhedrosis viruses. It has no reported toxicity *per se*. Fraser *et al.* (81) proposed a chimeric protein consisting of polyhedrin and *B.t.* δ-endotoxins for insecticidal use. Since extracellular viral particles are typically embedded in a polyhedrin matrix (i.e., the occluded form), they would now be embedded in a polyhedrin-*B.t.* δ-endotoxin matrix. Recombinant virus may serve as a vehicle for abundant production of insecticidal toxin.

Insect virus protein fusions to lytic proteins or pore-forming *B.t.* δ-endotoxins were proposed (82) as molecules with possible improved host-range or toxicity. Also proposed were chimeras with luteoviral proteins.

SUMMARY

Nature has provided potent insecticidal toxins as fermentation products of many *Bacillus thuringiensis* strains. Elucidation of structure-function relationships for this class of natural toxins is in its early stages. Both direct experimentation and application of theoretical structure-function principles emerging from the rapidly growing field of protein structure are accelerating understanding of these toxins. Coupled with the increasing demand for biologically sound pesticides, the benefits from engineering nature's toxins for improved performance set the stage for exciting, fast growth in discovery, characterization and commercialization of new active ingredients for biopesticides and transgenic plants.

REFERENCES

1 Feitelson, J.S., Payne, J. and Kim, L. (1992) Bio/Technology 10, 271-275.
2 Feitelson, J.S. (1993) in Advanced Engineered Pesticides (Kim, L., ed.) pp. 63-71, Marcel Dekker Inc., New York, NY.
3 Ojcius, D.M. and Young, J.D.-E. (1991) TIBS 16, 225-229.
4 Parker, M.W., Tucker, A.D., Tsernoglou, D. and Pattus, F. (1990) TIBS 15, 126-129.
5 Montecucco, C. and Schniavo, G. (1993) TIBS 18, 324-327.
6 Brunke, K.J. and Meeusen, R.L. (1991) TIBTECH 9, 197-200.
7 Gaertner, G.H., Quick, T.C. and Thompson, M.A. (1993) in Advanced Engineered Pesticides (Kim, L., ed.) pp.73-83, Marcel Dekker Inc., New York, NY.
8 Tabashnik, B.E. (1994) Annu. Rev. Entomol. 39, 47-79.
9 Höfte, H. and Whiteley, H.R. (1989) Microbiol. Rev. 53, 242-255.
10 Li, J., Carroll, J. and Ellar, D.J. (1991) Nature 353, 815-821.
11 Gill, S.S., Cowles, E.A. and Pietrantonio, P.V. (1992) Annu. Rev. Entomol. 37, 615-636.
12 Knowles, B.H. and Dow, J.A.T. (1993) BioEssays 15, 469-476.
13 English, L. and Slatin, S.L. (1992) Insect Biochem. Mol. Biol. 22, 1-7.
14 Yamamoto, T. and Powell, G.K. (1993) in Advanced Engineered Pesticides (Kim, L., ed.) pp. 3-42, Marcel Dekker Inc., New York, NY.
15 Visser, B., Bosch, D. and Honee, G. (1993) in *Bacillus thuringiensis*, an Environmental Biopesticide: Theory and Practice. (Entwistle, P.F., Corey, J.S., Bailey, M.J. and Higgs, S., eds.) pp. 71-88, John Wiley and Sons Ltd., Chichester.
16 Aronson, A.I. (1993) Molec. Microbiol. 7, 489-496.
17 Wabiko, H., Raymond, K.C. and Bulla, L.A., Jr. (1986) DNA 5, 305-314.
18 Hoefte, H., de Greve, H., Seurinck, J., Jansens, S., Mahillon, J., Ampe, C., Vandekerckhove, J., Vanderbruggen, H., Van Montagu, M., Zabeau, M. and Vaeck, M. (1986) Eur. J. Biochem. 161, 273-280.
19 Geiser, M., Schweitzer, S. and Grimm, C. (1986) Gene 48, 109-118.
20 Adang, M.J., Idler, K.F. and Rocheleau, T.A. (1987) in Biotechnology in Invertebrate Cell Culture (Maramorosch, K., ed.) pp. 85-99, Academic Press Inc., San Diego, CA.
21 Schnepf, H.E. and Whiteley, H.R. (1985) J. Biol. Chem. 260, 6273-6280.
22 Lilley, M., Ruffel, R.N. and Somerville, H.J. (1980) J. Gen. Microbiol. 118, 1-11.
23 Bulla, L.A., Jr., Kramer, K.J., Cox, D.J., Jones, B.L., Davidson, L.I. and Lookhart, G.L. (1981) J. Biol. Chem. 256, 3000-3004.
24 Chestukhina, G.G., Kostina, L.I., Mikhailova, A.L., Tyurin, S.A., Klepikova, F.S. and Stepanov, V.M. (1982) Arch. Microbiol. 132, 159-162.
25 Nagamatsu, Y., Itai, Y., Hatanaka, C., Fumatsu, G. and Hayashi, K. (1984) Agric. Biol. Chem. 48, 611-619.
26 Luthy, P. and Ebersold, H.R. (1981) in Pathogenesis of Invertebrate Microbial Diseases (E.W. Davidson, ed.) pp. 235-267, Allanheld, Osmun and Co., Totowa, NJ.

27 Jaquet, J., Hutter, R. and Luthy, P. (1987) Appl. Env. Microbiol. 53, 500-504.
28 Aronson, A.I., Han, E.S., McGaughey, W. and Johnson, D. (1991) Appl. Env. Microbiol. 57, 981-986.
29 Huber, H.E. and Luthy, P. (1981) in Pathogenesis of Invertebrate Microbial Diseases (E.W. Davidson, ed.) pp. 209-234, Allanheld, Osmun and Co., Totowa, NJ.
30 Johnson, D.E. and McGaughy, W.H. (1984) J. Invertebr. Pathol. 43, 156-159.
31 Höfte, H.R., Annys, K., Lambert, B., Jansens, S., Soetaert, P. and Peferoen, M. (1992) Appl. Env. Microbiol. 58, 2536-2542.
32 Bietlot, H.P., Vishnubhatla, I., Carey, P.R., Pozsgay, M. and Kaplan, H. (1990) Biochem. J. 267, 309-315.
33 Huber-Lukac, M., Jaquet, F., Luethy, P., Huetter, R. and Braun, D.G. (1986) Infect. Immun. 54, 228-232.
34 Choma, C.T., Surewicz., W.K., Carey, P.R., Pozsgay, M., Raynor, T. and Kaplan, H. (1990) Eur. J. Biochem. 189, 523-527.
35 Choma, C.T., Surewicz., W.K. and Kaplan, H. (1991) Biochem. Biophys, Res. Commun. 179, 933-938.
36 Kalman, S., Kiehne, K.L., Libs, J.L. and Yamamoto, T. (1993) Appl. Env. Microbiol. 59, 1131-1137.
37 Tailor, R., Tippet, J., Gibb, G., Pells, S., Pike, D., Jordan, L. and Ely, S. (1992) Molec. Microbiol. 6, 1211-1217.
38 Bradley, D., Biever, D., Bauer, L.S. and Harkey, M.A. (1992) Abstracts of Annual Meeting Society for Invertebrate Pathology, Heidelberg, Germany, 16-21 Aug. 1992.
39 Visser, B., Munsterman, E., Stoker, A. and Dirkse, W.G. (1990) J. Bacteriol. 172, 6783-6788.
40 Chou, K.-C., Maggiora, G.M., Némethy, G. and Scheraga, H.A. (1988) Proc. Nat. Acad. Sci. U.S.A. 85, 4295-4299.
41 Lear, J.D., Wasserman, Z.R. and DeGrado, W.F. (1988) Science 240, 1177-1181.
42 Grove, A., Tomich, J.M. and Montal, M. (1992) in Genetic Engineering (ed. Setlow, J.K.) pp. 163-184, Vol. 14, Plenum Press, New York, NY.
43 Grove, A., Rivier, J.E., Mutter, M. and Montal, M. (1993) Prot. Sci. 2 (suppl. 1), 79, Abstract 138-S.
44 Segrest, J.B., De Loof, H., Dohlman, J.G., Brouillette, C.G. and Anantharamaiah, G.M. (1990) Proteins: Struct. Funct. Genet. 8, 103-117.
45 Tytler, E.M., Segrest, J.P., Epand, R.M., Nie, S.-Q., Epand, R.F., Mishra, V.K., Venkatachalapathi, Y.V. and Anantharamaiah, G.M. (1993) J. Biol. Chem. 268, 22112-22118.
46 Ahmad, W. and Ellar, D.J. (1990) FEMS Microbiol. Lett. 68, 97-104.
47 Wu, D. and Aronson, A.I. (1992) J. Biol. Chem. 267, 2311-2317.
48 Gazit, E. and Shai, Y. (1993) Biochemistry 32: 3429-3436.
49 Shai, Y. and Gazit, E., Proceedings Sixth International Colloquium on Invertebrate Pathology and Microbial Control, Montpellier, France, 1994, Abstract SY6#3, pp. 211-213.
50 Lee, M.K., Milne, R.E., Ge, A.Z. and Dean, D.H. (1992) J. Biol. Chem. 267, 3115-3121.
51 van Frankenhuyzen, K., Gringorten, J.L., Milne, R.E., Gauthier, D., Pusztai, M., Brousseau, R. and Masson, L. (1991) Appl. Env. Microbiol. 57, 1650-1655.
52 Denolf, P., Jansens, S., Peferoen, M., Degheele, D. and Van Rie, J. (1993) Appl. Env. Microbiol. 59, 1828-1837.
53 Ge, A.Z., Shivarova, N.L. and Dean, D.H. (1989) Proc. Nat. Acad. Sci. U.S.A. 86, 4037-4041.

54 Ge, A.Z., Rivers, D., Milne, R. and Dean, D.H. (1991) J. Biol. Chem. 266, 17954-17958.
55 Schnepf, H.E., Tomczak, K., Ortega, J.P. and Whiteley, H.R. (1990) J. Biol. Chem. 265, 20923-20930.
56 Honee, G., Convents, D., Van Rie, J., Jansens, S., Peferoen, M. and Visser, B. (1991) Mol. Microbiol. 5, 2799-2806.
57 Lu, H., Rajamohan, F. and Dean, D.H. (1994) J. Bacteriol. 176, 5554-5559.
58 Almond, B.D. and Dean, D.H. (1993) Biochemistry 32, 1040-1046.
59 Whiteley, H.R. (1990) Lecture at the Annual Meeting of the American Society of Microbiology, Anaheim, CA.
60 Chen, X.J., Lee, M.K. and Dean, D.H. (1993) Proc. Nat. Acad. Sci. U.S.A. 90, 9041-9045.
61 Choe, S., Bennett, M.J., Fuji, G., Curmi, P.M.G., Kantardjieff, K.A., Collier, R.J. and Eisenberg, D. (1992) Nature 357, 216-222.
62 Allured, V.S., Collier, R.J., Carroll, S.F. and McKay, C.B. (1986) Proc. Nat. Acad. Sci U.S.A. 83, 1320-1324.
63 Bosch, D., Schipper, B., van der Kliej, H., de Maagd, R.A. and Stiekema, W.J. (1994) Bio/Technology 12, 915-918.
64 Knowles, B.H., Thomas, W.E. and Ellar, D.J. (1984) FEBS Lett. 168, 197-202.
65 Knowles, B.H. and Ellar, D.J. (1986) J. Cell Sci. 83, 89-101.
66 Garczynski, S.F., Crim, J.W. and Adang, M.J. (1991) Appl. Env. Microbiol. 57, 2816-2820.
67 Oddou, P., Hartman, H., Radecke, F. and Geiser, M. (1993) Eur. J. Biochem. 212, 145-150.
68 Sanchis, V. and Ellar, D.J. (1993) FEBS Letters 316, 264-268.
69 Vadlamundi, R.K., Ji, T.H. and Bulla, L.A.,Jr. (1993) J. Biol. Chem. 268, 12334-12340.
70 Sangadala, S., Walters, F.S., English, L.H. and Adang, M.J. (1994) J. Biol. Chem. 269, 10088-10092.
71 Knight, P.J., Crickmore, N. and Ellar, D.J. (1994) Molec. Microbiol. 11, 429-436.
72 Ellar, D.J., Proceedings Sixth International Colloquium on Invertebrate Pathology and Microbial Control, Montpellier, France, 1994, Abstract PS2#1, pp. 10-15.
73 Cramer, W.A., Cohen, F.S., Merrill, A.R. and Song, H.Y. (1990) Molec. Microbiol. 4, 519-526.
74 Shin,Y.-K., Levinthal, C., Levinthal, F. and Hubbell, W.L. (1993) Science 259, 960-963.
75 Barnes, W.M. (1985) Abstracts: First International Congress of Plant Molecular Biology, International Society for Plant Molecular Biology (Galau, G.A., ed.), University of Georgia Center for Continuing Education, Athens, Georgia, abstract OR-21-10.
76 Vaeck, M., Reynaerts, A., Höfte, H., Jansens, S., De Beuckeleer, M., Dean, C., Zabeau, M., Van Montagu, M. and Leemans, J. (1987) Nature 328, 33-37.
77 Honee, G., Vriezen, W. and Visser, B. (1990) Appl. Environ. Microbiol. 56, 823-825.
78 Murphy, J.R. and Williams, D.P. (1991) in Sourcebook of Bacterial Protein Toxins (Alouf, J.E. and Freer, J.H., eds.) pp. 491-505, Academic Press, London.
79 Pastan, I., Chaudhary, V. and Fitzgerald, K.J. (1992) Annu. Rev. Biochem. 61, 331-354.
80 Wilcox, E., Edwards, D., Schwab, G.E., Thompson, M. and Culver, P. (1989) European Patent Application EP 0 340 948 A1.

81 Fraser, M.J., Rosen, E.D. and Ploplis, V.A. (1989) U.S. Patent 4,870,023.
82 Sivasubramanian, N. (1991) PCT Patent Application WO91/17254.
83 Devereux, J., Haeberli, P. and Smithies, O. (1984) Nucl. Acids Research 12, 387-395.

USES FOR GAL4 EXPRESSION IN MAMMALIAN CELLS

Ivan Sadowski

Department of Biochemistry and Molecular Biology
University of British Columbia
2146 Health Sciences Mall
Vancouver, B.C. V6T 1Z3

INTRODUCTION

Research investigating regulation of the *GAL* genes in the yeast *Saccharomyces* has propelled this model system to a prototypical stature. The transcriptional activator GAL4 is, arguably, one of the most thoroughly characterized eukaryotic transcriptional regulatory proteins studied to date. This intense investigation over the past decade has spawned much of our current understanding of eukaryotic gene regulation, and additionally has provided an extensive toolbox of strategies and reagents for characterization of the functional and structural features of other eukaryotic transcriptional regulatory proteins. In this article I review the current knowledge of how GAL4 is regulated in yeast, and describe currently popular molecular analysis strategies which employ the functional domains of GAL4 in mammalian cells. To date these include - in addition to examining functional domains of transcription factors - novel approaches towards studying protein-protein interactions, site-specific proteinases, and proteinase inhibitors.

GAL4 FUNCTION IN *SACCHAROMYCES CEREVISIAE*

GAL4-regulated genes

The ability of yeast to utilize galactose efficiently requires the combined function of a group of genes whose products transport galactose into the cell and catalyze its entry into glycolysis by conversion into glucose-6-phosphate (see Table 1). Expression of most genes necessary for galactose catabolism is closely adjusted by a combination of mechanisms which ensure that galactose is utilized when available, but that glucose is used preferentially (1). Thus, their transcription is induced by galactose, but is repressed by the presence of glucose - this stringent regulation is conferred by upstream regulatory sequences which mediate the effects of galactose and glucose. GAL4 is responsible for

Genetic Engineering, Vol. 17, Edited by J.K. Setlow
Plenum Press, New York, 1995

Table 1. Genes and their products necessary for galactose catabolism in *Saccharomyces cerevisiae*.

A) Genes Encoding Regulatory Proteins

Gene	Function of the Encoded Gene Product:	Regulated By:[a] UAS$_G$	URS$_G$
GAL4	transcriptional activator protein, binds to UAS$_G$.	N	Y
GAL3	inducer synthesis, (inhibitor of GAL80p?).	Y	Y
GAL80	inhibitor of GAL4 protein.	Y	N
GAL11	co-activator for GAL4, RNA polymerase II associated protein.	N	N
GAL82	glucose repression, ??.	??	??
GAL83	glucose repression, ??.	??	??
GAL6	protease, binds to UAS$_G$, ??.	Y	??

B) Genes Encoding Catalytic Enzymes

	Function of Encoded Gene Product	Regulated By:[a] UAS$_G$	URS$_G$
GAL2	galactose permease	Y	Y
GAL1	galactokinase	Y	Y
GAL7	α-D-galactose-1-phosphate uridyltransferase	Y	Y
GAL10	uridine diphosphoglucose 4-epimerase	Y	Y
GAL5	phosphoglucomutase	N	N
MEL1	α-galactosidase, cleaves melibiose disaccharide into galactose and glucose.	Y	??

[a] Indicates whether the gene has an upstream activating sequence for galactose (UAS$_G$) and/or an upstream repression sequence (URS$_G$): Y, the gene has the regulatory element; N, the gene does not have the regulatory element; ??, not known.

activating transcription of the galactose-inducible genes; GAL4's activity is induced by galactose but is strongly inhibited by glucose.

Each galactose-inducible gene bears an upstream regulatory DNA element termed the upstream activating sequence for galactose (UAS$_G$) (see Table 1). Within each UAS$_G$ there are one or more 17-mer dyad symmetrical elements of the consensus 5′-C-G-G-A-G-G-A-C-T-G-T-C-C-T-C-C-G-3′, to which a dimer of the GAL4 protein binds (2). The DNA-recognition motif (see below) of each GAL4 monomer appears to form specific amino acid - nucleotide contacts with the outermost 5′-C-G-G-3′ sequence of the 17 base-pair binding site (3). A single consensus GAL4 DNA binding site can confer galactose-responsive regulation to heterologous yeast genes (4,5). However, most GAL4-regulated genes bear at least two upstream GAL4 binding sites (6-10). For example, the UAS$_G$ for the divergently transcribed *GAL1* and *GAL10* genes has four binding sites for GAL4 within a 365 nucleotide DNA fragment (9,10). Many transcription factors have been found to activate synergistically when presented with multiple upstream binding sites. This is true also for GAL4; transcriptional activation from a template bearing multiple sites is far greater than the sum of activation from individual sites (11,12). For GAL4 the synergistic effect observed *in vivo* likely results from the combined effect of cooperative DNA-binding to multiple adjacent sites by wild type GAL4 (see below, 13), and synergistic activation through multiple contacts with target molecules in the general initiation complex (12). The multiple binding sites within the *GAL1-10* divergent promoter, each having different affinities for a GAL4 dimer (2,11,13), confers sensitive regulation of *GAL1* and *GAL10* expression in response to small changes in the concentration of the GAL4 protein (14). Cooperative DNA-binding of wild type GAL4 therefore produces a sensitive "genetic switch" for *GAL1* and *GAL10* transcription similar to that controlling development of lambda phage (13,14).

Most genes regulated by GAL4 bear additional regulatory elements termed the upstream repression sequences (URS_G), which cause strong transcriptional repression when glucose is present (see Table 1, 14-17). The URS_G elements present in *GAL* gene promoters are bound by a complex of proteins which include the products of the *MIG1*, *SSN6* and *TUP1* genes (17,18). MIG1 protein binds specifically to the URS_G thereby directing the complex to DNA (17), while SSN6 and TUP1 appear to be involved in activator-independent upstream transcriptional repression (18,19). The *GAL4* gene itself has URS_G, which presumably results in depletion of GAL4 from cells grown in glucose (14), providing one effective mechanism for repression of the *GAL* genes by glucose (see below).

Regulation by Galactose

When galactose is absent, transcriptional activation by DNA-bound GAL4 is prevented by direct interaction with the negative regulatory protein GAL80 (see below and Figure 1, panel A) (20,21). Disruption of the *GAL80* gene, or mutations in GAL4 that prevent interaction with GAL80, result in constitutive expression of GAL4-regulated genes (22-25). Although GAL80's function has not been precisely determined, the simplest model would be that by binding to GAL4 it physically prevents interaction between the activation domains and targets within the general initiation complex (20,23,26,27). Alternatively, the GAL80 protein may possess a catalytic function which modifies GAL4 protein causing inhibition of its activity in the absence of galactose (28). Whatever the mechanism for inhibition, the negative effect of GAL80 is relieved upon addition of galactose. The precise mechanism for induction has not been defined either, but may involve dissociation of GAL80 from GAL4 (implied in Figure 1, panel B), or may result from an allosteric change in the GAL4-GAL80 complex. Several recent experiments seem to suggest that GAL4 and GAL80 may not dissociate upon induction (29,30). However, we believe this question has not been adequately resolved since immunoprecipitates of GAL4 prepared from galactose-grown yeast do not contain GAL80 (C. Costa and I. Sadowski, unpublished data).

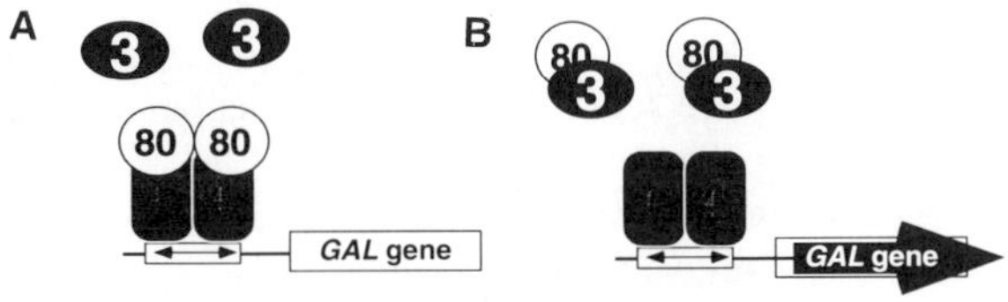

Figure 1. Schematic representation of GAL4 regulation. A) No Galactose: GAL4 (4) is bound to DNA but is inhibited by direct interaction with GAL80 (80). B) Galactose: GAL3 (3) inhibits GAL80.

Induction by galactose requires the product of *GAL3*, a protein that is structurally similar to the galactokinase (GAL1) and is thought to produce an inducer molecule from galactose (31-33). Disruption of *GAL3* significantly reduces the rate of induction, while disruption of both *GAL3* and *GAL1* prevents induction by galactose altogether (34). Furthermore, overexpression of *GAL1* in a *gal3* $^-$ strain can partially suppress the slow induction phenotype (34). These observations suggest that GAL3 performs a function that can at least partially be complemented by GAL1. Despite its similarity to GAL1 however, GAL3 has not been reported to have galactokinase activity (31). Experiments with a temperature-sensitive *gal3* allele have shown that GAL3 is required to maintain induction by galactose (35), a result that suggests the GAL3 protein itself, rather than a metabolic product, is required for induction. Perhaps a GAL3-galactose complex somehow forces an alteration in the interaction between GAL4 and GAL80 thus allowing GAL4 to activate transcription (see Figure 1, panel B).

Regulation by Glucose

Glucose, which enters glycolysis directly and thus is consumed more easily than galactose, reverses or prevents induction by galactose through several different mechanisms (14-17). As mentioned above, the *GAL4* gene, as well as most genes activated by GAL4 protein, has upstream repression sequences (URS_G) which inhibit transcription in glucose (Table 1). The protein complex binding to the URS_G presumably represses transcription in an activator-independent mechanism (19,36). Direct URS_G-mediated repression of *GAL1* occurs very quickly (15,36); this effect is compounded in the long term by repression of *GAL4* transcription resulting in depletion of the activator from the cell (14). The effect of reduced *GAL4* transcription in glucose would not be observed for hours since the half-life of GAL4 appears to be approximately 4 to 6 hours (I. Sadowski, unpublished data).

The activity of GAL2, the galactose permease, is directly inhibited by glucose; glucose appears to inhibit galactose import competitively (37). GAL80 also seems to mediate some glucose repression of *GAL* transcription, as deletion of *GAL80* causes 50-fold derepression of *GAL1* transcription in cells growing in glucose (16,38). The GAL80 effect appears to be much larger than can be accounted for by the combined effect of decreased *GAL2* and *GAL3* transcription and inhibition of GAL2 activity by the presence of glucose (17,25,36). Therefore, the function of GAL80 may be enhanced in cells growing in glucose. Alternatively, or additionally, the function of the GAL3 protein may be inhibited by glucose. Formation of the inducer may be competitively inhibited by glucose, similar to the effect on GAL2.

We have demonstrated that a large central region of GAL4 is responsible for inhibiting transcriptional activation function in response to glucose (39). Transcriptional activation by wild type GAL4 is inhibited at least 10-fold in cells exposed to glucose for 20 minutes, but activation by a mutant deleted of the central region is largely unaffected by glucose (39). Furthermore, fusion of the GAL4 central region to LexA-VP16 produces a transcription factor that is highly active in cells growing in the absence of glucose but is strongly inhibited in cells growing in glucose (39). It remains to be unambiguously determined how the central region effect relates to other mechanisms for glucose repression of the *GAL* genes. Eliminating the central region from GAL4 in an otherwise wild type yeast cell completely relieves repression of *GAL1* for at least two hours following addition of glucose (C. Perelli and I. Sadowski, unpublished data).

GAL4 Structural and Functional Features

An increasingly detailed map of GAL4's structural and functional features is emerging from numerous molecular and genetic dissection experiments over the past decade (Figure 2). Specific binding of GAL4 to DNA is conferred by a zinc-cluster motif located within the N-terminal 64 residues (3,40,41). GAL4 binds as a dimer to DNA (2), with the zinc-cluster motif of each monomer contacting the outer three nucleotides (C-G-G) of the 17-nucleotide dyad element (3). Dimerization is mediated at least in part by an α-helical region within residues 65-94 which form a coiled-coil interaction (2,3). At least one further region capable of directing dimerization is located within the C-terminal 400 residues (M. Hirst and I. Sadowski, unpublished data, 42). GAL4 also has its nuclear localization signal within its N-terminal 74 residues (43-45, not indicated on Figure 2).

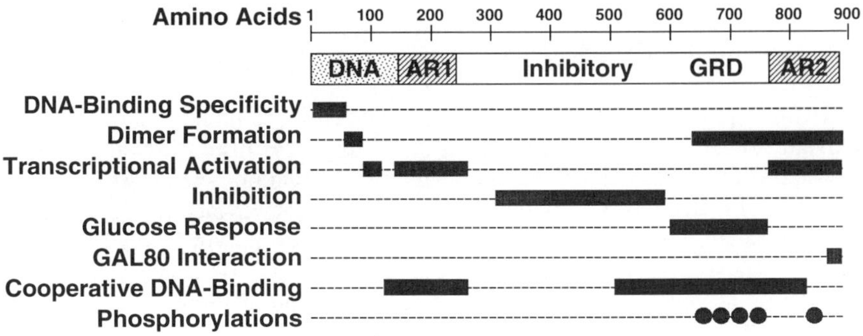

Figure 2. Structural and functional features of the GAL4 protein. Shown is a schematic representation of GAL4's functional domains. DNA, DNA-binding domain; AR1, Activating Region 1; Inhibitory, inhibitory domains; GRD, glucose response domain; AR2, activating region 2. Shaded horizontal bars indicate regions of the GAL4 protein which are necessary for the specified function.

N-terminal protein fragments containing only the zinc-cluster DNA recognition motif and dimerization domain are capable of specific and strong binding to DNA, but are unable to activate transcription either *in vitro* or *in vivo* (42,46,47). Transcriptional activation *in vivo* is dependent upon two major activating domains, termed activating region 1 (residues 148-238), and activating region 2 (residues 767-881) (40,42). Deletion of either activating region severely impairs GAL4's ability to induce expression of *GAL1-LacZ* reporter genes (39,42,48). Residues 94-106 also contain sequences that contribute to transcriptional activation by GAL4 derivatives *in vitro* (activating region 3) (46,47). All three of these activating regions contain a high proportion of negatively charged residues, a characteristic common amongst eukaryotic activating domains (see below, 49,50).

The C-terminal 30 residues of GAL4 are necessary for inhibition by the negative regulatory protein GAL80 (24). GAL80 has been demonstrated to bind to this region *in vitro* (21), and deletion of these residues causes constitutive, albeit weak, activation by GAL4 *in vivo* (24,42). GAL80 may also contact other regions of the GAL4 protein, since deletion of portions of the central region partially relieve repression by GAL80 in non-inducing conditions (I. Sadowski, unpublished data).

Our experiments have suggested that the large central region of GAL4, representing approximately 60% of its mass, is involved in regulating activity in response to glucose (39). Responsiveness to glucose appears to require two separate functional components. Residues 239-600 contain at least three separate segments which can strongly inhibit transcriptional activation by a variety of GAL4 and LexA fusion proteins; we have termed these segments the inhibitory domains (IDs). The inhibitory domains can inhibit transcriptional activation by the VP16 activating domain as effectively as GAL4 AR1 or AR2. GAL4 derivatives bearing the inhibitory domains are constitutively inactivated when the adjacent residues 600 to 768 are removed (39,42). This region has been denoted the glucose response domain (GRD). Our current model proposes that the inhibitory domains are somehow regulated by the glucose response domain which is itself regulated by the presence or absence of glucose (39). Needless to say, a lot of work remains to be done before the function of the central region can be fully appreciated.

Recombinant wild-type GAL4 produced by expression in insect cells binds cooperatively to multiple adjacent sites on DNA (13). In contrast however, a recombinant GAL4 N-terminal fragment consisting of the zinc-cluster recognition motif, N-terminal dimerization domain and AR3 (residues 1-147), produced by expression in *E. coli*, binds to a single site on DNA as efficiently as wild-type GAL4, but is incapable of binding cooperatively to multiple adjacent sites (2,13, 51). Cooperative DNA-binding must be mediated by interaction between GAL4 dimers, and therefore there are likely region(s) within the C-terminal 734 residues which mediate multimerization (Figure 2, M. Hirst and I. Sadowski, unpublished data).

GAL4 is phosphorylated on at least 4 serine residues under conditions in which it can activate transcription (28,48) (see Figure 2). We have demonstrated that the phosphorylations are unnecessary for transcriptional activation when GAL80 is absent (R. Dhanawansa and I. Sadowski, unpublished, 48). Preliminary evidence suggests that at least one major phosphorylation at serine-837, within activating region 2, regulates interaction between GAL4 and GAL80 (I. Sadowski, unpublished data). We have demonstrated that phosphorylation at this serine is not required for activation of transcription, but instead is likely acquired during interaction with the general initiation complex (48). The function of a cluster of three phosphorylations within the glucose response domain has not been identified (Figure 2); it is attractive to speculate that they are involved in regulation of GAL4 by glucose, but preliminary experiments tend to discount this hypothesis.

TRANSCRIPTIONAL ACTIVATION BY GAL4 IN METAZOAN CELLS

The Function of Transcriptional Activators is Highly Conserved in Eukaryotes

During the early part of the last decade, researchers studying control of gene expression in yeast began to recognize similarities between the upstream activating sequences (UASs) found within promoters of yeast genes and enhancer elements from genes of higher eukaryotes (52). Both types of DNA elements caused transcriptional activation in an orientation and positionally-independent manner. The only differences between them appeared to be that UASs generally could not cause activation from distances greater than 400 to 500 nucleotides from the site of transcriptional initiation, nor could they function downstream of the initiation site (52,53). In contrast, enhancer elements from higher eukaryotes could function at distances of a kilobase or greater, and were frequently observed within introns and downstream of the polyadenylation signal (54). Despite these differences, the reason(s) for which have yet to be identified, the similarities between yeast UAS elements and enhancers fueled the opinion that the basic mechanisms for transcriptional regulation are highly conserved in eukaryotes.

An important test of this prediction was to determine whether *trans*-acting factors which bound to yeast UASs could cause activation of transcription if expressed in mammalian cells when provided with an appropriate template bearing a cognate DNA recognition element. In 1988 several groups demonstrated that GAL4, when expressed transiently in either HeLa or CHO cells, stimulated transcription from reporter genes bearing upstream binding sites for GAL4 (55,56). In both experiments the UAS_G functioned precisely as would a mammalian enhancer, but in a GAL4-dependent fashion. The success of these experiments was facilitated by the fact that higher eukaryotes do not possess proteins capable of activating transcription from a GAL4 binding site. In fact, apart from *Saccharomyces*-related fungal species, other eukaryotes do not express proteins which can bind to the GAL4 consensus recognition element. A series of subsequent experiments demonstrated that GAL4 could activate transcription in insect (57), plant (58) and avian (59) cells in addition to mammalian cells, further demonstrating the high conservation of the basic eukaryotic transcriptional initiation machinery. In contrast, despite intense effort, GAL4 derivatives were found to be incapable of activating transcription of *E. coli* reporter genes (S. Liang and M. Ptashne, personal communication), suggesting that the conserved function of this class of transactivator does not extend to prokaryotes.

Conserved Function of GAL4 Domains in Mammalian Cells

Most deletion derivatives of GAL4 activate transcription of reporter genes in CHO cells to similar relative levels as when expressed in yeast, demonstrating a conservation of the function of the major domains. Importantly, the activating regions 1 and 2 are necessary for transcriptional activation in mammalian cells, and can activate transcription quite efficiently in mammalian cells when fused individually to the DNA-binding domain of GAL4 (see Figure 3, 55). This finding is not so surprising in retrospect because these activation domains contain a high proportion of negatively charged residues, a characteristic common amongst eukaryotic transcriptional activators (49,50,54). Similarly, the inhibitory regions we identified in yeast also cause strong inhibition of GAL4 derivatives when expressed in CHO cells: GAL4's DNA-binding domain fused to either AR1 or AR2 and the inhibitory domains are completely inactive in both yeast and CHO cells (Figure 3, unpublished data). This result implies that either the inhibitory domains prevent transcriptional activation without the aid of additional cellular proteins, or any additional proteins which mediate their effect must be present in both yeast and CHO cells. Experiments investigating these possibilities are currently under way. The N-terminal region of GAL4 which mediates nuclear localization in yeast must also function in mammalian cells since the DNA-binding domain appears capable of nuclear targeting of otherwise non-nuclear proteins when expressed in COS cells (60).

We have observed that not all GAL4 derivatives function in CHO cells as they do when expressed in yeast. GAL4 derivatives bearing short C-terminal truncations within activating region 2 are capable of activating transcription in yeast, at least when overexpressed (42), but are completely inactive in mammalian cells even when expressed at high levels (see Figure 3, unpublished data). Therefore at least one portion of GAL4 appears not to work in mammalian cells despite the conserved function of the DNA-binding, transcriptional activation, inhibitory domains and nuclear localization signal. Although we have not determined the mechanism for this discrepancy, we suspect it relates to the function of the central region. Our working hypothesis is that proteins necessary for the function of the glucose-response domain are absent in mammalian cells and therefore GAL4 derivatives may be constitutively inhibited by its central region. Therefore, elimination of portions of the major C-terminal activating region may be more devastating for GAL4 in mammalian cells because of the strong constitutive inhibitory effect of the central portion of the protein.

GAL80 appears to function when expressed in mammalian cells as well. Cotransfection of a plasmid expressing GAL80 prevents activation by GAL4 derivatives bearing the C-terminal 30 residues of GAL4 (I. Sadowski, unpublished data, P. Broad personal communication). However, attempts to induce GAL4 activity in these experiments by cotransfection of plasmids expressing GAL3 and addition of galactose have not been successful to date. Perhaps additional components are necessary for induction, or for overcoming the negative effect of the central region. Nevertheless, these initial experiments suggest that it may be possible to dissect the mechanisms for regulation of GAL4 *in vivo*,

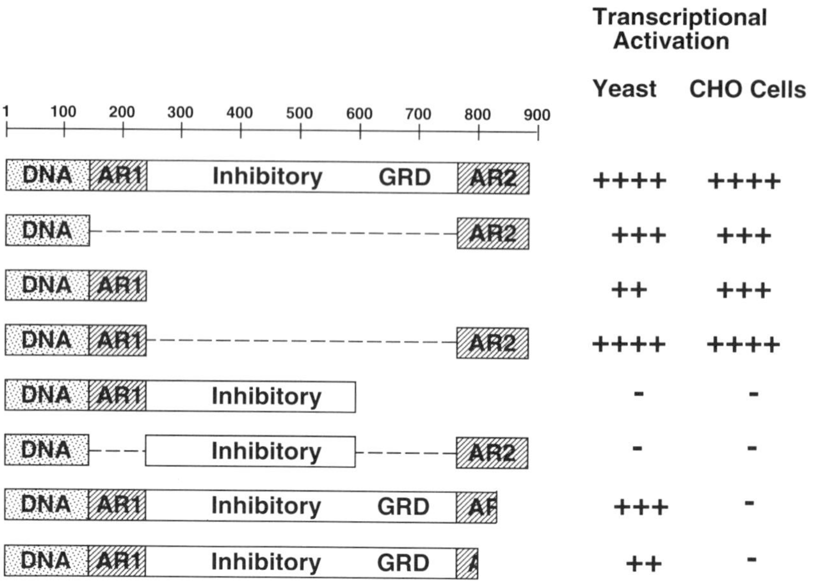

Figure 3. The function of GAL4's domains are conserved in mammalian cells. Shown is a schematic representation of wild-type GAL4 and a series of deletion derivatives (left). The relative transcriptional activation ability of each derivative in yeast and CHO cells is indicated at the right: ++++, 100% activation; +++, 75% activation; ++, 50% activation; + 25% activation; -, no transcriptional activation.

by further cotransfection of plasmids expressing the appropriate components into CHO cells.

STRATEGIES EMPLOYING GAL4 IN MAMMALIAN CELLS

Consistent with the notion that the function of eukaryotic *trans*-acting factors is highly conserved, it has become apparent that virtually every transcription factor studied in detail to date possesses modular functional domains which confer specific DNA-binding, transcriptional activation and in some cases regulation by a specific signal. Characterization of the functional organization of transcriptional regulatory proteins has relied heavily on strategies involving systematic deletion analysis, followed by fusion of putative functional domains to heterologous proteins. The invariable success of these experiments highlights the remarkable flexibility of transcription factor functional domains. GAL4 has proven to be useful for these strategies largely because of its well-characterized DNA-binding motif and *cis*-acting recognition element, and the fact that higher eukaryotes do not possess proteins capable of activating transcription from GAL4 binding sites (55,56). Many of the initial experiments involving expression of GAL4 in mammalian cells were directed at investigating the transcriptional activation functions of other eukaryotic transcription factors. However, numerous variations of strategies exploiting GAL4 fusions have evolved for other purposes since the initial experiments in 1988.

GAL4 Domains as Molecular Tools

Most strategies using GAL4 in mammalian cells employ fusions between a protein of interest and GAL4's DNA-binding domain, typically residues 1-147 which contains the zinc-cluster DNA recognition motif and dimerization region (see Figure 2). It is important to point out however, that this fragment also contains the activation domain identified in *in vitro* transcription experiments termed activating region 3 (12). It is not clear whether this region functions as a transcriptional activation domain *in vivo* since transcriptional activation by GAL4(1-147) is negligible even when expressed at very high levels in yeast or mammalian cells (24,55,56). Many experiments have demonstrated a synergistic effect of having multiple independently functioning activation domains present on the same activator protein (61-63, and see below); perhaps activating region 3 functions synergistically *in vivo* with GAL4's two major activating regions. If this were true then perhaps the sensitivity of assaying transcriptional activation function of heterologous activation domains is enhanced synergistically by the presence of AR3. Consequently, fusion to residues 1-147 of GAL4 may be a more sensitive assay for weaker transcriptional activation domains. To my knowledge a systematic study of this possibility has not been undertaken. On the other hand, GAL4(1-147) would be inappropriate to use in experiments which require that there be no interference in the transcriptional activation function arising from the DNA binding domain. In this regard, several groups have used smaller N-terminal fragments of GAL4 encompassing residues 1-96, in which activating region 3 is deleted (64-66). Both of these N-terminal fragments bind equally well to the GAL4 17-mer DNA binding site *in vitro* (67).

A variety of plasmid vectors have been constructed specifically for expressing fusion proteins containing the DNA-binding domain of GAL4 in mammalian cells (68-70), making the creation and expression of appropriate fusions a relatively simple matter. Most groups produce a fusion containing an N-terminal fragment of GAL4 fused to a C-terminal fragment of their protein of interest. However, the GAL4 DNA-binding domain has been demonstrated to function when fused either in the center of a protein or at the C-terminus; these alternatives are typically chosen in instances in which one wishes to replace the DNA-binding specificity of a protein with that of GAL4 without disrupting the overall structural organization of the molecule (60,63,71-74).

Similarly, many different reporter gene constructs bearing GAL4 binding sites have been constructed for assaying the regulatory effect of GAL4 fusion derivatives. Most of these utilize the *CAT* gene from Tn9 which encodes chloramphenicol acetyltransferase. The CAT enzyme is relatively stable when expressed in mammalian cells and is easily assayed by a variety of techniques (75,76). Alternatively, some reporter plasmids bear the gene for firefly luciferase as a reporter enzyme, which is also both easily assayed and extremely sensitive (77,78). Some groups use a β-globin reporter gene which produces a stable RNA transcript that can be measured by S1 or ribonuclease protection assays (79). Most reporter constructs for measuring the transcriptional activation function of GAL4 derivatives have minimal promoters with one or more 17-mer consensus GAL4 binding sites (see above) placed immediately upstream of a TATA-box, usually derived from the adenovirus E1B (69,80) or thymidine kinase (TK) genes (81). Experiments assaying the ability of GAL4-fusions to repress transcription from upstream binding sites typically use promoters bearing additional enhancer elements to generate a stronger basal signal; many groups have used the SV40 early enhancer and 21 base-pair repeats for these types of experiments (73,82), or the thymidine kinase promoter (83-85). The sensitivity of the assay can be adjusted in both cases by varying the number of upstream binding sites for GAL4. The effect of adding additional binding sites is generally found to be synergistic rather than additive (12,46). Many experements have involved the use of five closely-spaced strong consensus GAL4 binding sites (69), or alternatively a 365 nucleotide DNA fragment bearing the *GAL1-10* UAS_G (see above) which has four binding sites each with different affinities for the GAL4 DNA-binding domain. It should be noted that by using the latter fragment one is more likely to see effects produced by cooperative DNA-binding unless vast overexpression of fusion derivatives can be ensured; this possibility should be considered when interpretation is made of the differences in levels of transcriptional activation produced by various GAL4 fusions.

Some groups have constructed more specialized reporter genes to study context-specific interaction between GAL4 derivatives and other transcription factors. For example, the T-cell specific factor TCF-1α (hLEF) can activate transcription when fused to GAL4(1-147) and assayed on a T-cell receptor reporter gene bearing a GAL4 binding site, but is inactive when assayed on a reporter gene consisting of GAL4 sites positioned immediately upstream of a minimal TATA-box (74). Also, transactivation by fusions between HIV-1 Tat and the GAL4 DNA-binding domain require an SP1 site positioned adjacent to the GAL4 site (86,87). Reporter genes bearing GAL4 sites have been constructed from the MMTV (55,88), HTLV-1 (89) and HIV-1 (87,90-92) long terminal repeats as well as the phosphoglycerate kinase (93), thymidine kinase (94), β-globin (64), albumin (95), c-*fos* (96), CRBPII (97) and HLA-DRA (98) promoters, to name a few. Virtually any promoter can be made responsive to GAL4 fusion derivatives by simply inserting one or more of GAL4's cognate recognition elements (see above).

General Assay Procedure

A typical experiment to assay the transcriptional activation function of a GAL4 fusion derivative involves transient cotransfection of at least two plasmids into the mammalian cells of choice (Figure 4). The effector plasmid typically has a strong constitutive promoter expressing the GAL4 fusion derivative, and the reporter plasmid bears the reporter gene with DNA-binding sites for GAL4. There are as many transfection protocols as groups performing transfections, most of which involve calcium phosphate precipitation, DEAE-dextran, electroporation or lipofection. Transfected cells are typically harvested 48 to 72 hours post-transfection, extracts are prepared and assayed for the presence of CAT or luciferase activity, or RNA is extracted and assayed by ribonuclease protection or S1 analysis. Many groups include a third plasmid in the cotransfection which contains a constitutive promoter expressing a second reporter enzyme, such as β-galactosidase, for use as an internal control to standardize the transfection efficiency from sample to sample. However, since strong transcriptional activators produced by the effector plasmid can cause "squelching" of cotransfected constitutive reporter genes, an effect produced by sequestering of critical general initiation factors (49,50,99,100), one must be careful in interpreting the results of the internal control for assays in which the activators produced by different effector plasmids vary greatly in activity.

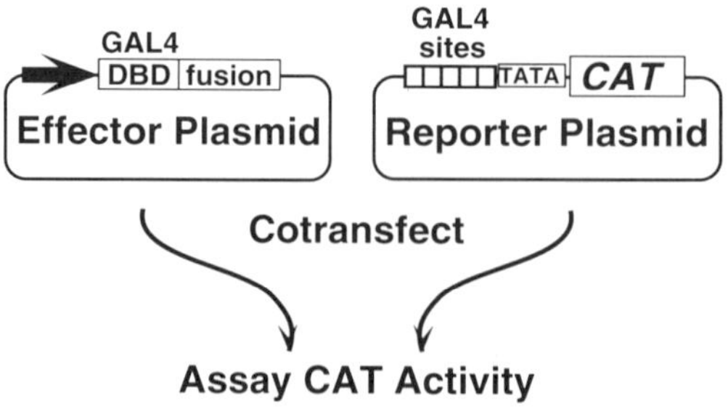

Figure 4. General assay for measuring the transcriptional activation ability of GAL4 fusions. GAL4 DNA-binding domain (DBD) fusions are expressed from a strong constitutive promoter (indicated with arrow) on the effector plasmid. The reporter plasmid contains consensus GAL4 DNA-binding sites (GAL4 sites) placed upstream of a TATA-box and CAT gene encoding chloramphenicol acetyl transferase.

Identification of Potential Transcriptional Activator Proteins, Mapping and Analysis of Transcriptional Activation Domains

Brent and Ptashne first demonstrated that *trans*-acting factors possessed separable DNA-binding and transcriptional activation functions by fusion of GAL4 to the prokaryotic repressor protein LexA; the resulting chimera was found to activate transcription of a yeast reporter gene bearing binding sites for LexA (101). This experiment demonstrated that transcriptional activation was not a function peculiar to the DNA-binding properties of GAL4 but rather that DNA-binding served simply to tether the transactivating components of GAL4 to a position adjacent GAL4-regulated genes. From this and a series of similar experiments was born the concept of the transcriptional activation domain (TAD) (49,54). GAL4's transcriptional activating domains were subsequently mapped in yeast with a series of deletion/fusion constructs (40,42), and similar experiments revealed transcriptional activation domains within transcription factors from many different eukaryotic species (53,102).

A common strategy for identification of transactivation domains emerged which typically involves deletion analysis of a transcriptional activator to define the regions necessary for transcriptional activation, and subsequently to confirm that these portions of the protein could function in that capacity by fusing it to a heterologous DNA-binding domain which itself lacked a transcriptional activation function (88,102,103). The DNA-binding domain of GAL4 is useful for these strategies because it is contained within a short N-terminal polypeptide which confers both DNA-binding specificity and dimer formation, and has a well defined and characterized *cis*-element which is not bound by mammalian factors (see above). GAL4(1-147), despite possessing activation region 3, activates transcription very inefficiently *in vivo* on its own. However, when fused in-frame to its own activation domains, or that of a heterologous transcription factor, it can activate transcription to a level proportional to the relative strength of the activating domain. The strategy has been so frequently applied, and with such success, that the functional test of an activation domain for all practical purposes is whether it can activate transcription of a reporter gene with GAL4 DNA-binding sites when fused in-frame to GAL4's DNA-binding domain.

Many laboratories have used fusions with GAL4's DNA-binding domain to assay whether a protein of unknown function may be a transcriptional activator (59,73,87,92,104-110). This approach can be used even in the absence of prior knowledge of the cognate DNA-recognition element or even if the protein binds directly to DNA. Providing the initial fusion is shown to activate transcription, the putative activation domains can be mapped by deletion analysis. It is not uncommon to find that a fusion of a full length protein to GAL4(1-147) is relatively inactive as a transactivator, while sub-fragments activate efficiently (60,111,112). This is usually interpreted to mean that the protein in question has domains that are capable of inhibiting transcriptional activation (see below). Demonstration that a protein fragment can activate transcription when artificially forced to bind DNA via GAL4(1-147) is by no means conclusive evidence that the protein is a transcriptional regulator, but nonetheless the approach is often a critical determinant in the initial analysis of suspected transcriptional regulators. Similarly, a negative result in such an assay may not be indicative of the function of the protein in question - some transcription factors are only capable of activating transcription in specific promoter contexts or in specific tissues (see below).

Tables 2, 3 and 4 provide a summary of some of the transcriptional regulatory proteins, oncogene and proto-oncogene products, and viral proteins whose potential transcriptional activation function has been assayed in fusion experiments with the DNA-binding domain of GAL4. Initial experiments to define the structural elements necessary

Table 2. Eukaryotic Transcription Factors Assayed for Transcriptional Activation Function with GAL4 DNA-Binding Domain Fusions

Transcription Factor	Predominant Transcriptional Activation Domain Structural Features[a]	References
AP-2	Proline-rich	(64, 180)
ATF-1	Regulated	(181)
ATF-α1	Putative metal-binding domain, requires E1A	(184)
ATF-2	N-terminal cryptic TAD, requires E1A	(106, 182, 183)
ATF-4	Acidic ??	(185)
BTEB2	Proline-rich	(172)
C/EBP	Multiple, proline-rich ??	(61, 95, 170)
CREB	Glutamine-rich, Regulated	(72, 163, 186)
CREM	Glutamine-rich	(187)
CTF-1/ NF-1	Proline-rich	(64, 171)
E2A	Loop - helix, ??	(117, 178)
E2F	Acidic	(130, 177)
Egr-1/ NGFI-A	Multiple, Acidic, Pro/Gln-rich, Ser/Thr-rich	(71, 168)
GF-1/ NF-E1/Eryf1	??	(175)
HOXD-4	Proline-rich	(194)
HNF1/LFB1	Glutamine/Proline/Serine-rich	(173)
ITF-1 and ITF-2	Serine, Threonine-rich	(64)
MyoD	Acidic	(112, 174)
Myogenin	Multiple- Serine/Threonine, acidic	(62)
Oct-1	Glutamine-rich	(64)
Oct-2	Glutamine-rich, Proline-rich	(64-66)
Oct-6/SCIP	??	(119)
Sox-4	Serine-rich	(179)
SP1	Glutamine-rich	(64, 87, 118)
SRF	Proline-Glutamine-rich	(96, 134)
TBP-1	??	(92)
TCF-1/hLEF	Proline/Tyrosine-Serine (requires TCF-2 *cis* site)	(74)
TEF-1	Proline/ Serine-threonine rich	(81)
TFE3	Acidic	(64, 169)
YY1/ NF-E1	??	(176)
Estrogen Receptor (ER)	Multiple	(188, 189)
Glucocorticoid Rec. (GR)	Multiple	(63, 190)
T3 Receptor/ c-ErbA	Multiple	(191, 192)

Transcription Factor	Predominant Transcriptional Activation Domain Structural Features[a]	References
v-ErbA	None	(191)
Progesterone Rec. (PR)	Multiple	(193)
Retanoic Acid Rec.	Multiple, ligand dependent and independent	(97)

[a]Indicates the predominant structural features thought to be responsible for activity of the transcriptional activation domain (TAD). None, does not activate transcription; ??, undetermined or unknown.

Table 3. Oncogene Products and Related Proteins Assayed for Transcriptional Activation Function with GAL4 DNA-Binding Domain Fusions

Gene Product/ Factor	Predominant TAD Feature[a]	References
c-Myc	Proline-rich	(131, 195, 196)
L-Myc	Weak, missing region homologous to c-Myc TAD	(132)
MAX	None	(197)
c-ets-1	Acidic	(198)
c-ets-2	Acidic	(198, 199)
v-Myb	??	(59, 200)
c-Myb	??	(59, 201)
B-Myb	None	(200)
v-Jun	Multiple, Proline-rich ??	(129, 202)
c-Jun	Multiple, Proline-rich ??	(129, 202-204)
JunB	Multiple	(205)
v-Fos	Proline-rich ??	(206)
c-Fos	Multiple, Proline-rich ??	(203, 206, 207)
FosB	Proline-rich ??	(206, 208)
FosB2	None, deleted in alternatively spliced form	(206, 208)
Fra1 and Fra2	None	(206)
v-Rel	Weak, missing most of c-rel TAD	(209, 210)
c-Rel	Acidic, Proline/Serine-rich	(60, 90, 209)
NF-kB p50	None (but activates in yeast)	(94)
NF-kB p65	Acidic; Proline-rich	(64, 94, 135)
I-kBα/ MAD-3	Acidic, ankyrin-like repeats, requires interaction with Rel	(104, 105)
I-kBβ	Acidic, ankyrin-like repeats	(105)
I-kBγ/ p105	Acidic	(104, 105)
I-Rel	None	(211)
Retinoblastoma	None, but can activate if co-expressed with Adenovirus E1A	(106)
p53	Acidic	(73, 107, 108)

[a]Indicates the predominant structural features thought to be responsible for activity of the transcriptional activation domain (TAD). None, does not activate transcription; ??, undetermined or unknown.

Table 4. Viral Gene Products Assayed for Transcriptional Activation Function with GAL4 DNA-Binding Domain Fusions

Gene Product	Virus	Predominant TAD Feature[a]	References
E1A	Adenovirus	Metal-binding motif ??	(80, 182, 183)
ZEBRA/BZLF1	EBV	Glutamine/Proline-rich ??	(109, 212)
Rta	EBV	Acidic, Hydrophobic, Proline-rich	(213) (79)
EBNA2	EBV	Acidic, Hydrophobic	(214)
EBNA 2	HPV	Acidic, Hydrophobic	(215)
VP16/ Vmw65	HSV-1	Acidic	(64, 88, 116)
ICP0	HSV - 1	??	(216)
ORF4	Varicella-Zoster	??	(217)
ORF62	Varicella-Zoster	Acidic	(218, 219)
T-antigen	SV40	None	(220)
IE	Pseudorabies	Acidic	(221)
Bel-1	H. foamy virus	Acidic	(91, 111, 222, 223)
Taf	S. foamy virus	Acidic, ??	(91, 224)
Tat	HIV-1, HIV-2	??, Require SP1	(86, 87, 110, 118)
Tax	HTLV-1	??	(89, 225-227)

[a]Indicates the predominant structural features thought to be responsible for activity of the transcriptional activation domain (TAD). None, does not activate transcription; ??, undetermined or unknown.

for activation by transcription factors indicated that most activating domains could be grouped into one of several classes based on their predominant amino acid compositions and predicted secondary structures (49,50,102). These were denoted the acidic, proline-rich, glutamine-rich and serine/threonine-rich activation domains. Experiments performed *in vitro* have demonstrated that various activating domains specifically contact proteins of the general initiation complex, namely the TATA-binding protein (TBP), its associated factors (TAFs) and/or TFIIB (reviewed in refs. 50,113,114). These interactions between sequence-specific DNA-binding proteins and the general transcriptional initiation components likely serve to catalyze the assembly of an active initiation complex. However, inspection of Tables 2, 3 and 4 indicates that many transcriptional activator proteins have activating domains that cannot be easily placed in one of the aforementioned categories. Since the target(s) for most activating domains have not been identified, and there are multiple categories of activating domains, it seems likely that there are at least as many different targets within the initiation complex, or that some transcriptional activators require adapter molecules or co-activators which can themselves contact the general initiation complex (115). Many groups who identify a transcriptional activation domain in mammalian cells go on to demonstrate that the same region can activate transcription in yeast (60,88,91,94,105,109,116,117). The ability of a transcriptional activator protein to activate in both yeast and mammalian cells indicates that the target(s) for protein-protein interactions of the activation domain are likely conserved as well. Consistent with this notion, the well characterized factor Sp-1 has been shown to be fully capable of activating transcription in mammalian cells when fused to GAL4 but the same fusion is inactive when expressed in yeast (64,87,118). Yeast apparently lack a protein that is functionally analogous to TAF110, which has been demonstrated to be a likely target for the Sp-1 activation domain (115). Similarly, some activation domains are only capable of activating transcription in specific cell types (61,62,74,112,119), suggesting either that their targets or adapters are cell-type-specific as well.

An additional feature common to transcriptional activator proteins that is revealed by Tables 2, 3 and 4 is that many have multiple regions that can contribute to transcriptional activation. The reiterated structure of multiple and varied activation domains on a single transactivating protein has been shown to contribute synergistically to activation *in vivo* (61-63,65,66,120) and *in vitro* (120,121).

Characterization of Regulatory Functions

The ability of many transcription factors to activate transcription is conditionally regulated by specific cellular signals. Mammalian signal-regulated transcription factors include the steroid hormone-receptor superfamily, as well as members of the ATF/CREB and Fos/Jun (AP1) families of transcriptional regulatory proteins. Fusions between members of all three of these families and the DNA-binding domain of GAL4 have resulted in appropriate signal-regulated transcriptional activation of reporter genes bearing GAL4 binding sites (see Table 5). This approach is particularly useful when one wishes to examine the regulatory function of these factors using mutagenesis without the interference of transcriptional activation from the corresponding endogenous protein.

Transcriptional regulatory proteins appear to be regulated either by phosphorylation, ligand binding, or interaction with regulatory proteins. Steroid hormone receptors, for example, are inactive as transcriptional activators until binding their cognate ligand. Ligand binding induces activity by causing dissociation of a heat shock protein complex, promoting nuclear localization and/or dimerization and/or unmasking of the activation domains (122,123). The activity of many transcription factors is controlled by phosphorylation (reviewed in refs. 124-126). Phosphorylation may directly control transcriptional activation function, DNA-binding ability, interaction with a regulatory molecule, or promote nuclear localization. Fusions of CREB/ ATF-1 and Jun with GAL4 were found to be appropriately regulated by phosphorylation in response to cAMP, or tyrosine kinase/ RAS activation, respectively (see Table 5).

Analogous to regulation of GAL4 in yeast by the negative regulator GAL80 (see above), some mammalian transcription factors are controlled by interaction with a regulatory protein. Prototypical members of this class of proteins include the NF-kB/Rel proteins whose activity is inhibited by I-kB/I-Rel (124,126). Similarly, c-Jun appears to be inhibited by a cell-specific regulator whose effect is relieved by src/RAS-induced signals (127-129). Experiments involving fusions of c-Jun with GAL4 were instrumental in characterization of this mode of regulation (see Tables 5 and 6). Similarly, the transcription factor E2F appears to be inhibited by direct interaction with the retinoblastoma protein (Rb) (130). Rb binds to a C-terminal segment of E2F and apparently blocks its ability to activate transcription (130).

Normal regulation of a transcription factor can sometimes be disrupted by fusion to GAL4's DNA-binding domain. For example, the ability of c-Myc to activate transcription is regulated throughout the cell cycle such that its ability to activate transcription occurs during G_1 and at the G_2M boundary (131). However, transcriptional activation by GAL4-c-Myc is constitutive throughout the cell cycle (132); the difference apparently relates to the fact that the GAL4 fusion is constitutively bound to DNA whereas the DNA-

Table 5. Regulated Transcription Factors Fused to GAL4's DNA Binding Domain

Transcription Factor	Positive Regulatory Signal(s)[a]	Notes:	References:
CREB	cAMP, Calcium Ionophores	Phosphorylated by PKA	(72, 163-166)
ATF-1	cAMP, Calcium Ionophores	Phosphorylated by PKA and CaM kinase II	(181)
c-Fos	Protein Kinase A	??	(228)
c-Jun	Src, Ras, UV-irradiation, Protein Kinase A	Phosphorylation by MAP/ERK-related protein kinase; inhibition of negative regulation	(127, 204, 228, 229)
Jun-B	Protein Kinase A	??	(228)
Glucocorticoid receptor	Glucocorticoid	Dissociation of HSPs, Multimerization	(63, 230)
Estrogen receptor	Estrogen Ligand	Nuclear Localization, Multimerization	(188, 230-232)
c-ErbA (T3-receptor)	T3		(85, 192)
Retanoic Acid receptor	All-*trans*-Retanoic Acid		(85, 97)
Retanoid-X-receptor	9-*cis*-Retanoic Acid		(97)

[a]Indicates ligand or stimulus which induces transcriptional activation by wild-type protein and GAL4 fusions.

binding activity of c-Myc is cell cycle regulated (131). Similarly, the serum response factor (SRF) is normally only capable of activating transcription in response to growth factors such as EGF and PDGF (133,134). However, when fused to GAL4(1-147), SRF activates transcription constitutively (96). Regulation of SRF involves formation of a ternary complex at the SRE; fusion to GAL4's DNA binding domain likely disrupts this normal mode of regulation (133). Transcription factors whose activity is normally controlled by regulating nuclear transport would also be constitutive transactivators when fused to GAL4 because of GAL4's functional and constitutive nuclear localization signal (64,94,135).

Many mammalian transcription factors have been found to possess regions similar to the inhibitory domains which we have observed in GAL4. These are typically identified as portions of the protein which when removed cause an increase in transcriptional activation. Table 6 displays a summary of some of the factors in which such inhibitory domains have been identified using GAL4 fusions. The effect of an inhibitory domain could be mediated by several mechanisms. A trivial possibility is that the putative inhibitory domain may simply make the fusion protein inherently unstable or inefficiently translated. This possibility can be examined directly by assaying the abundance of the fusion protein in Western blots or immunoprecipitations with antibodies against GAL4 or the fusion protein. This is an essential control to perform in any experiment involving GAL4 fusions. Providing the domain in question does not reduce the abundance of the GAL4 fusions, experiments can be performed to determine whether the function of the inhibitory domain is dependent upon a second *trans*-acting factor or whether the domain functions independently. This is typically addressed by determining if the effect of the inhibitory domain can be relieved by overexpression of protein fragments bearing the inhibitory domain but lacking a DNA-binding domain. If overexpression of a competitor protein fragment relieves inhibition then it could be argued that the inhibitory domain may

Table 6. Inhibitory Domains From Eukaryotic Transcription Factors Identified in GAL4 Fusions

Factor	*Cis-trans* Test[a]	Het. TAD[b]	Notes	Refs.
Bel-1	N.D.	N	Does not inhibit VP16	(111)
c-Rel	Overexpression of competitor is ineffective	N.D.	N-terminal inhibitory domain works similarly in yeast	(60)
SRF	N.D.	N.D.	Inhibitory region can be moved to various positions on the protein	(96)
c-Myb	Cotransfection of inhibitory fragment inhibits Myb activation *in trans*	Y	Can inhibit VP16; inhibits in both *cis* and *trans* independently of DNA-binding	(201)
C/EBP	N.D.	N.D.	Inhibits within GAL4 fusions as effectively as native C/EBP	(170)
Egr-1/ NGFI-A	Overexpression relieves inhibition.	N.D.	An overlapping domain can repress transcription when fused to GAL4 (see Table 7)	(71)
v-Jun, c-Jun	Overexpression relieves inhibition	Y	Can inhibit VP16; inhibitory domain of v-Jun is weaker; δ-domain	(127-129)

[a]Indicates whether the inhibitory effect could be relieved by overexpression of competitor inhibitory domain. N.D., not done.
[b]Indicates whether the inhibitory domain could inhibit a heterologous transcriptional activation domain (TAD). Y, the domain can inhibit heterologous activation domain; N, the domain cannot inhibit heterologous activation domain.

function by interaction with a second molecule whose effect can be titrated away. Such an approach led to the identification of the regulated δ-inhibitory motif within c-Jun (127-129). Interestingly, in yeast the effect of GAL4's inhibitory domains on both GAL4 and LexA derived activators cannot be relieved by independent overexpression of the inhibitory domains (G. Stone and I. Sadowski, unpublished data). This result suggests that either GAL4's inhibitory domains work independently, or alternatively that they require a protein whose abundance is untitratable. Another useful experiment which is often used to address the mechanism for inhibition is to determine whether the inhibitory domain can inhibit activation by heterologous transcriptional activation domains. Domains capable of inhibiting only specific activating regions may indicate a specific interaction between the two regions. Alternatively, indiscriminate inhibition of any transcriptional activating domain may suggest a more general mechanism, such as blocking binding to DNA for example.

Identifying and Assaying Transcriptional Repressors

An increasing number of transcriptional regulatory proteins have been identified which are capable of repressing transcription when bound to upstream sites. In many cases GAL4(1-147) has been converted into a transcriptional repressor by fusion with a specific "repression motif" - the resulting fusion is capable of repressing transcription of constitutive promoters bearing upstream GAL4 DNA-binding sites (see Table 7). GAL4's DNA-binding domain alone typically has no effect on these reporter genes, but when fused to a repression domain can cause significant repression of transcription, sometimes to undetectable levels. Many transcriptional repressor proteins contain a repression motif

Table 7. Factors Containing Domains Capable of Upstream Repression in GAL4 Fusions

Factor/ Protein	Reporter Construct[a]	Notes	Ref.
Egr-1	5 GAL4 sites upstream of TK promoter.	Deletion enhances Egr-1 transcriptional activation function	(168)
v-ErbA	1 GAL4 site upstream of TK promoter		(85)
c-ErbA (T3-receptor)	1 GAL4 site upstream of TK promoter	Represses transcription in absence of T3 but activates in presence of T3	(85)
Retanoic Acid Rec.	1 GAL4 site upstream of TK promoter	Represses in absence of RA but activates in presence of RA	(85)
YY1/ NF-E1	5 GAL4 sites upstream of TK promoter	Repression is cell-type-specific	(83, 84)
E4BP4	2 GAL4 sites upstream of GST-π promoter	Repression domain is highly charged	(233)
Adenovirus E1B	1 or 5 GAL4 sites upstream of TK promoter	W.T. E1B forms a complex with p53, represses p53-regulated genes	(234)
SCIP	5 GAL4 sites upstream of myelin P_0 promoter (β-globin reporter)	WT SCIP protein can repress from its own binding site in a position and orientation independent manner	(235)
WT-1 (Wilms' Tumor Suppressor)	5 GAL4 sites upstream of TK promoter (used a variety of constructs)	Can repress from at least 750 nucleotides upstream	(236)
Kid-1	5 GAL4 sites upstream of SV40 early enhancer and E1B TATA	Repression domain corresponds to Krüppel-associated Box (KRAB)	(82, 138)
ZNF2	5 GAL4 sites upstream of SV40 early enhancer and EIB TATA-CAT	Repression domain corresponds to Krüppel-associated Box (KRAB)	(82)
KOX-1	5 GAL4 sites upstream of TK promoter	Repression domain corresponds to Krüppel-associated Box (KRAB)	(137)

[a]Indicates the promoter used and how many binding sites for GAL4. All reporter genes are CAT unless otherwise specified.

related to that of the *Drosophila* Krüppel protein (136). In every case examined, the Krüppel-associated boxes (KRAB) of homology appear to be responsible for the upstream repression effect (82,137,138), as judged by the ability of GAL4 fusions with this domain to repress transcription. However, the KRAB domain is not the only motif which can mediate upstream repression, since Table 7 lists at least 6 factors whose repression domains have no obvious similarity to KRAB. The mechanism for repression of transcription by these mammalian *trans*-acting repressors are not well understood, but given the high evolutionary conservation of eukaryotic transcriptional regulatory mechanisms it is likely that many similarities will be found with the yeast upstream repressors SSN6 and TUP1 (18,19).

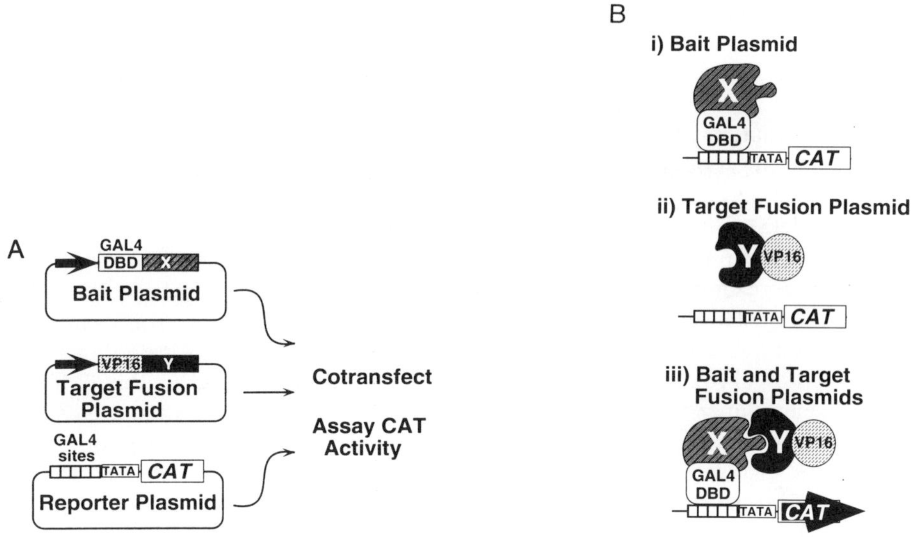

Figure 5. Mammalian two-hybrid strategy. A) The bait plasmid produces a GAL4 DNA-binding domain (DBD) bait (X) fusion from a strong constitutive promoter (indicated with arrow). The target plasmid produces the VP16 activation domain (VP16) fused to the target protein (Y). The expression plasmids are cotransfected with a reporter gene containing multiple consensus GAL4 DNA-binding sites (GAL4 sites) placed upstream of a TATA-box and CAT gene encoding chloramphenicol acetyltransferase. B) Transfection of either the bait plasmid (i) or the target plasmid (ii) alone do not produce a functional activator. Cotransfection of the bait and effector plasmids (iii) produces a functional activator providing the bait (X) and target (Y) proteins can form a productive interaction.

Characterization of Protein-Protein Interactions (Mammalian Two-Hybrid)

The modular properties of transcription factor DNA-binding and transcriptional activation domains have fostered an ingenious strategy for characterization and identification of specific protein-protein interactions. The strategy, termed the "two-hybrid" technique (reviewed in ref. 139), exploits the fact that the DNA-binding domain of GAL4 is incapable of activating transcription unless physically, but not necessarily covalently, associated with an activating domain. The impetus for the two-hybrid strategy came from experiments by Ma and Ptashne (140) who demonstrated that the GAL80 protein could be converted into a transcriptional activator by fusion to an activation domain; activation by the GAL80-TAD fusion was dependent upon the presence of a GAL4 derivative bearing the GAL80 binding region (C-terminal 30 amino acids, see Figure 2). In a two-hybrid experiment a protein fused to the DNA-binding domain of GAL4 is expressed in yeast bearing a GAL4-dependent reporter gene. Providing the fusion cannot activate transcription on its own, it can then be used as "bait" to screen a library of cDNA clones fused to an activation domain, usually activating region 2 of GAL4. cDNA clones within the library that encode proteins capable of forming protein-protein interactions with the bait fusion are identified by virtue of their ability to cause activation of the reporter gene. Variations of the strategy have also been developed which use LexA as a DNA-binding domain for fusion of the bait and an artificial acidic activating domain fused to the cDNA library (141).

In mammalian cells the two-hybrid strategy has been used primarily for characterization of the interaction between two proteins which are suspected to form specific contacts. The assay requires cotransfection of three plasmids (see Figure 5, panel A): a GAL4-bait fusion plasmid, a target fusion plasmid which produces the target protein fused to a transcriptional activation domain (that of VP16 (69) in Figure 5), and a reporter gene for measuring transcriptional activation by the two-hybrid complex. An interaction between the bait and target fusions is indicated by virtue of the fact that reporter gene expression is detected when the bait plasmid and target fusion plasmids are cotransfected (Figure 5B, iii), but not when transfected individually (Figure 5B, i and ii). This approach has been used to examine interaction in mammalian cells between RAP-30 and RAP74 (142,143), the nucleocapsid and phosphoproteins of VSV (144), and the transcription factors YY1 and SP1 (145). The technique has also been used to investigate dimerization of the HIV-1 Rev (146) and Tat proteins (147), dimerization of HTLV-1 Rex (146), and homo- and heterodimer formation of myogenin and MyoD with E12, Id and E47 (148). The advantages of using mammalian two-hybrid strategies are that it is easier to perform in laboratories that are unaccustomed to working with yeast, and also that it allows examination of protein-protein interactions which require modifications specific to higher eukaryotes.

A modified two-hybrid system has also been developed that would allow screening two-hybrid libraries in mammalian cells (149). In this system the TAD-target fusion library is expressed from a plasmid bearing the SV40 promoter/ replication origin. The reporter plasmid in turn contains a minimal promoter bearing binding sites for GAL4 upstream of the gene for SV40 large T antigen. Productive association of the GAL4-bait hybrid with a TAD-target hybrid results in expression of large T antigen which in turn can stimulate replication of the TAD-target expression plasmid. Replicated plasmids can be selectively enriched with *Dpn*I digestion (see below) of recovered plasmid DNA, and by several rescreening cycles (149).

Characterization of Proteases and Protease Inhibitors

A genetic technique for analysis of sequence-specific proteinases and proteinase inhibitors has evolved from the "two-hybrid" system which employs a diametrically opposed strategy (Figure 6). The scheme again relies upon the fact that transcriptional activators require physical association between their DNA-binding and transcriptional activation domains (150,151). A cleavage site for a sequence- or site-specific proteinase is inserted between the DNA-binding domain of GAL4 and a strong transcriptional activating domain (that of VP16 is shown in Figure 6). When expressed in a cell with an active proteinase function, the fusion protein will be specifically cleaved, thus separating the activation domain from the DNA-binding domain. Cleavage causes inactivation of the GAL4 activator and consequently loss of reporter gene expression (Figure 6, panel B). Mutations in either the proteinase (panel C) or the cleavage site (panel D), or agents that inhibit the proteinase reaction, can be identified by reporter gene expression (150). Initial experiments using this strategy employed fusions of GAL4 to the self-cleaving 3C protease from coxsackievirus (150) and the HIV-1 proteinase (151). However, the system should also work as diagrammed in Figure 6 by insertion of an appropriate specific cleavage site in the GAL4 activator and expression of the proteinase separately by cotransfection.

Since activation of transcription by the GAL4 fusion indicates inactivation of proteinase function, the strategy as described above and in Figure 6 is ideal for assaying

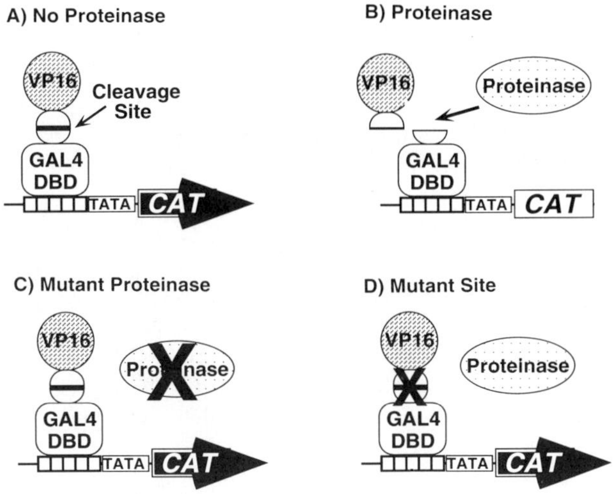

Figure 6. A genetic technique to assay sequence-specific proteinases. A) A fusion protein consisting of the GAL4 DNA-binding domain (GAL4 DBD) and VP16 activating domain (VP16) separated by a cleavage site for a sequence-specific proteinase can activate transcription in the absence of proteinase. B) Co-expression of the proteinase causes inactivation of the hybrid activator by site-specific cleavage. C) A mutant proteinase is unable to inactivate the hybrid activator. D) A mutation at the specific cleavage site on the hybrid activator prevents inactivation by the proteinase.

the effect of mutations in the proteinase or cleavage site, or the relative effect of specific inhibitors. However, if the goal is to assay relative proteinase activity the strategy is somewhat inappropriate because full activity would generate a negative result, i.e., no reporter gene expression (Figure 6, panel B). To accommodate such objectives it should be possible to modify the assay to measure repression of transcription by the GAL4 fusion rather than activation. In this variant strategy the cleavage site would be inserted between the GAL4 DNA-binding domain and a strong transcriptional repression domain (see Table 7). In the absence of proteinase function the GAL4 fusion should repress transcription of a constitutive promoter bearing upstream GAL4 sites (no reporter gene expression); cleavage of the fusion would inactivate the repressor and cause induction of reporter gene transcription.

Studying Factors that Regulate Replication of DNA Viruses

Initiation of replication has been shown to be closely linked to transcriptional regulatory elements in a variety of systems varying from prokaryotic plasmids to eukaryotic viruses, organelle DNA and chromosomes. The replication origins of DNA viruses are among the most thoroughly characterized, and typically contain both a "core" and "enhancer" or "auxiliary" component (152,153). The core component functions to determine where replication will be initiated and typically binds a specific major viral replication protein. For example, the polyoma and SV40 Large T-antigens each bind their respective ori-core elements. The auxiliary or enhancer elements, while usually not absolutely essential for replication, serve to stimulate the efficiency of the replication origin (153-156). The finding that the auxiliary elements of polyoma and SV40 viruses bind host cell transcription factors fuels the notion that transcriptional regulation and replication of these viruses is somehow closely linked. Similarly, factors encoded by Epstein-Barr virus which bind the auxiliary elements of the origin for lytic replication (oriLyt), namely ZEBRA (BZLF1/ EB1, zta, Z), possess all of the characteristics of a cellular transcriptional activator (154).

Several groups have investigated the spatial requirements of the auxiliary elements, and the structural requirements of their cognate binding factors, by substituting *cis*-acting elements at the origin of replication with GAL4 binding sites. The ability of various GAL4 derivatives to activate replication from these chimeric origins is then assayed in cotransfection experiments (154-157). Replication of the target plasmid is usually measured by extraction of the episomal DNA following the transfection and analyzed by Southern blotting following digestion with a diagnostic restriction enzyme in combination with *Dpn*I, which can only cleave the unreplicated DNA. These experiments have conclusively demonstrated that GAL4 derivatives which are capable of activating transcription can also stimulate replication of polyoma and SV40 (155-157). Similar to that observed for activation of transcription (see above), the effect of inserting additional GAL4 binding sites was found to stimulate replication synergistically rather than additively (156,157). In contrast to that observed with polyoma and SV40, the oriLyt of Epstein-Barr virus was found to be dependent upon specific activation by the factor ZEBRA; replication of a construct bearing GAL4 sites adjacent to oriLyt was found to be activated by a GAL4-ZEBRA fusion but not by GAL4-VP16 (154). These initial studies implicate common mechanisms for regulation of transcription and replication and suggest that strategies involving chimeric regulatory factors will be instrumental in elucidating these mechanisms.

Use of GAL4 in Transgenic Mice and *Drosophila*

Transgenic manipulation of model organisms has contributed significantly to our understanding of invertebrate and vertebrate development, and to the characterization of disease progression. Given the success and usefulness of GAL4 expression in mammalian cell culture experiments, it seems a logical progression to apply similar strategies to transgenic experiments with model organisms. The first such application was the development of a "binary transgenic mouse" system which allows expression of an otherwise transcriptionally inert transgene in the progeny of a genetic cross (158). Such an application is useful for many transgenic mouse experiments because of the lethal effects of some expressed transgenes. Oncogenes and proto-oncogenes, for example, sometimes produce embryonic lethality when expressed ectopically as transgenes. A binary system with the use of a GAL4 activator was performed by creating "transactivator" mouse lines which bore transgenes comprised of the MMTV long terminal repeat expressing the GAL4 DNA-binding domain fused to activating region 2 (see Figures 2 and 3). The "target" mouse lines contained transgenes in which the *int-2* gene was expressed from a GAL4-dependent promoter. The usefulness of the strategy was illustrated by the finding that both the transactivator and target mouse lines were phenotypically normal, but progeny resulting from crosses between the two lines produced phenotypes characteristic of transgenic mice bearing MMTV-*int-2* transgenes (158). This type of strategy seems promising for use in expressing transgenes which would otherwise be impossible because of lethality. GAL4-derived activators expressed in transgenic mice appear to be phenotypically silent, at least when using the MMTV-LTR, a finding that is critical for the usefulness of the strategy. Similar applications have been developed for use in *Drosophila*. In one experiment a GAL4 activator was used to express *even-skipped* inappropriately in a variety of temporal and spatial patterns (159), and in another the homeotic gene *abdominal-A* was ectopically expressed in the mesoderm (160). Both experiments used strategies analogous to that in transgenic mice; "transactivator" and "target" flies were generated separately and then cross-bred to force specific ectopic expression of GAL4-regulated promoters in the progeny.

FUTURE STRATEGIES

The true measure of a useful set of tools is their ability to be modified for application to new problems. As such, strategies derived from expressing GAL4 in mammalian cells continue to evolve and adapt in order to address new problems and also to simplify established techniques. For example, genetic analysis of mammalian somatic cells has suffered from the lack of tightly regulated inducible promoters, as are commonly available in prokaryotes or yeast. Among the new systems which are being developed to address this problem are several applications of the DNA-binding domain of GAL4. Fusions between GAL4's DNA-binding domain and steroid hormone receptors can confer steroid inducibility on promoters bearing GAL4 sites (see Table 5). This finding has prompted the development of systems that allow stringent regulation of specific transfected genes by steroid hormones (161). One such system involving the *Drosophila* ecdysone receptor may be of use for selectively regulating the temporal and spatial expression of genes in transgenic mouse experiments since mammalian cells do not normally respond to ecdysteroids (162). Related to these applications is the possibility that transcriptional activation by GAL4 derivatives can be used as a genetic marker for the activity of specific effector or regulatory proteins. For example, since some transcription factors are known to be regulated by specific protein kinases, the activity of the transcription factor can be used as a measure of protein kinase activity *in vivo*. It is now well established that CREB/ATF is controlled by direct phosphorylation from cAMP-dependent and calcium-dependent protein kinases, and furthermore that this phosphorylation-dependent regulation can be transferred to the DNA-binding domain of GAL4 (72, 163-166). An application of this would be to use transcriptional activation by GAL4-CREB chimeras as a marker for the analysis of mutants in the protein kinases (167). This would allow structure/function analysis of the protein kinases using genetic analysis.

New strategies utilizing expression of GAL4's DNA-binding domain in mammalian cells will undoubtedly evolve from those I have described. However, the DNA-binding domain fragment most frequently used in these strategies (residues 1-147) represents only 17% of the total mass of what may become the most thoroughly characterized eukaryotic transcription factor. GAL4 has a wealth of functional elements which could be used to confer dimerization, cooperative DNA-binding, specific protein-protein interactions and inhibition of transcriptional activation. As the molecular mechanisms conferring these functions to the GAL4 protein are elucidated I look forward to observing their use in newer and even more imaginative applications.

ACKNOWLEDGMENTS

I thank Neena Kuriakose, John Rohde and Martin Hirst for comments on the manuscript. Work in my laboratory is supported by grants from the Medical Research Council, the B.C. Health Research Foundation and the National Cancer Institute of Canada with funds from the Canadian Cancer Society. The author is a Research Scientist of the National Cancer Institute of Canada.

REFERENCES

1 Johnston, M. (1987) Microb. Rev. 51, 458-476.
2 Carey, M., Kakidani, H., Leatherwood, J., Mostashari, F. and Ptashne, M. (1989) J. Mol. Biol. 209, 423-432.
3 Marmorstein, R., Carey, M., Ptashne, M. and Harrison, S.C. (1992) Nature 356, 408-414.
4 Ruden, D.M., Ma, J. and Ptashne, M. (1988) Proc. Nat. Acad. Sci. U.S.A. 85, 4262-4266.
5 Giniger, E., Varnum, S.M. and Ptashne, M. (1985) Cell 40, 767-774.
6 Bajwa, W., Torchia, T.E. and Hopper, J.E. (1988) Mol. Cell. Biol. 8, 3439-3447.
7 Lorch, Y. and Kornberg, R.D. (1985) J. Mol. Biol. 186, 821-824.
8 Shimada, H. and Fukasawa, T. (1985) Gene 39, 1-9.
9 West, R., Yocum, R. and Ptashne, M. (1984) Mol. Cell. Biol. 4, 2467-2478.
10 Yocum, R.R., Hanley, S., West, R. and Ptashne, M. (1984) Mol. Cell. Biol. 4, 1985-1998.
11 Giniger, E. and Ptashne, M. (1988) Proc. Nat. Acad. Sci. U.S.A. 85, 382-386.
12 Carey, M., Lin, Y.S., Green, M.R. and Ptashne, M. (1990) Nature 345, 361-364.
13 Kang, T., Martins, T. and Sadowski, I. (1993) J. Biol. Chem. 268, 9629-9635.
14 Griggs, D.W. and Johnston, M. (1991) Proc. Nat. Acad. Sci. U.S.A. 88, 8597-8601.
15 Flick, J.S. and Johnston, M. (1992) Genetics 130, 295-304.
16 Flick, J.S. and Johnston, M. (1990) Mol. Cell. Biol. 10, 4757-4769.
17 Nehlin, J.O., Carlberg, M. and Ronne, H. (1991) EMBO J. 10, 3373-3377.
18 Keleher, C.A., Redd, M.J., Schultz, J., Carlson, M. and Johnson, A.D. (1992) Cell 68, 709-719.
19 Tzamarias, D. and Struhl, K. (1994) Nature 369, 758-761.
20 Chasman, D.I. and Kornberg, R.D. (1990) Mol. Cell. Biol. 10, 2916-2923.
21 Lue, N.F., Chasman, D.I., Buchman, A.R. and Kornberg, R.D. (1987) Mol. Cell. Biol. 7, 3446-3451.
22 Salmeron, J.J., Leuther, K.K. and Johnston, S.A. (1990) Genetics 125, 21-27.
23 Nogi, Y. and Fukasawa, T. (1989) Mol. Cell. Biol. 9, 3009-3017.
24 Ma, J. and Ptashne, M. (1987) Cell 50, 137-142.
25 Torchia, T.E., Hamilton, R.W., Cano, C.L. and Hopper, J.E. (1984) Mol. Cell. Biol. 4, 1521-1527.
26 Yun, S.-J., Hiraoka, Y., Nishizawa, M. and Fukasawa, T. (1990) Yeast 6, S311.
27 Leuther, K.K. and Johnston, S.A. (1990) Yeast 6, S250.
28 Mylin, L.M., Bhat, J.P. and Hopper, J.E. (1989) Genes and Dev. 3, 1157-1165.
29 Parthun, M.R. and Jaehning, J.A. (1990) J. Biol. Chem. 265, 209-213.
30 Leuther, K. and Johnston, S.A. (1992) Science 256, 1333-1335.
31 Bhat, P.J. and Hopper, J.E. (1991) Genetics 128, 233-239.
32 Bhat, P.J., Oh, D. and Hopper, J.E. (1990) Genetics 125, 281-291.
33 Torchia, T.E. and Hopper, J.E. (1986) Genetics 113, 229-246.
34 Bhat, P.J. and Hopper, J.E. (1992) Mol. Cell. Biol. 12, 2701-2707.
35 Nogi, Y. (1986) J. Bacteriol. 165, 101-106.
36 Johnston, M., Flick, J.S. and Pexton, T. (1994) Mol. Cell. Biol. 14, 3834-3852.
37 Matern, H. and Holzer, H. (1977) J. Biol. Chem. 252, 6399-6402.
38 Lamphier, M.S. and Ptashne, M. (1992) Proc. Nat. Acad. Sci. U.S.A. 89, 5922-5926.
39 Stone, G. and Sadowski, I. (1993) EMBO J. 12, 1375-1385.

40 Johnston, S.A., Zavortink, M.J., Debouck, C. and Hopper, J.E. (1986) Proc. Nat. Acad. Sci. U.S.A. 83, 6553-6557.
41 Keegan, L., Gill, G. and Ptashne, M. (1986) Science 231, 699-704.
42 Ma, J. and Ptashne, M. (1987) Cell 48, 847-853.
43 Silver, P.A., Keegan, L.P. and Ptashne, M. (1984) Proc. Nat. Acad. Sci. U.S.A. 81, 5951-5955.
44 Silver, P.A., Brent, R. and Ptashne, M. (1986) Mol. Cell. Biol. 6, 4763-4766.
45 Silver, P.A., Chiang, A. and Sadler, I. (1988) Genes and Dev. 2, 707-717.
46 Lin, Y.S., Carey, M.F., Ptashne, M. and Green, M.R. (1988) Cell 54, 659-664.
47 Carey, M., Lin, Y.S., Green, M.R. and Ptashne, M. (1990) Nature 345, 361-364.
48 Sadowski, I., Niedbala, D., Wood, K. and Ptashne, M. (1991) Proc. Nat. Acad. Sci. U.S.A. 88, 10510-10514.
49 Ptashne, M. (1989) Sci. Amer. January 1989, 23-29.
50 Ptashne, M. and Gann, A.A.F. (1990) Nature 346, 329-331.
51 Taylor, I.C.A., Workman, J.L., Schuetz, T.J. and Kingston, R.E. (1991) Genes and Dev. 5, 1285-1298.
52 Guarente, L. (1988) Cell 52, 303-305.
53 Struhl, K. (1989) Annu. Rev. Biochem. 58, 1051-1077.
54 Ptashne, M. (1986) Nature 322, 697-701.
55 Kakidani, H. and Ptashne, M. (1988) Cell 52, 161-167.
56 Webster, N., Jin, J.R., Green, S., Hollis, M. and Chambon, P. (1988) Cell 52, 169-178.
57 Fischer, J.A., Giniger, E., Maniatis, T. and Ptashne, M. (1988) Nature 332, 853-856.
58 Ma, J., Przibilla, E., Hu, J., Bogorad, L. and Ptashne, M. (1988) Nature 334, 631-633.
59 Weston, K. and Bishop, J.M. (1989) Cell 58, 85-93.
60 Bull, P., Morley, K.L., Hoekstra, M.F., Hunter, T. and Verma, I.M. (1990) Mol. Cell. Biol. 10, 5473-5485.
61 Nerlov, C. and Ziff, E.B. (1994) Genes and Dev. 8, 350-362.
62 Schwarz, J.J., Chakraborty, T., Martin, J., Zhou, J. and Olson, E.N. (1992) Mol. Cell. Biol. 12, 266-275.
63 Hollenberg, S. and Evans, R.M. (1988) Cell 55, 8899-8906.
64 Seipel, K., Georgiev, O. and Schaffner, W. (1992) EMBO J. 11, 4961-4968.
65 Tanaka, M., Clouston, W.M. and Herr, W. (1994) Mol. Cell. Biol. 14, 6046-6055.
66 Tanaka, M. and Herr, W. (1994) Mol. Cell. Biol. 14, 6056-6067.
67 Carey, M., Leatherwood, J. and Ptashne, M. (1990) Science 247, 710-712.
68 Sadowski, I. and Ptashne, M. (1989) Nucl. Acids Res. 17, 7539.
69 Sadowski, I., Bell, B., Broad, P. and Hollis, M. (1992) Gene 118, 137-141.
70 Raycroft, L. and Lozano, G. (1992) Gene 118, 143-144.
71 Russo, M.W., Matheny, C. and Milbrandt, J. (1993) Mol. Cell. Biol. 13, 6858-6865.
72 Quinn, P.G. (1993) J. Biol. Chem. 268, 16999-17009.
73 Raycroft, L., Wu, H. and Lozano, G. (1990) Science 249, 1049-1051.
74 Carlsson, P., Waterman, M.L. and Jones, K.A. (1993) Genes and Dev. 7, 2418-2430.
75 Gorman, C.M., Moffat, L.F. and Howard, B.H. (1982) Mol. Cell. Biol. 2, 1044-1051.
76 Seed, B. and Sheen, J.Y. (1988) Gene 67, 271-277.
77 de Wet, J.R., Wood, K.V., DeLuca, M., Helinski, D.R. and Subramani, S. (1987) Mol. Cell. Biol. 7, 725-737.

78 Brasier, A.R., Tate, J.E. and Habener, J.F. (1989) BioTechniques 7, 1116-1122.
79 Manet, E., Allera, C., Gruffat, H., Mikaelian, I., Rigolet, A. and Sergeant, A. (1993) Gene Expr. 3, 49-59.
80 Martin, K.J., Lillie, J.W. and Green, M.R. (1990) Nature 346, 147-52.
81 Hwang, J.-J., Chambon, P. and Davidson, I. (1993) EMBO J. 12, 2337-2348.
82 Witzgall, R., O'Leary, E., Leaf, A., Onaldi, D. and Bonaventre, J.V. (1994) Proc. Nat. Acad. Sci. U.S.A. 91, 4514-4518.
83 Park, K. and Atchison, M.L. (1991) Proc. Nat. Acad. Sci. U.S.A. 88, 9804-9808.
84 Shi, Y., Seto, E., Chang, L.S. and Shenk, T. (1991) Cell 67, 377-388.
85 Baniahmad, A., Kohne, A.C. and Renkawitz, R. (1992) EMBO J. 11, 1015-1023.
86 Kamine, J., Subramanian, T. and Chinnaduai, G. (1991) Proc. Nat. Acad. Sci. U.S.A. 88, 8510-8514.
87 Southgate, C., Zapp, M.L. and Green, M.R. (1990) Nature 345, 640-642.
88 Sadowski, I., Ma, J., Triezenberg, S. and Ptashne, M. (1988) Nature 335, 563-564.
89 Connor, L., Oxman, M.N., Brady, J.N. and Marriott, S.J. (1993) Virology 195, 569-577.
90 Muchardt, C., Seeler, J.-S., Nirula, A., Shurland, D.L. and Gaynor, R.B. (1992) J. Virol. 66, 244-250.
91 Garrett, E.D., He, F., Bogerd, H.P. and Cullen, B.R. (1993) J. Virol 67, 6824-6827.
92 Ohana, B., Moore, P.A., Ruben, S.M., Southgate, C.D., Green, M.R. and Rosen, C. (1993) Proc. Nat. Acad. Sci. U.S.A. 90, 138-142.
93 Michelson, A.M., Blake, C.C.F., Evans, S.T. and Orkin, S.H. (1985) Proc. Nat. Acad. Sci. U.S.A. 82, 6965-6969.
94 Moore, P.A., Ruben, S.M. and Rosen, C. (1993) Mol. Cell. Biol. 13, 1666-1674.
95 Friedman, A.D. and McKnight, S.L. (1990) Genes and Dev. 4, 1416-1426.
96 Johansen, F.-E. and Prywes, R. (1993) Mol. Cell. Biol. 13, 4640-4647.
97 Nagpal, S., Friant, S., Nakshatri, H. and Chambon, P. (1993) EMBO J. 12, 2349-2360.
98 Riley, J.L. and Boss, J.M. (1993) J. Immunol. 151, 6942-6953.
99 Gill, G. and Ptashne, M. (1988) Nature 334, 721-724.
100 Berger, S.L., Cress, W.D., Cress, A., Triezenberg, S.J. and Guarente, L. (1990) Cell 61, 1199-1208.
101 Brent, R. and Ptashne, M. (1985) Cell 43, 729-736.
102 Johnson, P.F. and McKnight, S.L. (1989) Annu. Rev. Biochem. 58, 799-839.
103 Triezenberg, S.J., Kingsbury, R.C. and McKnight, S.L. (1988) Genes and Dev. 2, 718-729.
104 Morin, P.J., Subramanian, G.S. and Gilmore, T.D. (1993) Nucl. Acids Res. 21, 2157-2163.
105 Morin, P.J. and Gilmore, T.D. (1992) Nucl. Acids Res. 20, 2453-2458.
106 Liu, F. and Green, M.R. (1990) Cell 61, 1217-1224.
107 Fields, S. and Jang, K. (1990) Science 249, 1046-1048.
108 Liu, X., Miller, C.W., Koeffler, P.H. and Berk, A.J. (1993) Mol. Cell. Biol. 13, 3291-3000.
109 Miller, G., Himmelfarb, H., Heston, L., Countryman, J., Gradoville, L., Baumann, R., Chi, T. and Carey, M. (1993) J. Virol. 67, 7472-7481.
110 Rhim, H. and Rice, A.P. (1993) J. Virol. 67, 1110-1121.
111 Lee, C.W., Chang, J., Lee, K.J. and Sung, Y.C. (1994) J. Virol. 68, 2708-2719.
112 Weintraub, H., Dwarki, V.J., Verma, I., Davis, R., Hollenberg, S., Snider, L., Lassar, A. and Tapscott, S.J. (1991) Genes and Dev. 5, 1377-1386.
113 Greenblatt, J. (1991) Cell 66, 1067-1070.

114 Carey, M. (1994) Nature 368, 402-403.
115 Hoey, T., Weinzierl, R.O.J., Gill, G., Chen, J.-L., Dynlacht, B.D. and Tjian, R. (1993) Cell 72, 247-260.
116 Cousens, D.J., Greaves, R., Goding, C.R. and O'Hare, P.O. (1989) EMBO J. 8, 2337-2342.
117 Quong, M.W., Massari, M.E., Zwart, R. and Murre, C. (1993) Mol. Cell. Biol. 13, 792-800.
118 Kamine, J. and Chinnadurai, G. (1992) J. Virol. 66, 3932-3936.
119 Monuki, E.S., Kuhn, R. and Lemke, G. (1993) Proc. Nat. Acad. Sci. U.S.A. 90, 9978-9982.
120 Gerber, H.-P., Seipel, K., Georgiev, O., Hofferer, M., Hug, M., Rusconi, S. and Schaffner, W. (1994) Science 263, 808-811.
121 Chi, T. and Carey, M. (1993) Mol. Cell. Biol. 13, 7045-7055.
122 Green, S. and Chambon, P. (1988) TIG 4, 309-314.
123 Picard, D., Salser, S.J. and Yamamoto, K.R. (1988) Cell 54, 1073-1080.
124 Karin, M. and Smeal, T. (1992) TIBS 17, 418-422.
125 Jackson, S.P. (1992) TICB 2, 104-108.
126 Hunter, T. and Karin, M. (1992) Cell 70, 375-387.
127 Baichwal, V.R., Park, A. and Tjian, R. (1991) Nature 352, 165-168.
128 Baichwal, V.R., Park, A. and Tjian, R. (1992) Genes and Dev. 6, 1493-1502.
129 Baichwal, V.R. and Tjian, R. (1990) Cell 63, 815-825.
130 Helin, K. and Fattaey, A. (1993) Mol. Cell. Biol. 13, 6501-6508.
131 Seth, A., Gupta, S. and Davis, R.J. (1993) Mol. Cell. Biol. 13, 4125-4136.
132 Barrett, J., Birrer, M.J., Kato, G.J., Dosaka-Akita, H. and Dang, C.V. (1992) Mol. Cell. Biol. 12, 3130-3137.
133 Rivera, V.M., Miranti, C.K., Misra, R.P., Ginty, D.D., Chen, R.-H., Blenis, J. and Greenberg, M.E. (1993) Mol. Cell. Biol. 13, 6260-6273.
134 Liu, S.H., Ma, J.T., Yueh, A.Y., Lees, M.S., Anderson, C.W. and Ng, S.Y. (1993) J. Biol. Chem. 268, 21147-21154.
135 Ruben, S.M., Narayanan, R., Klement, J.F., Chen, C.H. and Rosen, C.A. (1992) Mol. Cell. Biol. 12, 444-454.
136 Licht, J.D., Hanna-Rose, W., Reddy, J.C., English, M.A., Ro, M., Grossel, M., Shaknovich, R. and Hansen, U. (1994) Mol. Cell. Biol. 14, 4057-4066.
137 Margolin, J.F., Friedman, J.R., Meyer, W.K.H., Vissing, H., Thiesen, H.-J. and Rauscher, F.J.I. (1994) Proc. Nat. Acad. Sci. U.S.A. 91, 4509-4513.
138 Witzgall, R., O'Leary, E., Gessner, R., Ouellette, A.J. and Bonaventre, J.V. (1993) Mol. Cell. Biol. 13, 1933-1942.
139 Fields, S. and Sternglanz, R. (1994) TIG 10, 286-292.
140 Ma, J. and Ptashne, M. (1988) Cell 55, 443-446.
141 Zervos, A.S., Gyuris, J. and Brent, R. (1993) Cell 72, 223-232.
142 Yonaha, M., Aso, T., Kobayashi, Y., Vasavada, H., Yasukochi, Y., Weissman, S.M. and Kitajima, S. (1993) Nucl. Acids Res. 21,273-279.
143 Aso, T., Vasavada, H.A., Kawaguchi, T., Germino, F.J., Ganguly, S., Kitajima, S., Weissman, S.M. and Yasukochi, Y. (1992) Nature 355, 461-464.
144 Takacs, A.M., Das, T. and Banerjee, A.K. (1993) Proc. Nat. Acad. Sci. U.S.A. 90, 10375-10379.
145 Lee, J.-S., Galvin, K.M. and Shi, Y. (1993) Proc. Nat. Acad. Sci. U.S.A. 90, 6145-6149.
146 Bogerd, H. and Greene, W.C. (1993) J. Virol. 67, 2496-2502.
147 Bogerd, H.P., Fridell, R.A., Blair, W.S. and Cullen, B.R. (1993) J. Virol. 67, 5030-5034.

148 Finkel, T., Duc, J., Fearon, E.R., Dang, C.V. and Tomaselli, G.F. (1993) J. Biol. Chem. 268, 5-8.
149 Vasavada, H.A., Ganguly, S., Germino, F.J., Wang, Z.X. and Weissman, S.M. (1991) Proc. Nat. Acad. Sci. U.S.A. 88, 10686-10690.
150 Dasmahapatra, B., DiDomenico, B., Dwyer, S., Ma, J., Sadowski, I. and Schwartz, J. (1992) Proc. Nat. Acad. Sci. U.S.A. 89, 4159-4162.
151 Murray, M.G., Hung, W., Sadowski, I. and Das Mahapatra, B. (1993) Gene 134, 123-128.
152 DePamphilis, M.L. (1992) in Advances in Developmental Biochemistry (Wassarman, P.M., ed.), vol. 1, pp. 31-83, JAI Press, Greenwich, CT.
153 DePamphilis, M.L. (1988) Cell 52, 635-638.
154 Schepers, A., Pich, D. and Hammerschmidt, W. (1993) EMBO J. 12, 3921-3929.
155 Guo, Z.-S. and DePamphilis, M.L. (1992) Mol. Cell. Biol. 12, 2514-2524.
156 Bennet-Cook, E.R. and Hassell, J.A. (1991) EMBO J. 10, 959-969.
157 Baru, M., Shlissel, M. and Manor, H. (1991) Mol. Cell. Biol. 65, 3496-3503.
158 Ornitz, D.M., Moreadith, R.W. and Leder, P. (1991) Proc. Nat. Acad. Sci. U.S.A. 88, 698-702.
159 Brand, A.H. and Perrimon, N. (1993) Devel. 118, 401-415.
160 Greig, S. and Akam, M. (1993) Nature 362, 630-632.
161 Braselmann, S., Graninger, P. and Busslinger, M. (1993) Proc. Nat. Acad. Sci. U.S.A. 90, 1657-1661.
162 Christopherson, K.S., Mark, M.R., Bajaj, V. and Godowski, P.J. (1992) Proc. Nat. Acad. Sci. U.S.A. 89, 6314-6318.
163 Sheng, M., Thompson, M.A. and Greenberg, M.E. (1991) Science 252, 1427-1430.
164 Brindle, P., Linke, S. and Montminy, M. (1993) Nature 364, 821-824.
165 Quinn, P.G. (1994) J. Biol. Chem. 269, 14375-14378.
166 Leonard, J., Serup, P., Gonzalez, G., Edlund, T. and Montminy, M. (1992) Proc. Nat. Acad. Sci. U.S.A. 89, 6247-6251.
167 Enslen, H., Sun, P., Brickey, D., Soderling, S.H., Klamo, E. and Soderling, T.R. (1994) J. Biol. Chem. 269, 18-25.
168 Gashler, A.L., Swaminathan, S. and Sukhatme, V.P. (1993) Mol. Cell. Biol. 13, 4556-4571.
169 Beckmann, H., Su, L.-K. and Kadesh, T. (1990) Genes and Dev. 4, 167-179.
170 Pei, D.Q. and Shih, C.H. (1991) Mol. Cell. Biol. 11, 1480-1487.
171 Martinez, E., Dusserre, Y., Wahli, W. and Mermod, N. (1991) Mol. Cell. Biol. 11, 2937-2945.
172 Sogawa, K., Imataka, H., Yamasaki, Y., Kusume, H., Abe, H. and Fuji-Kuriyama, Y. (1993) Nucl. Acids. Res. 21, 1527-1532.
173 Toniatti, C., Monaci, P., Nicosia, A., Cortese, R. and Ciliberto, G. (1993) DNA Cell. Biol. 12, 199-208.
174 Johnston, L.A., Tapscott, S.J. and Eisen, H. (1992) Mol. Cell. Biol. 12, 5123-5130.
175 Martin, D. and Orkin, S.H. (1990) Genes and Dev. 4, 1886-1898.
176 Satyamoorthy, K., Park, K., Atchison, M.L. and Howe, C.C. (1993) Mol. Cell. Biol. 13, 6621-6628.
177 Shan, B., Zhu, X., Chen, P.L., Durfee, T., Yang, Y., Sharp, P. and Lee, W.H. (1992) Mol. Cell. Biol. 12, 5620-5631.
178 Aronhaeim, A., Shiran, R., Rosen, A. and Walker, M.D. (1993) Proc. Nat. Acad. Sci. U.S.A. 90, 8063-8067.
179 Wetering, M., Oosterwegel, M., Norren, K. and Clevers, H. (1993) EMBO J. 12, 3847-3854.

180 Williams, T. and Tjian, R. (1991) Genes and Dev. 5, 670-682.
181 Liu, F. (1993) J. Biol. Chem. 268, 6714-6720.
182 Liu, F. and Green, M.R. (1994) Nature 368, 520-525.
183 Bondesson, M., Svensson, C., Linder, S. and Akusjarvi, G. (1992) EMBO J. 11, 3347-3354.
184 Chatton, B., Bocco, J.L., Gaire, M., Hauss, C., Reimund, B., Goetz, J. and Kedinger, C. (1993) Mol. Cell. Biol. 13, 561-570.
185 Lemaigre, F.P., Ace, C.I. and Green, M.R. (1993) Nucl. Acids Res. 21, 2907-2911.
186 Sun, P., Schoderbek, W.E. and Maurer, R.A. (1992) Mol. Endocr. 6, 1858-1866.
187 Laoide, B.M., Foulkes, N.S., Schlotter, F. and Sassone-Corsi, P. (1993) EMBO J. 12, 1179-1191.
188 Meyer, M.-E., Gronemeyer, H., Turcotte, B., Bocquel, M.-T., Tasset, D. and Chambon, P. (1989) Cell 57, 433-442.
189 Tora, L., White, J., Brou, C., Tasset, D., Webster, N., Scheer, E. and Chambon, P. (1989) Cell 59, 477-487.
190 Danielian, P.S., White, R., Lees, J.A. and Parker, M.A. (1992) EMBO J. 11, 1025-1033.
191 Saatcioglu, F., Bartunek, P., Deng, T., Zenke, M. and Karin, M. (1993) Mol. Cell. Biol. 13, 3675-3685.
192 Katz, D. and Lazar, M.A. (1993) J. Biol. Chem. 268, 20904-20910.
193 Meyer, M.-E., Quirin-Stricker, C., Lerouge, T., Bocquel, M.-T. and Gronemeyer, H. (1992) J. Biol. Chem. 267, 10882-10887.
194 Rambaldi, I., Kovacs, E.N. and Featherstone, M.S. (1994) Nucl. Acids Res. 22, 376-382.
195 Kato, G.J., Barrett, J., Villa-Garcia, M. and Dang, C. (1990) Mol. Cell. Biol. 10, 5914-5920.
196 Seth, A., Alvarez, E., Gupta, S. and Davis, R.J. (1991) J. Biol. Chem. 266, 23521-23524.
197 Kato, G.J., Lee, W., Chen, L. and Dang, C.V. (1992) Genes and Dev. 6, 81-92.
198 Schneikert, J., Lutz, Y. and Wasylyk, B. (1992) Oncogene 7, 249-256.
199 Chumakov, A.M., Chen, D.L., Chumakova, E.A. and Koeffler, H.P. (1993) J. Virol. 67, 2421-2425.
200 Foos, G., Grimm, S. and Klempnauer, K.-H. (1992) EMBO J. 11, 4619-4629.
201 Dubendorff, J.W., Whittaker, L.J., Eltman, J.T. and Lipsick, J.S. (1992) Genes and Dev. 6, 2524-2535.
202 Angel, P., Smeal, T., Meek, J. and Karin, M. (1989) New Biol. 1, 35-43.
203 Sutherland, J.A., Cook, A., Bannister, A.J. and Kouzarides, T. (1992) Genes and Dev. 6, 1810-1819.
204 Hibi, M., Lin, A., Smeal, T., Minden, A. and Karin, M. (1993) Genes and Dev. 7, 2135-2148.
205 Deng, T. and Karin, M. (1993) Genes and Dev. 7, 479-490.
206 Wisdom, R. and Verma, I.M. (1993) Mol. Cell. Biol. 13, 7429-7438.
207 Metz, R., Bannister, A.J., Sutherland, J.A., Hagemeier, C., O'Rourke, E.C., Cook, A., Bravo, B. and Kouzarides, T. (1994) Mol. Cell. Biol. 14, 6021-6029.
208 Wisdom, R., Yen, J., Rashid, D. and Verma, I.M. (1992) Genes and Dev. 6, 667-675.
209 Sarkar, S. and Gilmore, T.D. (1993) Oncogene 88, 2243-2252.
210 Diehl, J.A. and Hannink, M. (1993) J. Virol. 67, 7161-7171.
211 Ruben, S.M., Klement, J.F., Coleman, T.A., Maher, M., Chen, C.-H. and Rosen, C. (1992) Genes and Dev. 6, 745-760.

212 Flemington, E.K., Borras, A.M., Lytle, J.P. and Speck, S.H. (1992) J. Virol. 66, 922-929.
213 Hardwick, J.M., Tse, L., Applegren, N., Nicholas, J. and Veliuona, M.A. (1992) J. Virol. 66, 5500-5508.
214 Cohen, J.I. and Kieff, E. (1991) J. Virol. 65, 5880-5885.
215 Ling, P.D., Ryon, J.J. and Hayward, S.D. (1993) J. Virol. 67, 2990-3003.
216 Weber, P. and Wigdahl, B. (1992) J. Virol. 66, 2261-2267.
217 Perera, L.P., Kaushal, S., Kinchington, P.R., Mosca, J.D., Hayward, G.S. and Straus, S.E. (1994) J. Virol. 68, 2468-2477.
218 Cohen, J.I., Heffel, D. and Seidel, K. (1993) J. Virol. 67, 4246-4251.
219 Perera, L.P., Mosca, J.D., Ruyechan, W.T., Hayward, G.S., Straus, S.E. and Hay, J. (1993) J. Virol. 67, 4474-4483.
220 Gruda, M.C., Zabolotny, J.M., Xiao, J.H., Davidson, I. and Alwine, J.C. (1993) Mol. Cell. Biol. 13, 961-969.
221 Martin, K.J., Lillie, J.W. and Green, M.R. (1990) Genes and Dev. 4, 2376-2382.
222 Blair, W.S., Bogerd, H. and Cullen, B.R. (1994) J. Virol. 68, 3803-3808.
223 Venkatesh, L.K. and Chinnadurai, G. (1993) J. Virol. 67, 3868-3876.
224 Mergia, A., Renshaw-Gegg, L.W., Stout, M.W., Renne, R. and Herchenroeder, O. (1993) J. Virol. 67, 4598-4604.
225 Fujisawa, J., Toita, M., Yoshimura, T. and Yoshida, M. (1991) J. Virol. 65, 4525-4528.
226 Fujii, M., Tsuchiya, H. and Seiki, M. (1991) Oncogene 6, 2349-2352.
227 Fujii, M., Tsuchiya, H., Chuhjo, T., Akizawa, T. and Seiki, M. (1992) Genes and Dev. 6, 2066-2076.
228 de Groot, R.P. and Sassone-Corsi, P. (1992) Oncogene 7, 2281-2286.
229 Binetruy, B., Smeal, T. and Karin, M. (1991) Nature 351, 122-127.
230 Webster, N.J.G., Green, S., Jin, J.R. and Chambon, P. (1988) Cell 52, 169-178.
231 Braselmann, S., Graninger, P. and Busslinger, M. (1993) Proc. Nat. Acad. Sci. U.S.A. 90, 1657-1661.
232 Berry, M., Metzger, D. and Chambon, P. (1990) EMBO J. 9, 2811-2818.
233 Cowell, I.G. and Hurst, H.C. (1994) Nucl. Acids Res. 22, 59-65.
234 Yew, P.R., Liu, X. and Berk, A.J. (1994) Genes and Dev. 8, 190-202.
235 Monuki, E.S., Kuhn, R. and Lemke, G. (1993) Mech. Devel. 42, 15-32.
236 Madden, S.L., Cook, D.M. and Rausher, F.J.I. (1993) Oncogene 8, 1713-1720.

PROTEIN THIOL MODIFICATION OF GLYCERALDEHYDE-3-PHOSPHATE DEHYDROGENASE AS A TARGET FOR NITRIC OXIDE SIGNALING

Bernhard Brüne[1] and Eduardo G. Lapetina[2]

[1]Faculty of Biology
University of Konstanz
P.O. Box 5560 M612
78434 Konstanz, Germany

[2]Burroughs Wellcome Co.
3030 Cornwallis Road
Research Triangle Park, NC 27709

INTRODUCTION

Nitric oxide (NO) serves as a messenger for a variety of physiological and patho-physiological processes in humans. The molecule, composed of only nitrogen and oxygen, is a key transducer of a vasodilator message from the endothelium to vascular smooth muscle, participates in central and peripheral neuronal transmission, and is a constitutent of the nonspecific immune defense system. As a radical, NO displays high reactivity in many reactions and facilitates a number of beneficial as well as harmful actions. A steady stream of reports continues to expand NO's sphere of action. Diverse and highly controlled signaling pathways in concert with a change in the ambient redox milieu may determine the physiological and patho-physiological actions of NO. The highly regulated constitutive and inducible NO output systems, a vast number of secondary metabolites stemming from the reaction of NO with oxygen or superoxide and a wide array of target groups all contribute to both beneficial and toxic reactions initiated by NO.

It is important that we know precisely how NO fulfills its fundamental role as one of the most important short-lived messengers. In this chapter, we briefly describe the generation of NO and the main physiologic effects of NO. Then we focus on the NO-induced covalent modification of the glycolytic enzyme glyceraldehyde-3-phosphate dehydrogenase (GAPDH), paying attention to the interaction of NO-derived redox species with protein thiols, S-nitrosothiol formation and NO-redox chemistry. Finally, NO toxicity due to the level of NAD^+ and cellular energy supply related to GAPDH modification are discussed.

Genetic Engineering, Vol. 17, Edited by J.K. Setlow
Plenum Press, New York, 1995

NITRIC OXIDE DONORS

Although nitric oxide releasing compounds have been used for more than 100 years in the treatment of congestive heart failure, angina pectoris, pulmonary hypertension, fibrinolytic therapy and blood pressure control, it was 1977 before the molecular mechanism of action of such compounds became apparent. NO released from these agents, generally termed nitrovasodilators, activates soluble guanylyl cyclase (1,2). Soluble guanylyl cyclase, a heterodimer heme protein with a molecular mass of 150 kD, consists of two immunologically distinct subunits and is present in most mammalian tissues and cells but virtually absent in intestinal epithelia. Activation of guanylyl cyclase is achieved by NO, or in some cases through S-nitrosothiol intermediates, released or generated by the nitrogen-oxide containing vasodilators. Mechanistically, enzyme activation is largely heme-dependent and occurs via the formation of a NO-heme interaction eliciting a conformational change that enhances the enzyme's catalytic activity (3-5).

The action of cGMP is controlled by enzymes that synthesize and degrade cGMP, like the soluble form of guanylyl cyclase and various phosphodiesterases, respectively. Moreover, the signaling of cGMP is mainly attributed to cGMP-dependent protein kinases, which become activated in response to increasing concentrations of the second messenger, cGMP. In addition to activating protein kinases, cGMP directly regulates cyclic nucleotide-gated ion channels, participating in visual as well as olfactory transduction which have been identified in the pineal gland and the kidney (6,7). In platelets and vascular smooth muscle, soluble guanylyl cyclase activators elevate tissue cGMP levels, causing distinct protein phosphorylation leading to vascular relaxation and inhibition of platelet aggregation in response to pharmacologically derived NO. Although exact mechanisms of this signaling system are unclear, recent findings suggest that phosphoprotein associated with actin filaments and focal contact areas serve as potential targets for regulation by cGMP-dependent protein kinases (8). Most NO actions can be mimicked experimentally by lipophilic cGMP analogs, diminished by inhibitors of guanylyl cyclase, and augmented by inhibiting cGMP-phosphodiesterases. This substantiates the suggestion that activation of cGMP-dependent protein kinases is a cardinal mechanism of the signal transduction pathway of NO.

Commonly used NO donors are 3-morpholinosydnonimine (SIN-1), S-nitrosothiols (i.e., S-nitrosoglutathione, GSNO), sodium nitroprusside (SNP), organic nitrates (i.e., nitroglycerin, NTG) and the recently introduced NONOates (compounds containing the $N(O)NO^-$ functional group). These prodrugs vary considerably in chemical nature, though all preseve NO in their molecular structure and all exhibit biological activity after NO liberation. Pharmacokinetic differences among NO donors arise in part from differences in bioactivation, chemical stability and enzymatic clearance (9). Two biotransformation systems--glutathione-S-transferase and the cytochrome P450 system--seem to be involved in the enzymatic release of NO from these prodrugs, although compounds like SNP spontaneously liberate NO.

Apart from differences in the bioactivation process, considerable variation arises from the NO-redoxspecies liberated from individual nitrovasodilators. SNP contains NO^+ (nitrosonium) bound to a ferrous ion coordinated with five cyanide anions. Spontaneous decomposition or a one-electron reduction process may support the release of the nitrogen species, with thiol or superoxide acting as reductant (10 and refs. therein). The release of cyanide prior to NO has to be considered a potential source of unwanted side effects (11). Decomposition of SIN-1 generates stoichiometric amounts of superoxide and NO, thereby giving rise to the formation of peroxynitrite ($ONOO^-$), which may lead to reactivities apart from those due to NO. The sulfur-nitrogen bond of S-nitrosothiols theoretically breaks up either homolytically or heterolytically, generating either nitrogen monoxide ($NO^•$) or the nitrosonium (NO^+) species. Thus, the group collectively known as "NO donors" generates an array of compounds with distinctive chemistries. This array of chemical species with unique properties and reactivities may be a clue to understanding the physiologic and pathophysiologic faces of NO (12).

NITRIC OXIDE GENERATING ENZYMES

Observations made in the early 1900s suggested that mammals produce oxides of nitrogen, as detected in the urine (13). These observations derived from studies of dietary nitrates as a source of carcinogenic nitrosamines. Attempts to delineate nitrosamine metabolism led to the observation that germ-free rats excrete more nitrate than they ingest and that this production was enhanced by endotoxin treatment (14,15). In the mid-1980s, evidence was found for induced nitrite and nitrate synthesis in murine macrophages (16). Next, L-arginine was revealed as the substrate for these metabolites; investigators noted that conversion of the substrate to nitrogen oxides was required for macrophage-induced cytotoxicity (17). In the meantime, the role of the endothelium as a source of an unidentified relaxing factor, endothelium-derived relaxing factor (EDRF), was advanced; the vasorelaxant action of acetylcholine in blood vessels was observed to depend on a highly labile, diffusible messenger released from endothelial cells (18). However, it was not until 1987/1988 that the biological properties of EDRF and those of NO synthesized from arginine and released from endothelial cells were found to be identical (19,20). A nearly complete overlap between the chemical and biological properties of EDRF and NO is now well established (21) and forms the basis for recent claims that EDRF and NO are identical (22).

Major insights into endogenous NO formation came from biological and pharmacological characterization of the release process, purification and molecular cloning of individual NOS (nitric oxide synthase) isoenzymes. A number of reviews published on this subject in recent years serve as general references for more detailed information (23-29).

The expanding family of NOS isoforms generally falls into two categories: (i) a constitutive form regulated by calcium and calmodulin, which leads to the release of NO over several minutes after an intracellular calcium increase, and (ii) a cytokine-inducible form expressed in many cell types after challenge with immunologic or inflammatory stimuli, generating large amounts of NO for several days. Apparently, there are two constitutive isoforms. One seems restricted to endothelium, whereas the other form is found in central as well as peripheral neurons, neuroblastomas, human skeletal muscle, ß cells, and several other cell types. In relation to the cytokine-inducible enzyme, macrophages represent a classical example for cells expressing the inducible isoform.

All NOS isoenzymes are NADPH-dependent, P450-like hemoproteins; all contain binding sites for FAD (flavin adenine dinucleotide) and FMN (flavin mononucleotide) and carry out a five-electron oxidation of their substrate, L-arginine, in the presence of tetrahydrobiopterin to produce NO and stoichiometric amounts of citrulline. The various isoforms are subjected to numerous transcriptional as well as translational control mechanisms, opening multiple paths to augment or suppress NO production (29). This may help to adjust the level of demand so that the molecule can carry out its specific functions. Several inhibitors of NOS are used to intervene pharmacologically with NO production and to trace back individual action to the NO-signaling pathway. Most of these inhibitors lack specificity for individual NOS-isoforms. Structurally they are analogs of the naturally occurring substrate L-arginine. N^G-monomethy-L-argine (NMMA) is just one classical example. The combinations of NOS inhibitors with NO-liberating drugs are valuable biochemical tools because they permit the investigation of the role of NO in signaling pathways without interfering with second messenger systems involved in NOS activation.

Activation of endothelial constitutive NOS by acetylcholine, bradykinin and adenosine triphosphate accounts for the release of EDRF. Both the normal formation of NO in the arterial side of the circulation and stimulated NO production in response to endothelium-dependent vasodilators efficiently control vascular tone, blood pressure and hemodynamic properties in the circulation, including platelet function. Any pharmacologically or patho-physicologically induced interference with the biosynthesis, release, or action of NO could provoke or lead to the development of localized vasospasm and platelet aggregation.

The existence of NO in the brain was first demonstrated in cerebellar slices that, upon activation, released a factor with properties resembling those of authentic NO, and by observations of high amounts of NOS activity in the cerebellum (30 and references therein; 31-33). NOS occurs in discrete neuronal populations in the brain. In the cerebral cortex, hippocampus, and corpus striatum, it is localized in small interneurons only, whereas in the cerebellum, all granule cells and basket cells contain NOS. The enzyme is absent from Purkinje cells; in contrast, cGMP formation is quite prominent in Purkinje cells. This pattern of localization provides a rational basis for the regulation of NOS and cGMP by glutamate, the major excitatory neurotransmitter in the brain. Cyclic nucleotide levels increase several-fold in response to activation of the N-methyl-D-aspartate subtype of the glutamate receptor on granule cells or basket cells. An intracellular calcium increase activates the constitutive NOS, leading to the formation of NO, which diffuses to the Purkinje cells to activate soluble guanylyl cyclase. Attention has now shifted to identification of possible roles of NO production in response to stimulation of excitatory amino acid receptors. One notion is that NO, as a freely diffusible molecule, may play a role in synaptic plasticity, such as long-term potentiation (LPT) and long-term synaptic depression (LTD) (34 and refs. therein; 35). A function of NO in the peripheral nervous system may be in the myenteric plexus and throughout the whole gastrointestinal pathway. In the latter, NO is released by stimulation of nonadrenergic-noncholinergic (NANC) neurons mediating relaxation of the intestines--one component of peristalsis. In this respect, NO posesses several characteristics of an unusual neurotransmitter: it is synthesized only on demand, diffusing to the postsynaptic cell and binding to its intracellular receptor, soluble guanylyl cyclase, all causing cGMP-mediated downstream effects.

OVERPRODUCTION OF NITRIC OXIDE

Although the numerous physiological roles of NO in the nervous, cardiovascular, pulmonary, renal and endocrine systems (mainly associated with activation of constitutive NOS) are well substantiated, excessive production of NO in response to inflammatory cytokines or continuously elevated glutamate levels suggests that a potentially toxic role also exists for NO.

In the central nervous system, NO has been implicated in the cytotoxicity following excess stimulation of neurons by glutamate (36-38). Excessive glutamate release underlies neurodegenerative disorders, including ischemic hypoxic insults and Alzheimer's and Huntington's diseases. Although the precise mechanism by which NO overproduction kills neurons remains to be clarified, exessive activation of the glutamate-responsive NMDA receptor, followed by pronounced NO production, apparently constitute the initiating process. Toxic insults are likely to overlap with the mechanisms by which macrophage--derived NO kills bacteria or neoplastic cells. Free radical formation and energy depletion as a result of mitochondrial iron-sulfur enzyme inhibition, and NAD^+ depeletion resulting from activation of poly-ADP-ribosyltransferase, are likely to be involved in the toxic process (39). Planned experiments will use NO donors like SNP to mimic NMDA-induced neurotoxicity; in the reverse experiment, an NMDA receptor antagonist and NOS inhibitor block neuronal destruction. An as yet unresolved paradox is the fact that NOS-containing neurons are selectively resistant to various neurotoxic insults, including NO.

It follows that if EDRF formation and function constitute the main physiologic regulator of vascular tone, induction of the high-output system for NO causes severe life-treatening hypotension in systemic inflammatory responses (sepsis). Moreover, it is plausible that NO production is a mechanism of the nonspecific defense system. The antimicrobial capacity of NO in the defense against pathogens invading mammalian tissues has been known for a long time (40) and is essentially linked to NOS induction. The cytostatic role of macrophage-derived NO applies to lymphocytes as well as tumor cells and microbes (26,41). The mechanisms by which NO inhibits microbial growth or generates a cytostatic effect include inhibition of iron-sulfur proteins of mitochondrial respiration, inhibition of aconitase, inhibition of ribonucleotide reductase, modification of GAPDH, DNA deamination reactions, inhibition of protein synthesis, toxic radical formation when NO combines with superoxide, and alterations of the intracellular iron status (24 and refs. therein; 42). Similar potentially toxic mechanisms involving NO may operate during type I diabetes mellitus when pancreatic ß cells become irreversibly damaged (43-46) or during cardiac dysfunction (47).

The unique chemical properties of NO cause a variety of effects in biological systems, conferring on the molecule the ability to act as a physiologic regulator molecule or as a toxic agent. The L-arginine:NO pathway in macrophages as a primary defense mechanism against tumor cells, intracellular micro-organisms and pathogens seems at the same time to be involved in turning a beneficial effect into a potentially harmful pathway directed against vital cells. As NO turns out to be cytotoxic in some cells and cytostatic in others, the sensitivity to NO varies considerably from one cell to another. The vulnerability of a given cell may depend on cellular characteristics, including dependence on iron-sulfur enzymes, total energy stores, free radical defense systems, repair mechanisms and the ambient redox milieu that can convert $NO^{\bullet}$ to NO^+ or NO^-.

THE ROLE OF NAD^+ IN INTRACELLULAR SIGNALING

Nicotinamide adenine dinucleotide (NAD^+), a pyridine coenzyme composed of ADP (adenosine-diphosphate), ribose and nicotinamide, is involved in several biochemical signaling pathways (48 and refs. therein; 49,50). NAD^+ participates either as a redox cofactor, or a substrate cofactor, or a precursor of an intracellular messenger in the following reactions:

- NAD^+ as a redox-cofactor of dehydrogenase reactions
- NAD^+ as a precursor for cyclic ADP-ribose
- NAD^+ as a cofactor for NAD^+ glycohydrolases
- NAD^+ as a cofactor for mono-ADP-ribosylation reactions
- NAD^+ as a cofactor for poly-ADP-ribosylation reactions
- NAD^+ in the covalent NO-dependent modification of GAPDH

The predominant role for NAD^+ is as a cofactor for oxidation-reduction reactions utilized by a variety of dehydrogenases and reductases that occupy critical positions in metabolic pathways. During reaction catalysis, the transfer of hydrogen and electrons reduces NAD^+, producing NADH. Reversible formation of NADH thereby couples dehydration of primary and secondary alcohols to the formation of ATP in the mitochondrial respiratory chain. However, under anaerobic glycolytic conditions, glyceraldehyde-3-phosphate dehydrogenase produces ATP. Again, NAD^+, acting as a redox cofactor, is reduced to NADH while glyceraldehyde-3-phosphate is converted to 3-phosphoglycerate.

Glycohydrolases use NAD^+, breaking up the ß-glycosidic bond, producing free ADP-ribose, nicotinamide, and a proton. Free ADP-ribose is known to ADP-ribosylate proteins in a non-enzymatic fashion (51). Known examples are mitochondrial proteins that become labeled in response to oxidants (52), and cytoplasmic actin for which non-enzymatic ADP-ribosylation at a cysteine residue inhibits actin polymerization (53). Another group of NAD^+-consuming enzymes are those that convert NAD^+ into cyclic ADP-ribose. Cyclic ADP-ribose is known to interact with the ryanodine receptor at the sarcoplasmic reticulum, releasing calcium from intracellular stores in a manner analogous to inositol 1,4,5-trisphosphate (50).

Beyond these functions, mono- and poly-ADP-ribosyltransferases utilize the ADP-ribose portion of NAD^+ to catalyze the post-translational modification of proteins. ADP-ribosylations are ubiquitous, and the high group transfer potential of the ß-glycosidic linkage between nicotinamide and ribose drives this enzymatic reaction. Nuclear poly-ADP-ribose synthase, also referred to as poly-ADP-ribose polymerase, transfers ADP-ribose to histones, chromatin, or the enzyme itself, and then elongates the ADP-ribose chain in a linear or branched polymer. The physiological functions of the nuclear enzyme include the regulation of DNA repair and probably regulation of the conformation of the chromatin. In addition, a role during transcription and cell differentiation has been reported. The enzyme has a specific requirement for ends of DNA strands; intact DNA does not activate the poly-ADP-ribose polymerase. Nicotinamide, benzamide, and its derivative, 3-aminobenzamide, are valuable pharmacological inhibitors (49,54).

Mono-ADP-ribosylation reactions are ubiquitous in nature, occurring in bacteria, viruses, plants, animals and humans. Such reactions are catalyzed by bacterial toxins and endogenous mammalian enzymes. All enzymes seem to work by transferring ADP-ribose from NAD^+ to individual amino acid residues like arginine, histidine, asparagine and cysteine. A large number of bacterial toxins, for example cholera toxin, pertussis toxin, diphteria toxin, pseudomonas toxin and botulinum toxins, use heterotrimeric G protein, low molecular weight G proteins, cytoplasmic actin or other specific proteins like the eukaryotic elongation factor 2 as natural substrates. Alterations brought about by bacterial toxin are well documented, because their pathological actions affect cell function, closely related to severe illness symptoms. Endogenous transferases have been characterized and isolated. However, detailed physiological functions still remain obscure (49).

COVALENT MODIFICATION OF GAPDH BY NITRIC OXIDE

Five years ago we showed that NO induces an NAD-dependent modification of GAPDH (55). In our initial studies we incubated human platelet cytosol with $[^{32}P]NAD^+$ and observed the labeling of a 39 kD protein. Incorporation of radioactivity from $[^{32}P]NAD^+$ increased dramatically in the presences of NO donors like SNP or SIN-1 (55). Protein modification withstood SDS gel electrophoresis (SDS-PAGE) performed under reducing conditions and occurred in a dose-dependent manner. Maximal labeling was reached after 90 to 120 min. Similar experiments with rat tissues (brain, liver, small intestine, heart and lung) in place of platelet cytosol resulted in the radioactive modification of the same protein. Although a few other proteins were modified as well, the 39 kD band was the most prominent one in SDS-PAGE. This protein showed the greatest stimulatory response toward NO donors when we employed cytosolic fractions of brain and heart. To determine whether NO generated from an active NOS would cause the radioactive labeling of the 39 kD protein, we used rat cerebellum cytosol fortified with all cofactors needed to continously stimulate the brain's constitutive NO-generating system (56). NO produced from the natural substrate L-arginine significantly stimulated incorporation of radioactivity into the 39 kD protein, while NOS inhibitors blocked protein labeling. Thus, an active NOS and several chemically distinct NO donors share the ability to cause similar post-translational protein modification of a 39 kD protein in the presence of $[^{32}P]NAD^+$.

An interesting, though unexpected, observation was that the reaction was independent of a cGMP response. Activation of soluble guanylyl cyclase by NO, endogenously produced or exogenously applied, was not involved. Moreover, cGMP or lipophilic cGMP analogs did not substitute for NO-liberating agents in stimulating this reaction. This NO action represents a new type of NO signaling. Since it was known from early observations that hemoglobin-constricted arteries inhibit nitrovasodilators and block the action of EDRF, we used hemoglobin for the purpose of capturing NO, thereby again linking the 39 kD protein modification to NO actions. Hemoglobin dose-dependently suppressed the NO-donor-caused modification to the 39 kD protein. We considered this further indication that NO directly drives this NAD^+-dependent reaction. Confirmation of our initial studies came from other groups working with brain tissue (57,58). Williams and coworkers observed the predominant labeling of a 39 kD protein in response to SNP and an active NOS in homogenates and cytosol from rat brain. Duman et al. reported the endogenous labeling of several proteins in homogenates of rat cerebral cortex, with SNP mainly stimulating the incorporation of radioactivity from $[^{32}P]NAD^+$ into a 36 kD protein band. Moreover, NO-releasing compounds stimulated the reversible ADP-ribosylation of a 41 kD cytosolic protein in *Dictyostelium discoideum* (59).

To reveal the identity of the 39 kD protein, we purified the protein from human platelet cytosol. The homogeneous protein was subjected to N-terminal sequence analysis. The first 29 amino acids revealed that the protein is identical to glyceraldehyde-3-phosphate dehydrogenase (GAPDH) (60). It was interesting to note that the modification of the homogeneous purified protein occurred in a time- and dose-dependent manner in the presence of NO donors only, indicative of an automodification process. A loss of enzyme activity is associated with GAPDH modification in the presence of NAD^+ and NO-releasing compounds. The identity of the 39 kD protein as GAPDH was established by two other groups using brain and erythrocyte cytosolic fractions, respectively (61,62).

Initial characterization of the reaction leading to increased incorporation of $[^{32}P]NAD^+$ radioactivity into GAPDH seemed consistent with an ADP-ribosylation mechanism (55,59,61-63). Support for this notion came from the following observations: (1) compared to NAD^+, free ADP-ribose is not a cofactor for this reaction; (2) snake venom phosphodiesterase cleavage of the radioactive protein releases only 5'-AMP; (3) labeling of GAPDH is observed using $[^{32}P]NAD^+$ or $[U\text{-}^{14}C]NAD^+$ but not when $[carbonyl\text{-}^{14}C]NAD^+$ is employed as a substrate (62,63). Although GAPDH is considered the predominant target of this NO-driven automodification in various cells and tissues, the exact mechanism behind the covalent modification with $[^{32}P]NAD^+$ is still a matter of debate. A recent study reported that NO-releasing compounds can stimulate some radiolabeling of GAPDH with $[nicotinamide\text{-}^{14}C]NAD^+$; this is consistent with the covalent binding of the whole molecule of NAD^+ (64). In contrast, other reports indicate that GAPDH as a major protein on the surface of group A streptococci transfers the ADP-ribose moiety of NAD^+ to free cysteine (65). The reaction product, resulting in a true thioglycosidic linkage, would indicate an ADP-ribosylation reaction catalyzed by GAPDH. Moreover, a close correlation between GAPDH ADP-ribosylation and enzyme inhibition is also suggested for the enzyme from *Dictyostelium discoideum* (66). Subsequent reports demonstrated that GAPDH functions as an NAD glycohydrolase generating free ADP-ribose, which is used to modifiy different amino acid residues, depending on assay conditions such as pH value and cofactor addition (66). Assay conditions may direct the transfer of ADP-ribose from NAD^+ to either cysteine residues or other amino acid residues. Moreover, incorporation of radioactivity from [^{32}P]ADP-ribose into GAPDH occurs, although in a NO-independent fashion (66,67).

GAPDH modification can be studied easily using purified enzyme or employing cytosolic fractions of cells. Although the reaction resembles known ADP-ribosylation reactions, conditions for optimal enzyme modification are different from those used in the pertussis toxin or cholera toxin labeling assays. ATP, thymidine, and high phosphate concentrations alone or in combination do not favor GAPDH modification, whereas NADPH drastically enhances the reaction carried out using human platelet cytosol (68). For the modification of GAPDH, we usually use a 100 mM Hepes buffer, pH 7.4, and 1 to 10 μM NAD^+ in a volume of 50 to 100 μl. Thiols and NO donors are added as required. Reactions are run for 30 to 60 min at 37°C, and then protein is precipitated with trichloroacetic acid. Pellets are washed two times with ether, and proteins are separated by 11% SDS-PAGE. Radioactivity in individual bands is detected with the phosphor imager system.

Nitric oxide required for the reaction can be applied exogenously through NO donors or can be endogenously delivered by an active constitutive or inducible NO synthase. Commercially available NO-releasing compounds like sodium nitroprusside (SNP), 3-morpholinosydnonimine (SIN-1; Cassella-Pharma, Frankfurt am Main, Germany), S-nitrosoglutathione (GSNO) or spermine-NO may be employed. Aqueous, pH-buffered fresh solutions may be used in a concentration range from 10 to 1000 μM. As mentioned, care should be taken to optimize conditions, for example, thiol should be added to encourage maximal release of NO from these compounds.

DTT (2.5 mM)	-	+	-	+	-	+
SNP (200 μM)	-	-	+	+	-	-
SIN-1 (200 μM)	-	-	-	-	+	+

Figure 1. Covalent Modification of GAPDH by Two Different NO-Donors.
GAPDH (10 μg/assay) was incubated at 37°C for 30 min. in 100 mM Hepes buffer, pH 7.5, containing 10 μM NAD^+ and 0.5 μCi[^{32}P]NAD^+ in a total volume of 100 μl, with dithiothreitol (DTT) and two NO-donors (concentration in parentheses as indicated). Preparation of samples and detection of radioactivity was essentially performed as described (63). The following NO donors were used: SNP, sodium nitroprusside; SIN-1, 3-morpholinosydnonimine.

Alternatively, a permanently active constitutive NO synthase may generate sufficient steady-state levels of NO to enhance radioactive modification of GAPDH. This has been achieved in rat cerebellum cytosol (56) and in rat brain with the addition of purified brain NO synthase (61). In this experimental setup, the addition of NO-synthase inhibitors blocked radioactive GAPDH labeling (56). In addition, similar experiments can be performed by activating the inducible NO synthase. Pancreatic ß-cells or the ß-cell line RINm5F respond to the cytokine IL-1 with an increased output of nitrite, stemming from the inducible form of the NO synthase. In cells treated with IL-1, GAPDH enzyme activity decreases to roughly 50% compared to controls, whereas addition of a NO synthase inhibitor restores enzymatic catalysis (69). Endogenous production of NO, besides affecting GAPDH enzyme activity, also prevents incorporation of [^{32}P]NAD^+ into GAPDH in the cytosolic fraction of IL-1-treated cells. In recent experiments with homogenates from stimulated hippocampus, SNP-induced ADP-ribosylation was reduced at preconvulsive stage II and stage V (tonic-clonic seizures) of dentate gyrus kindling, compared with controls (70). As suggested, this probably reflects enhanced endogenous ADP-ribosylation, depending on the progressive activation of the NMDA receptors and the generation of NO. Influence of GAPDH enzyme activity and ADP-ribosylation as a consequece of exogenous application of SNP has been reported for *Dictyostelium discoideum* as well (66). In an advanced system, we probed for GAPDH alterations under *in vivo* conditions. Mice were injected with bacterial endotoxin known to generate NO by an active cytokine inducible NO synthase. GAPDH enzyme activity was significantly reduced in cytosolic fractions of heart and spleen, and incorporation of radioactivity from [^{32}P]NAD$^+$ into the protein in the cytosol was similarly decreased (71). Exposure of cytosolic fractions to [^{32}P]NAD$^+$, which results in less incorporation of radioactivity into GAPDH of treated animal compared to controls, established the association of GAPDH modification, *in vivo*, with enhanced NO production.

Nitrogen monoxide (NO$^\bullet$) is extremely susceptible to both oxidation and reduction, which produce nitrosonium (NO$^+$) or nitroxyl anion (NO$^-$), respectively (12). Moreover, secondary reactions may led to the interaction of NO with thiol groups, generating S-nitrosothiols, or result in the formation of peroxynitrite (ONOO$^-$), the reaction product of NO and superoxide (O_2^-). During our studies we observed that the efficacy of various NO donors at inducing GAPDH modification seemed independent of rates of NO$^\bullet$ release. For example, upon decomposition, spermine releases NO with a half-life of roughly 40 min while producing moderate GAPDH modification. In contrast, SNP releases no detectable NO$^\bullet$ but always causes efficient GAPDH labeling. More interesting, SIN-1 releases very little or no NO$^\bullet$ but gives rise to very efficient GAPDH modulation. Overall, results from these studies seem to be compatible with a mechanism involving NO$^+$ inducing S-nitrosylation of active-site thiol. We further probed the chemical mechanism of post-translational modification of GAPDH using nitrosonium tetrafluoroborate (BF_4NO) (72). BF_4NO is unique because it hydrolyzes under controlled conditions in aqueous solution, producing NO$^+$ only. However, in contrast to SNP, where thiol addition promotes release of NO and GAPDH modification, application of thiol (DTT) together with BF_4NO significantly attenuates GAPDH labeling. Thus, the competing S-nitrosating reaction with DTT, present in high excess over enzyme, predictably hinders the reaction with the active-site thiol of the protein. These data implicate NO$^+$ as the species transferred to the active-site thiol in the mechanism of covalent modification of the enzyme. Our findings are in keeping with existing indirect evidence that protein S-nitrosylation precedes, and thereby initiates, subsequent covalent [^{32}P]NAD$^+$-dependent protein modification (73). Figure 2 shows a scheme of GAPDH modification induced by NO$^+$.

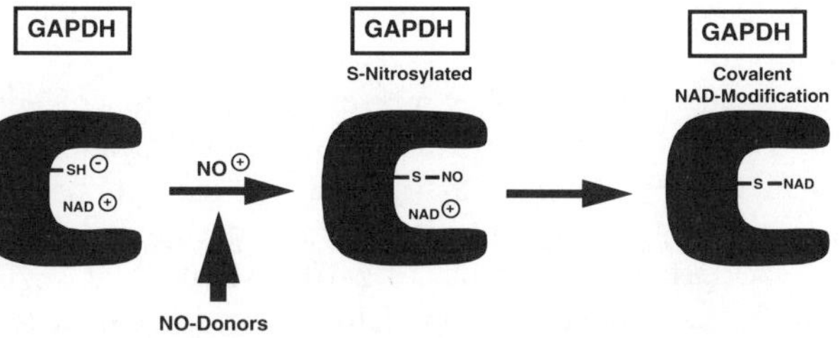

Figure 2. Schematic Modification of GAPDH.
Modification of GAPDH by NO-donors with the proposed sequence of S-nitrosylation, followed by the post-translational, covalent modification in the presenece of NAD$^+$. As indicated, a nitrosonium (NO$^+$) ion is required for S-nitrosylation of GAPDH.

Studies with the NO donor SIN-1 have led to conflicting results regarding the mechanism of GAPDH modification. SIN-1 generates NO$^\bullet$ and superoxide stoichiometrically and at the same time. The reaction between these two species results in the formation of peroxynitrite. Concluding that $ONOO^-$ mediates the action of SIN-1, we examined the effect of peroxynitrite in greater detail. $ONOO^-$ causes radioactive labeling of GAPDH (72). Notably, the action of peroxynitrite is critically dependent on the presence of thiol. With thiols (DTT) in relative excess over peroxynitrite, conditions are conducive to formation of RS-NO. Therefore, the likely scenario employing SIN-1 or peroxynitrite directly probably involves S-nitrosation of DTT followed by NO^+ donation to the reactive thiol of GAPDH. In contrast, when peroxynitrite is present in relative excess to thiols (DTT), it oxidizes protein thiols to sulfenic or sulfonic acids (74), thereby inhibiting subsequent covalent modification of GAPDH by $[^{32}P]NAD^+$. The unifying concept by which nitric oxide and related NO donors initiate non-enzymatic ADP-ribosylation (like) reactions thus favors a role for nitrosonium (NO^+) or nitronium (NO_2^+)-like species and emphasizes the role of NO group transfer chemistry in the initial step of this GAPDH modification.

PARTICIPATION OF THIOLS IN NO-MEDIATED GAPDH MODIFICATION

Free, low molecular weight thiols and protein thiol groups are critical factors influencing NO-mediated GAPDH modification (75). Thiols influence basal and NO-stimulated labeling reactions, they facilitate the release of NO from some NO donors, they scavenge NO by means of S-nitrosothiol formation, and they affect the redox chemistry of peroxynitrite and critical GAPDH thiol groups. For GAPDH modification induced by SNP, thiols are ultimately required because they promote the one-electron-driven release of NO from the NO donor. In contrast, when BF_4NO is used, thiols blunt the reaction: NO^+ immediately reacts with thiols, leading to S-nitrosothiol formation, thereby scavenging the driving redox species for GAPDH-modification (72). In the case of peroxynitrite, thiols are required to supress the oxidation of critical protein thiol groups. Reduced protein thiol groups, especially cysteine 149 of GAPDH, are a prerequisite for any modification leading to enzyme inhibition. Concentrations of 1 to 4 mM thiol (DTT being preferred over reduced glutathione or cysteine) suppress basal GAPDH modification carried out in the absence of any NO-releasing compound (64,66,75). In contrast, the NO-stimulated reaction is dramatically enhanced by thiols, again with DTT being most effective (75,76). However, DTT concentrations of 5 mM or higher reverse the stimulatory effect of NO-releasing compounds in modifying GAPDH (64). Explanations may be sought in S-nitrosothiol formation, a reaction eventually competing for NO, or, alternatively, one can speculate about the reversibility of the reaction in the presence of high thiol concentrations.

Protein thiol groups also are key players during the covalent modification of GAPDH induced by NO and NAD^+. Post-translational modification of GAPDH involves thiol groups, as indicated by the fact that radiolabel incorporated into the protein is sensitive to $HgCl_2$ cleavage (63,64). Cleavage experiments are carried out with $[^{32}P]NAD^+$-labeled GAPDH in the presence of 2-10 mM $HgCl_2$ at 37°C for up to 60 min. This treatment results in the removal of more than 90% of the radioactivity associated with GAPDH. The sensitivity toward $HgCl_2$, compared to acid or neutral hydroxylamine, is indicative of a linkage with a sulfhydryl group. However, ADP-ribosylated cysteine residues modified by pertussis toxin seem much more labile toward $HgCl_2$ compared to the situation seen with GAPDH. The requirement of reduced protein thiol groups for NO-induced GAPDH modification is further substantiated by employing thiol-blocking agents like dithionitrobenzene or N-ethylmaleimide (75). Any modification of the thiol group prior to NO-induced covalent GAPDH modification efficiently hinders incorporation of radioactivity later on (63).

Compatible with these results are observations made when the sesquiterpene antibiotic koningic acid, which binds to GAPDH at the Cys-149 of the active site and thereby inhibits NO-stimulated GAPDH modification, is used (77). Although other dehydrogenases share with GAPDH the active-site thiol group and the NAD^+ binding site, they do not respond to NO-releasing compounds with covalent enzyme modification (61,78). According to our experiments, GAPDH has a remarkably more sensitive critical thiol group than several other enzymes. Thus, GAPDH is an exceptional target for this NO-mediated covalent protein modification mechanism. The reactive thiol group, probably Cys-149 in the active site of the enzyme, is important for enzymatic activity during glycolysis as well as for NO-induced enzyme modification. Moreover, the thiol group is known for its extreme sensitivity toward all kinds of oxidants as well as autoxidation. Oxidation of this group with organic or inorganic peroxides or generally oxidative stress conditions may lead to the generation of intramolecular disulfides, sulfenic acids, or oxidant-specific S-thiolation (79,80). Oxidative modification of GAPDH is associated with inhibition of enzyme catalysis during glycolysis, although in most cases removing the oxidant largely restores enzyme activity.

The Cofactor NAD^+

Incorporation of radioactivity from $[^{32}P]NAD^+$, in the presence of increased levels of NO, into a 39 kD protein, later identified as GAPDH, has been extensively reported. Although initially addressed as an ADP-ribosylation reaction other observations assume that NO enhances the binding of one molecule of $[^{32}P]NAD^+$, instead of ADP-ribose only. In contrast to the proposed NAD^+ binding mechanism, the transfer of ADP-ribose from NAD^+ to free cysteine in the presence of GAPDH (65) suggests that enzymatic activity is shared by ADP-ribosylating reactions. Moreover, time kinetic studies revealed an ideal correlation between ADP-ribosylation and enzyme inhibition (66). Although different assay conditions may explain variations in GAPDH modification, eventually allowing for more than one mechanism to operate during this post-translational enzyme modification, there is a general agreement that NO or a related redox species drives this reaction. Moreover, there are consistent reports (60,64,66) that the presence of NO alone is not sufficient to inhibit GAPDH enzyme activity; NAD^+ or a NAD^+-like cofactor appear to be an ultimate requirement. Thus, the underlying mechanism for the inhibition of GAPDH is not likely to be a simple reaction of NO with protein thiol groups and concomitant formation of an S-nitrosothiol. Interestingly, a bell-shaped curve has been reported for the concentration of NAD^+ required to efficiently modify the protein (66). Optimal concentrations of roughly 10 μM NAD^+ are required to cause maximal ADP-ribosylation. This observation may help to explain difficulties experienced during examination of potent inhibition in the presence of NO (more than 80% inhibition of enzymatic activity compared to controls), compared to much lower incorporation of radioactivity into the protein under identical conditions. The amount of radioactivity incorporated into GAPDH, calculated on a molar basis, ranges from 1% up to 40% (63,64,66,81). Moreover, for stoichiometric considerations, one has to take into account the portion of NAD^+ tightly bound to the enzyme. If a large portion of cold NAD^+ in the presence of NO reacts with GAPDH, the amount of $[^{32}P]NAD^+$ bound to the enzyme is underestimated when radioactivity is measured. This is also complicated by the fact that the number of GAPDH binding sites already modified is unknown. As detected by polyethyleneimine TLC analysis, cleavage of the electroeluted, radioactive $[^{32}P]NAD^+$-modified protein with snake venom phosphodiesterase releases 5′ AMP (55). These results, obtained from a procedure that discriminates between mono- and poly-ADP-ribosylation, suggest that modification of GAPDH involves monomers of ADP-ribose only, irrespective of the transfer mechanism.

NAD^+-dependent, NO-stimulated reactions have been reported in several biological signaling pathways. In human neutrophils, actin becomes ADP-ribosylated when $[^{32}P]NAD^+$ is used in the presence of NO (82), whereas in an NG108-15 (mouse neuroblastoma x rat glioma somatic hybrid cell) cell homogenate (83), NO promotes the NAD^+ labeling of poly-ADP-ribose polymerase. In brain, NO-stimulated endogenous ADP-ribosylation has been linked to long-term potentiation (84), and changes in the ADP-ribosylation status of some hippocampal proteins have been linked to kindling progression (70).

CYTOTOXICITY OF NO

Nitric oxide is a patho-physiological mediator, and, as a potentially toxic molecule, NO can cause cell death. Some cells like neuronal cells and pancreatic islets are highly susceptible to endogenously formed or chemically generated NO (38,43,85,86). Mechanistically, NO may cause DNA-deaminating reactions (87) or may cause toxicity, at least in part, by reaction with superoxide anion, apparently leading to the formation of peroxynitrite ($ONOO^-$). Classically, cell death is believed to occur by one of two mechanisms--necrosis or apoptosis. Whereas apoptosis is characterized by distinct morphology and precise biochemical markers--i.e., chromatin condensation and DNA fragmentation into oligonucleosomal fragments of 180 bp and multiples of this, known as the "DNA ladder", necrosis is considered an accidental cell death leading to cell swelling, membrane rupture, and randomly digested DNA. For at least some cells, such as RINm5F cells and macrophages, NO-mediated cell death occurs by apoptosis (88,89; Messmer, Lapetina and Brüne, unpublished data). Moreover, the expression of the tumor supressor gene p53 seems to be involved (our observations). Furthermore, NO activates poly-ADP--ribosyltransferase (39), a nuclear enzyme that becomes activated following breakage of the DNA strand. The activation of this enzyme may play a role in NO-induced cell death because it has been linked to DNA damage followed by cellular NAD^+ and ATP depletion. This might have important implications for NO-mediated GAPDH modification as well. NO-induced DNA damage consumes NAD^+ during the poly-ADP-ribosylation reaction, thereby disturbing the intracellular concentration and redox balance of its nicotinamide cofactors. As a result, oxidative stress or enhanced NO generation may inactivate glycolysis, affecting the enzymatic activity of GAPDH by independent mechanisms: (i) direct inactivation of the enzyme by an NAD^+-dependent mechanism, (ii) oxidative inhibition, and (iii) reduction of the intracellular concentration of NAD^+. The net effect of these factors reduces the theoretical intracellular activity of GAPDH. In parallel with the described inhibition of mitochondrial activity, several alterations would diminish the amount of ATP formed, thus decreasing the overall energy supply.

CONCULSIONS/SUMMARY

Nitric oxide signaling is achieved through cGMP-dependent and -independent mechanisms. The latter are exemplified by the NAD^+-dependent automodification of the glycolytic enzyme glyceraldehyde-3-phosphate dehydrogenase (GAPDH). The experimental post-translational, covalent modification of the enzyme by $[^{32}P]NAD^+$ is achieved using NO-releasing compounds and an active constitutive or inducible NO-synthase. Potential roles for NO in this covalent enzyme modification can be grouped as follows:

- S-Nitrosylation of GAPDH by NO^+
- NAD^+-dependent, post-translational covalent automodification of GAPDH
- Oxidative modification of GAPDH by NO-related compounds, probably $ONOO^-$

GAPDH modification by one of the proposed mechanisms would lead to inhibition of enzyme catalysis. It is likely that the NAD^+-dependent automodification process occurs *in vitro*, in intact cells, and in whole animals.

Besides its normal function in glycolysis, GAPDH not only is a target for NO-mediated direct and indirect modifications but also is ADP-ribosylated in the presence of brefeldin A (90). The relation of such ADP-ribosylation to enzyme activity is so far unknown. GAPDH also may be involved in one of the following functions unrelated to its glycolytic activity (81 and refs. therein; 90):

- binding and transport of tRNA associated with nuclear localization of GAPDH
- DNA-repair activity, i.e., uracil DNA glycosylase
- activation of transcription in neurons
- interaction with tubulin and microtubules
- the transport of nitric oxide
- serves as a substrate for brefeldin A stimulated ADP-ribosylation

Because some of these alternative functions of GAPDH, just like NO-mediated modification of the enzyme, are related to the NAD^+ binding site of the protein, we are interested in searching for the significance of these activities in relation to NO actions.

In recent years, several functions of NO have been linked to direct, cGMP-independent actions. Modification of GAPDH is probably just one interesting target related to NO-redox chemistry and active-site thiol modification. It will be challenging to investigate NO biochemistry in closer detail and to elucidate how NO targets biological systems, especially in relation to the patho-physiological role of NO in medically related conditions.

REFERENCES

1 Schultz, K-D., Schultz, K. and Schultz, G. (1977) Nature 265, 750-751.
2 Arnold, W.P., Mittal, C.K, Katsuki, S. and Murad, F. (1977) Proc. Nat. Acad. Sci. U.S.A. 74, 3203-3207.
3 Waldman, S.A. and Murad, F. (1987) Pharmacol Rev. 39, 163-196.
4 Tremblay, J., Gerzer, R. and Hamet, P. (1988) Adv. Sec. Messenger and Phosphoprot. Res. 22, 319-383.
5 Schmidt, H.H.H.W. (1992) FEBS Lett. 307, 102-107.
6 Walter, U. (1989) Rev. Physiol. Biochem. Pharmacol. 113, 41-88.
7 Yau, K-W. (1994) Proc. Nat. Acad. Sci. U.S.A. 91, 3481-3483.
8 Reinhard, M., Halbrügge, M., Scheer, U., Wiegand, C., Jockusch, B.M. and Walter, U. (1992) EMBO J. 11, 2063-2070.
9 Bennett, B.M., McDonald, B.J., Nigam, R. and Simon, W.C. (1994) TIPS 15, 245-249.
10 Noack, E. and Murphy, M. (1991) in Oxidative Sress: Oxidants and Antioxidants (H. Sies, ed.) pp. 445-489, Academic Press, San Diego, CA.
11 Bates, J.N., Baker, M.T., Guerra, R. Jr. and Harrison, D.G. (1991) Biochem. Pharmacol 42, S157-S165.
12 Stamler, J.S., Singel, D.J. and Loscalzo, J. (1992) Science 258, 1898-1902.
13 Mitchell, H.H., Shonle, H.A. and Grindley, H.S. (1916) J. Biol. Chem. 24, 461-490.
14 Tannenbaum, S.R., Fett, D., Young V.R., Land, P.D. and Bruce, W.R. (1978) Science 200, 1487-1489.

15 Green, L.C., Tannenbaum, S.R. and Goldman, P. (1981) Science 212, 56-58.
16 Stuehr, D.J. and Marletta, M.A. (1985) Proc. Nat. Acad. Sci. U.S.A. 82, 7738-7742.
17 Hibbs, J.B. Jr., Taintor, R.R. and Vavrin, A. (1987) Science 235, 473-476.
18 Furchgott, R.F. and Zawadzki, J.V. (1980) Nature 288, 373-376.
19 Palmer, R.M.J., Ferrige, A.G. and Moncada, S. (1987) Nature 327, 524-526.
20 Palmer, R.M.J., Ashton, D.S. and Moncada, S. (1988) Nature 333, 664-666.
21 Ignarro, L.J. (1990) Annu. Rev. Pharmacol. Toxicol. 30, 535-560.
22 Feelisch, M., te Poel, M., Zamora, R., Deussen, A. and Moncada, S. (1994) Nature 368, 62-65.
23 Moncada, S., Palmer, R.M.J. and Higgs, E.A. (1991) Pharmacol. Rev. 43, 109-142.
24 Nathan, C. (1992) FASEB J. 6, 3051-3064.
25 Snyder, S.H. (1992) Science 257, 494-496.
26 Nussler, A.K. and Billiar, T.R. (1993) J. Leukocyte Biol. 54, 171-178.
27 Marletta, M.A. (1993) J. Biol. Chem. 268, 12231-12234.
28 Knowles, R.G. and Moncada, S. (1994) Biochem. J. 298, 249-258.
29 Nathan, C. and Xie, Q. (1994) Biochem. J. 298, 249-258.
30 Garthwaite, J., Charles, S.L. and Chess-Williams, R. (1988) Nature 336, 385-388.
31 Garthwaite, J. (1991) TINS 14, 60-67.
32 Bredt, D.S. and Snyder, S.H. (1992) Neuron 8, 3-11.
33 Garthwaite, J. (1993) The Neurosciences 5, 171-180.
34 Shibuki, K. (1993) Semin. Neurosci. 5, 217-223.
35 Zorumski, C.F. and Izumi, Y. (1993) Biochem. Pharmacol. 46, 777-785.
36 Lowenstein, C.J. and Snyder, S.H. (1992) Cell 70, 705-707.
37 Choi, D.W. (1993) Proc. Nat. Acad. Sci. U.S.A. 90, 9741-9743.
38 Dawson, V.L., Dawson, T.M., Bartley, D.A, Uhl, G.R. and Snyder, S.H. (1993) J. Neuroscience 13, 2651-2661.
39 Zhang, J., Dawson, V.L., Dawson, T.M. and Snyder, S.H. (1994) Science 263, 687-689.
40 Reddy, D., Lancaster, J.R. Jr. and Cornforth, D.P. (1983) Science 221, 769-770.
41 Moncada, S. (1992) Acta Physiol. Scand. 145, 201-227.
42 Henry, Y., Lepoivre, M., Drapier, J-C., Ducrocq, C., Boucher, J-L. and Guissani, A. (1993) FASEB J. 7, 1124-1134.
43 Kröncke, K.D., Brenner, H.H., Rodriguez, M.L., Etzkorn, K., Noack, E.A., Kolb, H. and Kolb-Bachofen, V. (1993) Biochim. Biophys. Acta 1182, 221-229.
44 Fehsel, K., Jalowy, A., Qi, S., Burkart, V., Hartmann, B. and Kolb, H. (1993) Diabetes 42, 496-500.
45 Corbett, J.A. and McDaniel, M.L. (1992) Diabetes 41, 897-903.
46 Eizirik, D.L., Sandler, S. and Palmer, J.P. (1993) Diabetes 42, 1383-1391.
47 Schulz, R., Nava, E. and Moncada, S. (1992) Brit. J. Pharmacol. 105, 575-580.
48 Everse, J., Anderson, B.M. and You, K.-S, eds. (1982) The Pyridine Nucleotide Coenzymes. Academic Press, New York, NY.
49 Jacobson, M.K. and Jackson, E.L., eds. (1989) ADP-Ribose Transfer Reactions: Mechanisms and Biological Significance. Springer-Verlag, New York, NY.
50 Berridge, M.J. (1993) Nature 365, 388-389.
51 Hilz, H., Koch, R., Fanick, W., Klapproth, K. and Adamietz, P. (1984) Proc. Nat. Acad. Sci. U.S.A. 81, 3929-3933.
52 Richter, C. and Frei, B. (1988) Free Rad. Biol. Med. 4, 365-375.
53 Just, I., Wollenberg, P., Moss, J. and Aktories, K. (1994) Eur. J. Biochem. 221, 1047-1054.
54 Althaus, F.R. and Richter, C., eds. (1987) ADP-Ribosylation of Proteins: Enzymology and Biological Significance. Springer-Verlag, Berlin, Germany.

55 Brüne, B. and Lapetina, E.G. (1989) J. Biol. Chem. 264, 8455-8458.
56 Dimmeler, S. and Brüne, B. (1991) Biochem. Biophys. Res. Commun. 178, 848-855.
57 Duman, R.S., Terwilliger, R.Z. and Nestler, E.J. (1991) J. Neurochem. 57, 2124-2132.
58 Williams, M.B., Li, X., Gu, X. and Jope, R.S. (1992) Brain Res. 592, 49-52.
59 Tao, Y., Howlett, A. and Klein, C. (1992) Proc. Nat. Acad. Sci. U.S.A. 89, 5902-5906.
60 Dimmeler, S., Lottspeich, F. and Brüne, B. (1992) J. Biol. Chem. 267, 16771-16774.
61 Zhang, J. and Snyder, S.H. (1992) Proc. Nat. Acad. Sci. U.S.A. 89, 9382-9385.
62 Kots, A.Y., Skurat, A.V., Sergienko, E.A., Bulargina, T.V. and Severin, E.S. (1992) FEBS Lett. 300, 9-12.
63 Dimmeler, S. and Brüne, B. (1992) Eur. J. Biochem. 210, 305-310.
64 McDonald, L.J. and Moss, J. (1993) Proc. Nat. Acad. Sci. U.S.A. 90, 6238-6241.
65 Pancholi, V. and Fischetti, V.A. (1993) Proc. Nat. Acad. Sci. U.S.A. 90, 8154-8158.
66 Tao, Y., Howlett, A. and Klein, C. (1994) Eur. J. Biochem. 224, 447-454.
67 Zocchi, E., Guida, L., Franco, L., Silvestro, L., Guerrini, M., Benatti, U. and de Flora, A. (1993) Biochem. J. 295, 121-130.
68 Brüne, B., Dimmeler, S. and Lapetina, E.G. (1992) Biochem. Biophys. Res. Commun. 182, 1166-1171.
69 Dimmeler, S., Ankarcrona, M., Nicotera, P. and Brüne, B. (1993) J. Immunol. 150, 2964-2971.
70 Vezzani, A., Sparvoli, S. Rizzi, M. Zinetti, M. and Fratelli, M. (1994) Neuro. Report 5, 1217-1220.
71 Dimmeler, S., Messmer, U.K., Tiegs, G. and Brüne, B. (1994) Eur. J. Pharmacol. 267, 105-112.
72 Mohr, S., Stamler, J.S. and Brüne, B. (1994) FEBS Lett. 348, 223-227.
73 Molina y Vedia, L., McDonald, B., Reep, B. Brüne, B., DiSilvio, M., Billiar, T.R. and Lapetina, E.G. (1992) J. Biol. Chem. 267, 24929-24932.
74 Radi, R., Beckman, J.S., Bush, K.M. and Freeman, B.A. (1991) J. Biol. Chem. 266, 4244-4250.
75 Brüne, B. and Lapetina, E.G. (1990) Arch. Biochem. Biophys. 279, 286-290.
76 Vaidyanathan, V.V., Sastry, P.S. and Ramasarma, T. (1993) Biochim. Biophys. Acta 1203, 36-44.
77 McDonald, B., Reep, B., Lapetina, E.G. and Molina y Vedia, L. (1993) Proc. Nat. Acad. Sci. U.S.A. 90, 11122-11126.
78 Dimmeler, S. and Brüne, B. (1993) FEBS Lett. 315, 21-24.
79 Claiborne, A.L., Miller, H., Parsonage, D. and Ross. R.P. (1993) FASEB J. 7, 1483-1490.
80 Schuppe-Koistinen, I., Moldeus, P., Bergman, T. and Cotgrave, I.A. (1994) Eur. J. Biochem. 221, 1033-1037.
81 Brüne, B., Dimmeler, S., Molina y Vedia, L. and Lapetina, E.G. (1994) Life Sciences 54, 61-70.
82 Clancy, R.M., Leszczynska-Piziak, J. and Abramson, S.B. (1993) Biochem. Biophys. Res. Commun. 191, 847-852.
83 Boyd, R.S., Donnelly, L.E., Allport, J.R. and MacDermot, J. (1993) Biochem. Biophys. Res. Commun. 197, 1277-1282.
84 Duman, R.S., Terwilliger, R.Z. and Nestler, E.J. (1993) J. Neurochem. 61, 1542-1545.
85 Delaney, C.A., Green, M.H.L., Lowe, J.E. and Green, I.C. (1993) FEBS Lett. 333, 291-295.

86 Lipton, S.A., Choi, Y.-B., Pan, Z.-H., Lei, S.Z., Chen, H.-S.V., Sucher, N.J., Loscalzo, J., Singel, D.J. and Stamler, J.S. (1993) Nature 364, 626-632.
87 Wink, D.A., Kasprzak, K.S., Maragos, C.M. Elespuru, R.K., Misra, M., Dunams, T.M., Cebula, T.A., Koch, W.H., Andrews, A.W., Allen, J.S. and Keefer, L.K. (1991) Science 254, 1001-1003.
88 Albina, J.E., Cui, S., Mateo, R.B. and Reichner, J.S. (1993) J. Immunol. 150, 5080-5085.
89 Ankarcrona, M., Dypbukt, J.M., Brüne, B. and Nicotera, P. (1994) Exp. Cell Res. 213, 172-177.
90 De Matteis, M.A., Di Girolamo, M., Colanzi, A., Pallas, M., Di Tullio, G., McDonald, L.J., Moss, J., Santini, G., Bannykh, S. Corda, D. and Luini, A. (1994) Proc. Nat. Acad. Sci. U.S.A. 91, 1114-1118.

THE GENETICS OF NUCLEAR MIGRATION IN FUNGI

Susan M. Beckwith, Christian H. Roghi and N. Ronald Morris

Department of Pharmacology
UMDNJ-Robert Wood Johnson Medical School
Piscataway, NJ 08854

INTRODUCTION

Nuclear migration occurs in a wide range of processes including fertilization, spindle localization, embryogenesis, brain development, maintenance of acetylcholine receptors in neuromuscular junctions, virus infection and cancer cell migration. Genetic analyses from fungal systems illuminate the complexity of the nuclear migration process. The cytoskeleton, motor molecules, cell wall, spindle pole bodies, nucleoporins and heterotrimeric G proteins have all been implicated in nuclear movement.

SIGNIFICANCE OF NUCLEAR MIGRATION IN EUKARYOTES

During fertilization, the egg and sperm pronuclei migrate toward each other. This movement is followed by a migration of the pronuclei to the center of the egg where their nuclear membranes fuse to achieve syngamy. Without nuclear migration, the egg is doomed and has no chance to develop into a new individual (reviewed in 1). Migration of nuclei is also necessary for mating (karyogamy) in lower eukaryotes such as yeast (reviewed in 2).

The positioning of the nucleus determines the location of the mitotic spindle which specifies the cleavage plane of the cell. Migration of the nucleus to an eccentric position within a cell results in the formation of unequal daughter cells, an essential aspect of normal eukaryotic development (3,4, reviewed in 5). The asymmetric distribution of the nucleus in neuroepithelial cells causes the cells to become wedge shaped, a requirement for the proper bending of the brain epithelium during development (6,7).

Nuclear migration is also a distinct step in cell migration. During central nervous system development, neuronal cells migrate to their appropriate position in the brain. Studies of migrating neuroblasts *in vitro* and *in situ* demonstrate that neuronal migration is achieved by the extension of a leading process. The nucleus then migrates through the extended process (8-10). Some brain disorders may result from defects in neuronal migration such as lissencephaly (smooth brain; reviewed in 11) and perhaps some cases of schizophrenia (reviewed in 12). Nuclear movement may also be required for tumor cell migration. Human lung adenocarcinoma cells extend a long process, the nucleus migrates through the extension, and the cell body then retracts behind the nucleus (13). The formation of extended processes may be a first step of tumor cell invasion *in vivo* (14,15).

There are many examples of nuclear migration in multinucleate cells as well. The developing egg of *Drosophila melanogaster* is a multinucleate syncytium. The nuclei migrate from the center of the egg to the periphery where they become cellularized to form the embryo (16). In multinucleated muscle cells, a subset of migrating nuclei actively cluster beneath acetylcholine receptors at neuromuscular junctions, thus providing a source of newly synthesized acetylcholine receptors to replace the degraded ones (17). When tissue culture cells are infected with the parainfluenza virus, these cells fuse to form a multinucleated syncytium. The nuclei migrate in the cytoplasm and align in tightly packed rows (18,19). In summary, nuclear migration is crucial from the beginning of life through the continued growth and development of an organism. Ironically, nuclear migration may also be involved in processes that end life, such as metastasis of cancer cells.

COMPONENTS OF NUCLEAR MIGRATION

Despite the importance of nuclear motility, little is known about the mechanism and regulation of nuclear migration. This process appears to involve 1) motor molecules to power the nuclear movement, 2) structural elements on which the motor and nucleus travel, 3) components to attach the nucleus to the motor and 4) molecules that initiate and regulate nuclear migration.

FUNGI AS MODEL SYSTEMS FOR GENETIC STUDIES OF NUCLEAR MIGRATION

Cytological, pharmacological, immunological and biochemical studies have implicated all three components of the cytoskeleton - microtubules, actin and intermediate filaments as participants in nuclear migration. In addition, inhibitors of the microtubule motor protein, dynein, and the regulatory molecule, protein kinase C, prevent nuclear migration. Fungal genetics has played an important role by confirming the involvement of the above mentioned proteins in nuclear migration. Genetics also opens the way to identify additional components of this complex process and to test the function of postulated components in the living cell (reviewed in 20-22).

One powerful way to study the function of proteins involved in nuclear migration is by mutational analysis. One can also study the effect of overexpression or depletion of the gene product *in vivo* (reviewed in 22). Functional interactions between two proteins can be inferred through the identification of extragenic suppressors, secondary mutations that reverse the mutant phenotype. However direct proof always requires substantiation through biochemistry (22). A combination of genetics and biochemistry should lead to the identification of most of the components of the nuclear migration process, establish their function *in vivo*, and define protein-protein interactions among the components (reviewed in 22).

The first step in genetic analysis is the isolation of mutants (21) that are defective in nuclear migration. Mutations provide the surest way to connect the function of proteins *in vivo* with their activities *in vitro* (22). The higher the eukaryote, the more difficult are genetic studies. Thus, the use of mutants from lower eukaryotes to investigate cellular processes has become a fruitful and well-known technique (20-22). The fungi *Aspergillus nidulans*, *Neurospora crassa*, *Saccharomyces cerevisiae* and *Schizosaccharomyces pombe* are helping to unravel the mystery of nuclear migration.

FILAMENTOUS FUNGI

Aspergillus nidulans

A. nidulans is one of the best developed genetic systems for studies of nuclear movement. Upon germination of conidia (uninucleate asexual spores), the nucleus in the spore divides and the daughter nuclei move into the developing germ tube. The cell becomes multinucleate as the nuclei continue to undergo multiple rounds of division (Figure 1A). A uniform distribution of nuclei along the growing hypha is needed for proper growth and development (23). Multiple temperature-sensitive mutants have been isolated in *A. nidulans* that affect nuclear migration.

The "*nud*" (nuclear distribution) mutants are unable to transport nuclei into the growing mycelium at restrictive temperature (23-25). The *nud* phenotype is dramatic: multiple nuclei are present in the conidial spore end of the germ tube followed by a long tube devoid of nuclei (Figure 1B). The *nud* mutants demonstrate that there is an active mechanism for nuclear migration in *A. nidulans* (23). The *nud* mutants are phenotypically similar, but genetic linkage and complementation analyses reveal the presence of at least five genes: *nudA*, *nudC*, *nudF*, *nudG* and *nudH*. Three of the *nud* genes have been cloned and sequenced (see below).

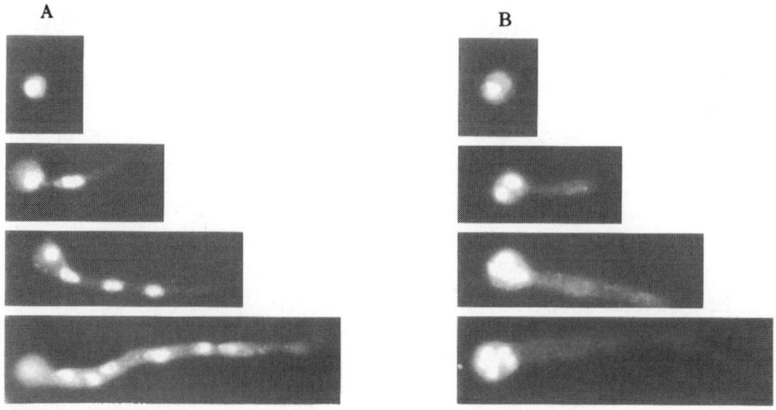

Figure 1. The effect of *nud* mutations on nuclear migration in *Aspergillus nidulans*. (A) Wild-type and (B) *nud* phenotypes during a time course incubation at restrictive temperature. Growth of conidia (asexual spores) was done at 42°C. The cells were fixed and nuclei were visualized by staining with the DNA-specific dye, DAPI.

Other mutants of *A. nidulans* are partially blocked in conidiation (asexual sporulation) because the nuclei do not migrate into certain conidiophore structures, called the primary sterigmata. These mutants, called *apsA* and *apsB* for anucleate primary sterigmata, also exhibit an irregular distribution of nuclei in the vegetative hyphae (26,27; Fischer and Timberlake, J. Cell Biol., in press). However, unlike the *nud* mutants, the nuclei are able to move out of the spore upon germination. The *nudA*, *nudC*, *nudF* and *apsA* mutants will be discussed below as will the *benA33* and *mipA1* tubulin mutants which also affect nuclear migration.

Neurospora crassa

As in *A. nidulans*, the nuclei are more or less evenly distributed in the growing hyphae of *N. crassa*. Eight genes have been identified that cause an asymmetric distribution of nuclei in which nuclear distribution is highly asymmetric. In these mutants, hyphal tips are curled and distorted. These nuclear migration mutants have been termed *ro* mutants based on their ropy appearance (28,29). Only the sequences of the RO1 and RO4 genes have been published. These will be discussed in detail below. Homologues of some of the NUD genes and RO genes have been found in higher eukaryotes. Thus, the results of nuclear migration studies in filamentous fungi may be broadly applicable to eukaryotes in general.

YEAST

Saccharomyces cerevisiae

Nuclear migration during cell division and conjugation (mating) are well characterized cellular processes in *S. cerevisiae*. During conjugation, the nuclei from two parents migrate toward each other and fuse to form a zygote with a single diploid nucleus (reviewed in 30). This process of nuclear fusion, karyogamy, is similar to fertilization in higher eukaryotes. Cell division in *S. cerevisiae* occurs by budding (reviewed in 2,22,31; Figure 2). A small bud emerges during the beginning of the cell cycle and grows during the cell cycle. It remains attached to the mother cell by a short neck. Near the end of DNA synthesis, the late G_2 nucleus migrates from the center of the mother cell to the bud neck. Upon completion of DNA synthesis, the nucleus elongates symmetrically into the mother and the daughter bud and then divides. The migration of the nucleus to the bud neck positions the spindle properly: upon elongation of the spindle, a complete set of chromosomes will move into the mother cell and the daughter cell. The cell cycle is completed following cytokinesis and cell separation.

Yeast geneticists have isolated numerous mutants which demonstrate defects in nuclear migration during karyogamy and/or cell division. The following *S. cerevisiae* mutants will be discussed below: *act1-4*, *act3*, *act5*, *bfr1-1*, *bik1*, *cdc20*, *chs2*, *cin1*, *cin2*, *cin4*, *dhc1*, *dyn1*, *jnm1*, *kar3-1*, *kar4*, *kar9*, *mdm1*, *myo1*, *num1*, *nup1*, *sis1*, *spa1-1*, *tcp1-1*, *tub2-401* and *ypt1*.

Schizosaccharomyces pombe

During cell division in the fission yeast, *S. pombe*, separation of the daughter nuclei proceeds until the spindle extends the length of the cell. Upon spindle depolymerization, the nuclei reverse direction and a centripetal postmitotic nuclear migration occurs. Each nucleus migrates to the center of the new daughter cell (reviewed in 32). During *S. pombe* mating, the premeiotic nucleus becomes elongated and moves around the cell. The fact that telomeres are attached to the elongated tip of the nucleus during this migration suggests a possible role for telomeres in nuclear movement during certain stages of the life cycle (33;

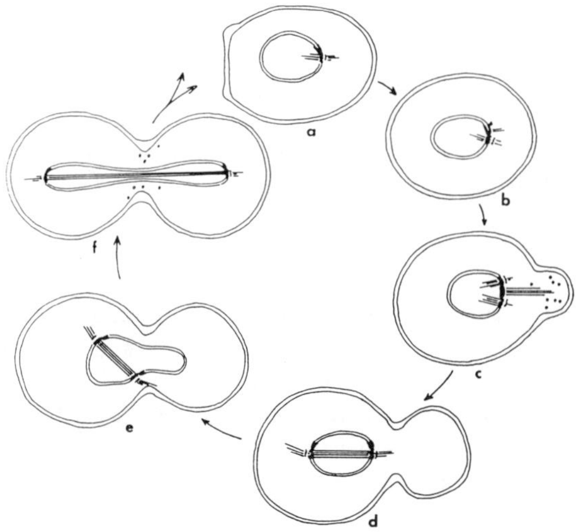

Figure 2. Nuclear organization and migration during the cell division cycle of the budding yeast, *Saccharomyces cerevisiae*. Reproduced with permission from reference 2, Figure 1, a-f, p. 61.

reviewed in 34). The following mutants with defects in nuclear migration will be discussed below: *nda2* and *ppb1*.

1) MUTATIONS IN STRUCTURAL ELEMENTS

Tubulin Gene Mutations

Microtubules (MTs) are long, cylindrical, multimeric tubes composed of α- and β-tubulin heterodimers (reviewed in 35). Studies in a variety of organisms have suggested that nuclear migration is a MT-mediated event. In addition to cytological observations, anti-MT drugs have provided much of the evidence for the role of MTs in nuclear movement.

For example, in *A. nidulans* nuclear migration is disrupted by benomyl, an inhibitor of MT polymerization (36). β-tubulin mutants with altered benomyl sensitivities were used to demonstrate that the inhibition of nuclear migration by benomyl was due to its effect on MTs and not to some nonspecific effect on the cell (36). Nuclear movement was supersensitive to benomyl in a benomyl supersensitive strain and was not affected by benomyl in a benomyl resistant strain (36). Thus, the nuclear migration defect in the presence of benomyl is due to a direct effect of benomyl on MTs.

A. nidulans benA33

In *A. nidulans*, a temperature-sensitive β-tubulin mutation, *benA33*, inhibits nuclear migration and nuclear division at the restrictive temperature (37). The *benA33* mutation confers benomyl resistance by causing the MTs to be hyperstable to depolymerization. This provides evidence that MT disassembly is required for nuclear migration and chromosomal movement. The blockage of nuclear movement and division is suppressed by *A. nidulans tubA1*, an α-tubulin mutation. Thus, α-tubulin must also be involved in these motility events (37).

S. pombe nda2-KM52

In *S. pombe*, the *nda2-KM52* (nuclear division arrest) mutant, which also confers supersensitivity to anti-MT drugs, displays a nuclear localization defect - the nucleus is diplaced from the center of the cell (38). NDA2 is an α-tubulin gene (39); thus α-tubulin is also required for nuclear migration in *S. pombe*.

S. cerevisiae tub2-401

As in other eukaryotic cells, *S. cerevisiae* MTs are found in both the mitotic spindle and the cytoplasm (reviewed in 22). All the MTs have one end associated with the spindle pole body (SPB), the fungal equivalent of the centrosome in higher eukaryotes. The spindle MTs are involved in chromosome segregation and nuclear division; the role of the cytoplasmic MTs (astral MTs) is less clear (reviewed in 22).

A cold-sensitive mutation in the *S. cerevisiae* β-tubulin gene, TUB2, demonstrates a role for cytoplasmic MTs during cell division by budding. At the semi-restrictive cold temperature, the *tub2-401* (tubulin defective; 40) mutant strain is incapable of assembling cytoplasmic (astral) MTs even though the spindle MTs assemble properly (42). This mutation results in a defect in nuclear migration to the bud neck prior to bud separation. Consequently there are an abnormally high number of binucleate and anucleate cells in the *tub2-401* mutant strains. These are rarely seen in the wild type (41,42).

A. nidulans mipA1

In *A. nidulans*, the MIPA gene encodes γ-tubulin, a new member of the tubulin superfamily. To determine the function of γ-tubulin *in vivo*, a disruption in the MIPA gene was created *in vitro*. The disrupted gene was transformed into *A. nidulans* in place of the wild-type gene. The *mipA1* (microtubule-interacting protein) mutation is lethal and partially inhibits nuclear migration and blocks nuclear division (43). Cytoplasmic MTs in the *mipA1* strain are decreased in number and in length, and mitotic spindles are totally absent. Therefore, γ-tubulin is essential for MT polymerization (43). Immunofluorescence microscopy with γ-tubulin-specific antibodies has shown that γ-tubulin is a component of the SPB (43).

Actin Gene Mutations

S. cerevisiae act1-4

Actin is a 43 kD cytoskeletal component that self-associates to form long polymers that perform numerous functions related to cell structure and motility (reviewed in 44-46). During most of the *S. cerevisiae* cell cycle, actin is unevenly distributed (47, reviewed in 22). When the bud begins to form, a ring of actin dots forms on the mother cell's side of the neck. In the mother cell, actin cables are directed towards the bud neck; just below the surface of the bud, there are actin patches. These asymmetries exist until the completion of the cell cycle at cytokinesis.

When actin function is disrupted in *S. cerevisiae* cells by the temperature-sensitive mutation, *act1-4* (actin defective), many of the cells become binucleate (41). Therefore, the wild-type actin protein is necessary for maintenance of proper nuclear distribution. Interestingly, actin-capping protein mutants also exhibit a slight defect in nuclear migration (noted in 48). Thus, in *S. cerevisiae*, both actin and MTs are required for proper nuclear orientation.

Intermediate Filament Gene Mutations

S. cerevesiae mdm1

Intermediate filaments (IFs) are 10 nm fibers that are components of the cytoskeleton. IF proteins have the intrinsic ability to self-assemble into filaments containing thousands of individual IF polypeptide chains (reviewed in 49). Immunolocalization studies in neuroepithelial cells have suggested that IFs may play a functional role in nuclear migration (50). There have been several other reports of IFs that colocalize with nuclei during nuclear migration events (19,51).

MDM1 encodes an essential 51 kD protein in *S. cerevisiae* that self-assembles into 10 nm filaments characteristic of IFs (52). *mdm1* mutant strains display defects in the movement of mitochondria and nuclei into buds. The positioning of the mitotic spindle is also affected (52,53). The MDM1 gene product was localized by indirect immunofluorescence to a unique pattern of spots and punctate arrays throughout the cytoplasm of yeast cells (54). The anti-MDM1 antibody stains animal cells in a filamentous pattern that is characteristic of previously described IF immunofluorescence patterns (54).

Cell Wall Gene Mutations

S. cerevisiae chs2

Chitin plays a structural role in the cell walls of fungi (55). During cell division in *S. cerevisiae*, a chitin collar is formed in the cell wall surrounding the bud neck. This collar colocalizes with actin patches which have been hypothesized to localize the synthesis of chitin to the neck (reviewed in 22). A mutation in the CHS2 chitin synthase II gene results in a nuclear migation defect. The *chs2* mutation leads to the production of multinucleate and anucleate cells (56). Perhaps the chitin ring provides an attachment site for cytoplasmic (astral) MTs needed to transport the nucleus to the bud site.

2) MUTATIONS IN MOTOR MOLECULES

Dynein Gene Mutations

A. nidulans nudA

In *A. nidulans*, the nuclei of the *nudA* (nuclear distribution) mutant do not move into the growing mycelium at restrictive temperatures (23). The NUDA gene of *A. nidulans* encodes the heavy chain of cytoplasmic dynein (25). Cytoplasmic dynein is a large multisubunit ATPase that transports cargo along MTs (reviewed in 57, 60,61). Dynein was first discovered as the motor that powers the beating of cilia and flagella; this dynein is called axonemal dynein (reviewed in 58,59). Although cilia and flagella contain several different axonemal dyneins, only a single cytoplasmic dynein has been discovered in the cytoplasm of cells so far. MTs are required for nuclear migration in *A. nidulans*, and cytoplasmic dynein appears to be the primary MT dependent motor that powers this movement.

N. crassa ro-1

In *N. crassa*, mutations in the cytoplasmic dynein heavy chain RO-1 gene also disrupt nuclear migration. The *ro-1* (ropy appearance) mutants fail to maintain a uniform nuclear distribution in fungal hyphae (29). Instead, a highly asymmetric distribution of nuclei results in clumps of nuclei and long anucleate hyphal regions. These mutations were called "*ropy*" because the hyphae are curled and distorted (29).

S. cerevisiae dhc1/dyn1

In *S. cerevisiae*, the cytoplasmic dynein heavy chain gene has also been identified as a component of the nuclear migration system (66,67). The *dhc1* (dynein heavy chain) mutation (67), also named *dyn1* (dynein defective) (66), was created by replacing the wild-type cytoplasmic dynein heavy chain with a disupted gene. These disruption mutants are defective in nuclear migration and have misaligned spindles. These mutants produce budded cells with a large binucleated mother cell and an anucleate bud (66,67). The dynein disruption is not lethal for vegetative growth, and cytoplasmic (astral) MTs are visible. Yeast and filamentous fungi share a conserved mechanism of nuclear migration in which cytoplasmic dynein mediates nuclear movement along MTs.

Kinesin Gene Mutations

S. cerevisiae kar3

Kinesin is a multisubunit ATPase that transports cargo along MTs (reviewed in 57, 62-64). The *S. cerevisiae* KAR3 gene encodes a kinesin-like protein (68). The *kar3* (karyogamy deficient) mutation disrupts nuclear movement during karyogamy in *S. cerevisiae* (68) and confers a sensitivity to anti-MT drugs such as benomyl. The sensitivity to benomyl in the *kar3* mutants correlates with the fact that kinesins interact with MTs. The KAR3 gene product is believed to function during karyogamy by pulling the two parental nuclei together *via* the cytoplasmic MTs (68).

Myosin Gene Mutations

S. cerevisiae myo1

The *S. cerevisiae* MYO1 gene encodes a myosin 1-like heavy chain (69). Myosin 1 is an actin-activated ATPase (reviewed in 65). The *myo1* (myosin 1 defective) mutants display an abnormal segregation of nuclei and have defects in cytokinesis (69). Thus actin motors and MT motors are involved in nuclear migration in yeast.

3) MUTATIONS IN THE NUCLEAR ENVELOPE

Mutations that affect Nuclear Pore Complex Proteins

S. cerevisiae nup1

The *S. cerevisiae* NUP1 gene is a member of a highly conserved family of nuclear pore complex proteins called nucleoporins (70). The nuclear pore complex is required for protein and RNA nucleocytoplasmic transport (reviewed in 71) and has been shown to be required for nuclear migration in budding yeast (70). Many of the *nup1* (nucleoporin deficient) mutant cells become multinucleate at all temperatures due to a defect in nuclear migration and spindle orientation (70). Experimental results suggest that the *nup1* mutants may have a defect in the structural connections between the nuclear envelope and the nuclear scaffold. This in turn may prevent the reorientation of the nucleus in response to cytoskeletal forces that normally move the nucleus towards the neck of the budding cell (70).

Mutations that Affect Nuclear Membrane Proteins

S. cerevisiae bfr1

BFR1 is a multicopy suppressor of brefeldin A-induced lethality in *S. cerevisiae* (72). A *bfr1* mutant strain exhibits cells with two nuclei in one cell body due to a nuclear mispositioning during mitosis (72). Brefeldin A can block anterograde vesicular traffic from the endoplasmic reticulum (ER) to the Golgi (reviewed in 73). It is hypothesized that the BFR1 gene product is involved in vesicle targeting and/or fusion of nuclear membrane components (72). Perhaps one of these components is required for the attachment of the nucleus to a motor or cytoskeletal element required for nuclear migration.

SPB Mutations

S. cerevisiae spa1

The spindle pole body (SPB) is embedded in the nuclear envelope throughout the cell cycle. The *S. cerevisiae* SPA1 gene product (spindle pole antigen) is a putative spindle pole body component (74). Mutations in the SPA1 gene affect cell growth, chromosome segregation and spindle morphology. A strain carrying a disruption of SPA1, *spa1-1*, displays missegregated nuclei due to a defect in nuclear migration (74). Because the SPB is required to nucleate both spindle and cytoplasmic MTs, it is not surprising that both chromosome movement and nuclear movement are affected when a mutation is introduced into a SPB component.

4) MISCELLANEOUS MUTATIONS

Mutations that Affect MTs

S. cerevisiae bik1

The *S. cerevisiae* BIK1 gene encodes a 58 kD protein that associates with MTs (75). A *bik1* (bilateral karyogamy defect) deletion affects a variety of MT-dependent functions. The nuclei frequently fail to move to the bud neck and often divide wholly within the mother cell. The *bik1* mutant is also defective in karyogamy (75). The pleiotropic defects imply that wild-type BIK1 is required for MT function during cell division and karyogamy.

S. cerevisiae cin1, cin2 and cin4

Three different mutants, *cin1*, *cin2* and *cin4* (chromosome instability) cause defects in MT-mediated nuclear migration events in *S. cerevisiae* including both nuclear migration and karyogamy (76). The *cin1*, *cin2* and *cin4* mutants affect MT stability. They are also sensitive to benomyl. Overexpression of the wild-type CIN1 gene product also causes benomyl sensitivity. CIN1, CIN2 and CIN4 may contribute to MT stability by regulating the activity of a MT component, or they may serve as structural components of MTs (76). The nuclear mispositioning presumably arises from a failure of the cytoplasmic MTs to draw the dividing nucleus into the neck.

S. cerevisiae jnm1

The *S. cerevisiae jnm1* (just nuclear migration) mutants are viable and phenotypically similar to the viable *S. cerevisiae dyn1* (dynein) mutants (92). Many of the *jnm1* mutant cells accumulate with a large bud and two nuclei in the mother cell. The mitotic spindles are often misaligned with respect to the mother/bud axis, and the astral MTs are long and often wind around the periphery of the cell (92). Like *dyn1* mutant cells, the *jnm1* mutant cells do not have many anucleate cells, only binucleate cells. The *jnm1* mutant cells exhibit a delay in anaphase that is not apparent in strains defective in astral MTs such as the *S. cerevisiae tub2-401* (tubulin) mutant. The presence of astral MTs may help prevent the completion of mitosis in binucleate mother cells (92).

The predicted JNM1 gene product is a 44 kD protein with three coiled coil domains and no significant homology to other proteins. A JNM1:lacZ hybrid protein localizes to a single spot often near the SPB in unbudded cells. In cells with a short spindle, the protein localizes asymmetrically to the SPB nearest to the bud; in cells with long spindles, the spot is in the bud and not always associated with the SPB (92). The JNM1 protein may tether astral MTs to the bud (92).

S. cerevisiae kar4 and *kar9*

The *S. cerevisiae kar3* (see above), *kar4* and *kar9* (karyogamy defective) mutants were identified in a screen to select 1) mutants whose nuclei do not migrate properly during mating and 2) mutants whose nuclei move properly towards each other, but are unable to undergo nuclear membrane fusion (77). The *kar3, kar4* and *kar9* mutants fall into the first class and have altered cytoplasmic MTs (77). The *kar4* mutant zygotes have abnormally long cytoplasmic MTs which fail to interconnect the nuclei. The *kar9* mutants display much shorter cytoplasmic MTs which are misoriented with respect to one another. The *kar9* mutant is hypersensitive to benomyl. KAR9 is believed to be important for other MT functions such as nuclear migration and spindle orientation during mitosis (77).

S. cerevisiae num1

A novel gene in *S. cerevisiae*, NUM1, may be involved in the MT-interaction between the G_2 nucleus and the bud neck (78). In *num1* (nuclear migration) disrupted cells the G_2 nucleus is often unable to migrate into the neck of the budded cells. Thus, the G_2 nuclei divide in the mother and then migrate to the neck. The normal order of events is reversed; nuclear division occurs before nuclear migration (78). The NUM1 gene encodes a hydrophilic 313 kD protein with a potential Ca^{2+} binding site and internal repeats characteristic of Ca^{2+}- and phospholipid-binding proteins that interact with both membranes and the cytoskeleton (78). The NUM1 gene protein sequence also has several potential phosphorylation sites for protein kinase A (PKA) and PKC (78).

Interestingly, experiments with protein kinase C (PKC) activators and inhibitors suggest an important role for PKC in initiation of nuclear migration during mouse embryo development (79). Inhibition of PKC prevents the calmodulin-mediated assembly and disassembly of the MT networks that move peripherally located nuclei to the center of the embryo during compaction (79).

A. nidulans apsA

In *A. nidulans*, *apsA* (anucleate primary sterigmata) mutants are defective in nuclear positioning in both hyphae and in the specialized, multicellular spore-producing structures called conidiophores (29; Fischer and Timberlake, J. Cell Biol., in press). APSA encodes a 183 kD coiled-coil protein that shares similarity with the *S. cerevisiae* NUM1 gene product. The carboxyl termini of the APSA protein has a 460 amino acid residue region that shares 72% similarity and 25% identity with the NUM1 protein. Additionally, the two proteins share structural similarities in their charge distribution and heptad repeats. The similarity between a yeast and filmentous fungal protein required for nuclear migration indicates a conserved role for these proteins in nuclear migration.

S. pombe ppb1

A calcineurin-like gene, PPB1, that is 57% identical to rat type 2B protein phosphatase has been identified in *S. pombe* (80). Calcineurin is a Ca^{2+}/calmodulin-regulated type 2B protein phosphatase. Pleiotropic effects are produced by null and overexpression mutations of PPB1, which include a defect in nuclear migration (80). Interestingly, the SPB is frequently mispositioned and associated with strange pointed tips in the chromatin region (80). Among other functions, calcineurin is involved in the T-cell mediated immune response (81,82). Interestingly, when T-cells interact with target cells, the centrosomes (higher eukaryotic version of SPBs) and cytoplasmic MTs orient towards the target cell (reviewed in 83). Perhaps the defect in nuclear migration in *ppb1* mutants is due to a lack of a signal required to orient the SPB and the cytoplasmic MTs properly towards the bud site.

S. cerevisiae tcp1

In *S. cerevisiae*, the TCP1 (84) gene is a homologue to the mouse TCP1 gene that affects embryogenesis and sperm differentiation (85). The *S. cerevisiae tcp1-1* (tailless complex polypeptide 1) mutation produces multinucleate and anucleate cells (84). In addition to the abnormal nuclear positioning in *tcp1-1* strains, the MT structures are deformed and longer than in wild-type. Overexpression of wild-type TCP1 gene leads to an increased sensitivity to the anti-MT drug benomyl, as does the *tcp1-1* mutation, suggesting that the TCP1 gene product affects MT-mediated processes required for migration of the nucleus to the bud neck (84).

S. cerevisiae ypt1

The *S. cerevisiae* ras-related YPT1 gene encodes a 23.5 kD ras-related GTP-binding protein. A YPT1 gene with a regulatable promoter was used to replace the wild-type gene. Down regulation of the promoter leads to loss of YPT protein function and results in the formation of abnormal MT structures (86). *ypt1* dominant-lethal strains were also constructed and these strains produce cells with two nuclei in the mother cell. The nuclei are unable to move through the neck between the mother and daughter bud and spindles are misaligned (86).

Dynein Activator Gene Mutations

S. cerevisiae act3 and *act5*

The ACT3 (87) and ACT5 (48) genes in *S. cerevisiae* share sequence identity with ARP1. ARP1 (actin related protein) was previously referred to as centractin (89,90) or actin-related protein of vertebrates (91). ARP1 is the major component of the glued/dynactin complex that has been shown to be essential for activating cytoplasmic dynein-driven motility *in vitro* (reviewed in 88). The *act 3* and *act5* (actin defective) mutant strains are viable and have defects in nuclear migration and spindle orientation that are similar to the viable *S. cerevisiae dyn1* (dynein) mutants (48,87). If dynein is not active *in vivo* unless the glued/dynactin complex is present, it makes sense that a similar phenotype would be observed for both mutations.

N. crassa ro-4

The *N.crassa* RO-4 gene is another member of the ARP1 gene family (29). The *ro-4* (ropy appearance) mutant is phenotypically identical to the *N. crassa ro-1* (dynein) mutant (29). This suggests that dynein requires the presence of the glued/dynactin complex to maintain proper uniform nuclear distribution in fungal hyphae (29). The conservation of this actin-related protein function between yeast and filamentous fungi provides evidence that the elements involved in nuclear migration are conserved.

Mutations in β-transducin-like Proteins

S. cerevisiae cdc20

S. cerevisiae *cdc20* (cell division cycle defects) mutants affect chromosome separation during mitosis and nuclear migration during karyogamy (93). The *cdc20* mutants show aberrant MT staining with an increase in the intensity of spindle staining, although the intracellular tubulin levels do not increase. CDC20 may therefore modulate MT structure (93). The CDC20 protein is a 57 kD protein with homology to the β-transducin family. CDC20 may function as a G-protein in intracellular signaling events that regulate MTs indirectly, or it could play a more direct role in MT dynamics by somehow modulating GTP binding to tubulin. Experiments to test both possibilities are under way (93).

A. nidulans nudF

The *A. nidulans* NUDF gene encodes a protein with the characteristics of a heterotrimeric G protein β subunit (Xiang et al., unpublished data). The *nudF* (nuclear distribution defect) mutants are unable to transport their nuclei into the growing germ tube (23). NUDF shares a high level of identity at the amino acid level to the human LIS1 (94) gene product (Xiang et al., unpublished data). Humans with a defect in the LIS1 gene are born with Miller-Dieker lissencephaly, a severe brain malformation. Victims of this disease have a smooth cerebral surface instead of a normal highly convoluted cortex because of a defect in neuronal migration (reviewed in 11, 94). If nuclear migration is essential to neuronal migration, it is not unreasonable to propose that a defect in nuclear migration may cause defects in neuronal migration such as in lissencephaly.

NUDF was cloned as an extra copy suppressor of the *nudC3* mutation (Xiang et al., unpublished data). The NUDC gene encodes a 23 kD protein in *A. nidulans* (24). The function of this protein is not known; however, homologues have been found in *Drosophila melanogaster* (95), *Caenorhabditis elegans* (expressed sequence tag in GenBank database) and rat (Y.-L. Li, personal communication). Thus, NUDC encodes an evolutionarily conserved protein. The *nudC3* mutation decreases the level of NUDF protein in the cell (Xiang et al., unpublished data). This explains why an extra copy of NUDF suppresses the ts *nudC3* mutation. The exact functions of NUDF and NUDC are still being explored, as is the nature of the interaction between these proteins.

Heat Shock Protein Gene Mutations

S. cerevisiae sis1

The *S. cerevisiae* SIS1 gene (96) is a high copy number suppressor of strains defective in SIT4, a serine/threonine protein phosphatase involved in transcription and cell cycle progression (97). The depletion of the wild-type SIS1 (*sit4* suppressor) gene product is achieved through the use of a regulatable promoter in the *sis1-2* strain. The lack of SIS1 protein results in cells that are unable to transport the nucleus to the bud neck (96). SIS1 is an essential gene that encodes a 37 kD protein with substantial homology to the *E.coli* bacterial DnaJ protein (96), a member of the heat shock family (reviewed in 98). The SIS1 gene has a heat shock consensus element in its promoter, and SIS1 protein increases in response to heat shock (96). SIS1 protein is localized throughout the cell; however it is more concentrated in the region of the nucleus (96). Further genetic and biochemical analysis is required to determine the cellular function of SIS1 (98).

CONCLUSION

Nuclear migration is a crucial process that occurs in many cellular and developmental events as diverse as fertilization and brain development. Due to the significance of nuclear migration, it is clearly important to determine the components involved in the mechanism and regulation of this process. By a combination of genetic approaches including mutant analysis and extragenic suppression, fungal geneticists have begun to dissect the process.

Nuclear migration occurs during yeast mating and cell division and during the growth of filamentous fungi. These movements provide models for nuclear migration during fertilization, and the growth and development of higher eukaryotes. Nuclear movement in fungi is mediated by many cytoskeletal elements. At least two MT motors, dynein and kinesin, have been shown to power nuclear movements. The actin related protein, ARP1, appears to be involved in nuclear migration, and IFs and myosin also participate in nuclear migration; however, their exact roles are unclear.

The structural integrity of the nuclear envelope, SPB and cell wall is critical for proper nuclear migration. In addition, the nuclear migration process involves G-proteins, heat shock proteins and mediators of MT structure and dynamics. Several of the components identified in fungal nuclear migration are also found in higher eukaryotes. Thus, the understanding of nuclear migration in fungi may be an important step forward in the comprehension of this process in higher eukaryotes.

ACKNOWLEDGMENTS

We thank Drs. X. Xiang and Y.-L. Li for sharing their unpublished data with us. The preparation of this manuscript was supported by grants from the NIH - National Institute of General Medical Sciences (GM-29228 and GM-34711).

REFERENCES

1 Schatten, G. (1982) Int. Rev. Cytol. 79, 59-163.
2 Byers, B. (1981) in The Molecular Biology of the Yeast *Saccharomyces*: Life Cycle and Inheritance (Strathern, J.N., Jones, E.W. and Broach, J.R., eds.), pp. 59-96, Cold Spring Harbor Laboratory, Cold Spring Harbor, NY.
3 Dan, K. (1979) Dev. Growth Differ. 21, 507-535.
4 Reeve, W.J. and Kelly, F.P. (1983) J. Embryol. Exp. Morph. 75, 117-139.
5 Sardet, C., McDougall, A. and Houliston, E. (1994) Trends Cell Biol. 4, 166-172.
6 Sauer, F.C. (1935) J. Comp. Neurol. 62, 377-405.
7 Schoenwolf, G.C. and Smith, J.L. (1990) Devel. 109, 243-270.
8 Rakic, P. (1971) J. Comp. Neur. 141, 283-312.
9 Book, K.J. and Morest, D.K. (1990) J. Comp. Neur. 297, 55-76.
10 Book, K.J., Howard, R. and Morest, D.K. (1991) Exp. Neur. 111, 228-243.
11 Barth, P.G. (1987) Can. J. Neurol. Sci. 14, 1-16.
12 Bloom, F.E. (1993) Arch. Gen. Psychiatry 50, 224-227.
13 Klominek, J., Sundqvist, K.-G. and Robert, K.-H. (1991) Proc. Nat. Acad. Sci. U.S.A. 88, 3902-3906.
14 Roos, E., Dingemans, K.P., van de Pavert, I.V. and van den Bergh-Weerman, M. (1977) J. Nat. Cancer Inst. 58, 399-407.
15 Roos, E., van de Pavert, I.V. and Middelkroop, O.P. (1981) J. Cell Sci. 47, 385-397.
16 Zolaker, M. and Erk, I. (1976) J. Microscopie Biol. Cell. 25, 97-106.
17 Englander, L.L. and Rubin, L.L. (1987) J. Cell Biol. 104, 87-95.
18 Holmes, K.V. and Choppin, P.W. (1968) J. Cell Biol. 39, 526-543.
19 Wang, E., Cross, R.K. and Choppin, P.W. (1979) J. Cell Biol. 83, 320-337.
20 Morris, N.R. (1989) Cell Motil. Cytoskel. 14, 58-61.
21 Botstein, D. and Maurer, R. (1982) Annu. Rev. Genet. 16, 61-83.
22 Huffaker, T.C., Hoyt, M.A. and Botstein, D. (1987) Annu. Rev. Genet. 21, 259-284.
23 Morris, N.R. (1976) Genet. Res. 26, 237-254.
24 Osmani, A.H., Osmani, S.A. and Morris, N.R. (1990) J. Cell Biol. 111, 543-551.
25 Xiang, X., Beckwith, S.M. and Morris, N.R. (1994) Proc. Nat. Acad. Sci. U.S.A. 91, 2100-2104.
26 Clutterbuck, A.J. (1969) Genetics 63, 317-327.
27 Clutterbuck, A.J. (1994) Microbiol. 140, 1169-1174.
28 Garnjobst, L. and Tatum, E.L. (1967) Genetics 57, 579-604.
29 Plamann, M., Minke, P.F., Tinsley, J.H. and Bruno, K.S. (1994) J. Cell Biol. 127, 139-149.
30 Rose, M.D. (1991) Annu. Rev. Microbiol. 45, 539-567.
31 Pringle, J.R. and Hartwell, L.H. (1981) in The Molecular Biology of the Yeast *Saccharomyces*: Life Cycle and Inheritance (Strathern, J.N., Jones, E.W. and Broach, J.R., eds.), pp. 97-142, Cold Spring Harbor Laboratory, Cold Spring Harbor, NY.
32 Robinow, C.F. and Hyams, J.S. (1989) in Molecular Biology of the Fission Yeast (Nasim, A., Young, P. and Johnson, B.F., eds.), pp. 273-330, Academic Press, Inc., San Diego, CA.
33 Chikashige, Y., Ding, D.-Q., Funabiki, H., Haraguchi, T., Mashiko, S., Yanagida, M. and Hiraoka, Y. (1994) Science 264, 270-273.
34 Blackburn, E.H. (1994) Cell 77, 621-623.
35 Burns, R.G. and Surridge, C.D. (1994) in Microtubules (Hyams, J.S. and Lloyd, C.W., eds.), pp. 3-31, Wiley-Liss, Inc., New York, NY.

36 Oakley, B.R. and Morris, N.R. (1980) Cell 19, 255-262.
37 Oakley, B.R. and Morris, N.R. (1981) Cell 24, 837-845.
38 Umesono, K., Toda, T., Hayashi, S. and Yanagida, M. (1983) J. Mol. Biol. 168, 271-284.
39 Toda, T., Adachi, Y., Hiraoka, Y. and Yanagida, M. (1984) Cell 37, 233-242.
40 Huffaker, T.C., Thomas, J.H. and Botstein, D. (1988) J. Cell Biol. 106, 1997-2010.
41 Palmer, R.E., Sullivan, D.S., Huffaker, T. and Koshland, D. (1992) J. Cell Biol. 119, 583-593.
42 Sullivan, D. and Huffaker, T. (1992) J. Cell Biol. 119, 379-388.
43 Oakley, B.R., Oakley, C. E., Yoon, Y. and Jung, M.K. (1990) Cell 61, 1289-1301.
44 Pollard, T.D. (1990) Curr. Opin. Cell Biol. 2, 33-40.
45 Mitchison, T.J. (1992) Mol. Biol. Cell 3, 1309-1315.
46 Bretscher, A., Drees, B., Harsay, E., Schott, D. and Wang, T. (1994) J. Cell Biol. 126, 821-825.
47 Adams, A.E.M. and Pringle, J.R. (1984) J. Cell Biol. 98, 934-945.
48 Muhua, L., Karpova, T.S. and Cooper, J.A. (1994) Cell 78, 669-679.
49 Fuchs, E. and Weber, K. (1994) Annu. Rev. Biochem. 63, 345-382.
50 Stagaard, M and Møllgård, K. (1989) Anat. Embryol. 180, 17-28.
51 Numata, O., Sugai, T. and Watanabe, Y. (1985) Nature 314, 192-194.
52 McConnell, S.J. and Yaffe, M.P. (1993) Science 260, 687-689.
53 McConnell, S.J., Stewart, L.C., Talin, A. and Yaffe, M.P. (1990) J. Cell Biol. 111, 967-976.
54 McConnell, S.J. and Yaffe, M.P. (1992) J. Cell Biol. 118, 385-395.
55 Cabib, E., Roberts, R. and Bowers, B. (1982) Annu. Rev. Biochem. 51, 763-793.
56 Bulawa, C.E. and Osmond, B.C. (1990) Proc. Nat. Acad. Sci. U.S.A. 87, 7424-7428.
57 Walker, R.A. and Sheetz, M.P. (1993) Annu. Rev. Biochem. 62, 429-451.
58 Gibbons, I.R. (1989) in Cell Movement, Volume 1: The Dynein ATPases (Warner, F.D., Satir, P. and Gibbons, I.R., eds.), pp. 3-22, Alan R. Liss, Inc., New York, NY.
59 Asai, D.J. and Brokaw, C.J. (1993) Trends Cell Biol. 3, 398-402.
60 Vallee, R. (1993) Proc. Nat. Acad. Sci. U.S.A. 90, 8769-8772.
61 Schroer, T.A. (1994) Curr. Opin. Cell Biol. 6, 69-73.
62 Goldstein, L.S.B. (1993) Annu. Rev. Genet. 27, 319-351.
63 Goodson, H.V., Kang, S.J. and Endow, S.A. (1994) J. Cell Sci. 107, 1875-1884.
64 Hoyt, M.A. (1994) Curr. Opin. Cell Biol. 6, 63-68.
65 Cheney, R.E. and Mooseker, M.S. (1992) Curr. Opin. Cell Biol. 4, 27-35.
66 Eshel, D., Urrestarazu, L.A., Vissers, S., Jauniaux, J.-C., van Vliet-Reedijk, J.C., Planta, R.J. and Gibbons, I.R. (1993) Proc. Nat. Acad. Sci. U.S.A. 90, 11172-11176.
67 Li, Y.-Y., Yeh, E., Hays,T. and Bloom, K. (1993) Proc. Nat. Acad. Sci. U.S.A. 90, 10096-10100.
68 Meluh, P.B. and Rose, M.D. (1990) Cell 60, 1029-1041.
69 Watts, F.Z., Shiels, G. and Orr, E. (1987) EMBO J. 6, 3499-3505.
70 Bogerd, A.M., Hoffman, J.A., Amberg, D.C., Fink, G.R. and Davis, L.I. (1994) J. Cell Biol. 127, 319-332.
71 Forbes, D.J. (1992) Annu. Rev. Cell Biol. 8, 495-527.
72 Jackson, C.L. and Képès, F. (1994) Genetics 137, 423-437.
73 Lippincott-Schwartz, J. (1993) Trends Cell Biol. 3, 81-88.
74 Snyder, M. and Davis, R.W. (1988) Cell 54, 743-754.
75 Berlin, V., Styles, C.A. and Fink, G.R. (1990) J. Cell Biol. 111, 2573-2596.
76 Hoyt, M.A., Stearns, T. and Botstein, D. (1990) Mol. Cell. Biol. 10, 223-234.
77 Kurihara, L.J., Beh, C.T., Latterich, M., Schekman, R. and Rose, M.D. (1994) J. Cell Biol. 126, 911-923.

78 Kormanec, J., Schaaff-Gerstenschlager, I., Zimmermann, F.K., Perecko, D. and Kuntzel, H. (1991) Mol. Gen. Genet. 230, 277-287.
79 Ohsugi, M., Ohsawa, T. and Yamamura, H. (1993) Devel. Biol. 156, 146-154.
80 Yoshida, T., Toda, T. and Yanagida, M. (1994) J. Cell Sci. 107, 1725-1735.
81 McKeon, F. (1991) Cell 66, 823-826.
82 Liu, J., Farmer, J.D., Lane, W.S., Friedman, J., Weissman, I. and Schreiber, S.L. (1991) Cell 66, 807-815.
83 Kupfer, A. and Singer, S.J. (1989) Annu. Rev. Immunol. 7, 309-337.
84 Ursic, D. and Culbertson, M.R. (1991) Mol. Cell. Biol. 11, 2629-2640.
85 Silver, L.M. (1981) Genet. Res. 38, 115-123.
86 Schmitt, H.D., Wagner, P., Pfaff, E. and Gallwitz, D. (1986) Cell 47, 401-412.
87 Clark, S.W. and Meyer, D.I. (1994) J. Cell Biol. 127, 129-138.
88 Schroer, T.A. (1994) J. Cell Biol. 127, 1-4.
89 Clark, S.W. and Meyer, D.I. (1992) Nature 359, 246-250.
90 Paschal, B.M., Holzbaur, E.L.F., Pfister, K.K., Clark, S., Meyer, D.I. and Vallee, R.B. (1993) J. Biol. Chem. 268, 15318-15323.
91 Lees-Miller, J.P., Helfman, D.M. and Schroer, T.A. (1992) Nature 359, 244-246.
92 McMillan, J.N. and Tatchell, K. (1994) J. Cell Biol. 125, 143-158.
93 Sethi, N., Monteagudo, C., Koshland, D., Hogan, E. and Burke, D.J. (1991) Mol. Cell. Biol. 11, 5592-5602.
94 Reiner, O., Carrozzo, R., Shen, Y., Wehnert, M., Faustinella, F., Dobyns, W.B., Caskey, C.T. and Ledbetter, D.H. (1993) Nature 364, 717-721.
95 Cunniff, J. and Warrior, R. (1993) Mol. Biol.Cell 4, 402a.
96 Luke, M.M., Sutton, A. and Arndt, K.T. (1991) J. Cell Biol. 623-638.
97 Sutton, A., Immanuel, D. and Arndt, K.T. (1991) Mol. Cell. Biol. 11, 2133-2148.
98 Lindquist, S. and Craig, E.A. (1988) Annu. Rev. Genet. 22, 631-677.

STRUCTURE AND FUNCTION OF THE PLATELET-DERIVED GROWTH FACTOR FAMILY AND THEIR RECEPTORS

Kristen C. Hart[3], Brendan D. Galvin[1] and Daniel J. Donoghue[1,2]

[1]Department of Chemistry and Biochemistry, [2]Center for Molecular Genetics and [3]Molecular Pathology Program, School of Medicine
University of California, San Diego
La Jolla, CA 92093-0322

INTRODUCTION

Growth factors play a pivotal role in the processes of cell proliferation and differentiation. Platelet-derived growth factor (PDGF) and related growth factors are involved in many normal and pathological processes. Of particular interest is the involvement of growth factors and their receptors in signal transduction, cell cycle control and differentiation. The oncogene v-*sis* encodes a protein, designated p28sis or SIS, that is homologous to the B-chain of PDGF. The role of PDGF-related growth factors in autocrine transformation (1) has been a controversial subject. The study of SIS and related proteins has been very important in elucidating the mechanisms of signal transduction and transformation; however, much remains to be discovered. This paper will review the structure of the PDGF family of growth factors and their receptors, summarize current knowledge about the proteins that are involved downstream in mediating signal transduction, and discuss the role of PDGF and related growth factors in various pathologic processes.

THE PDGF FAMILY OF GROWTH FACTORS

The growth factor family, consisting of platelet-derived growth factor (PDGF), vascular endothelial growth factor/vascular permeability factor (VEGF/VPF) and placental growth factor (PlGF), mediates a variety of biological responses. PDGF was the first member of this family to be cloned and consequently is the most thoroughly characterized. PDGF elicits a variety of responses including growth stimulation of fibroblasts, osteoblasts, arterial smooth muscle cells and glial cells; stimulation of cholesterol synthesis, amino acid transport and protein synthesis (2,3); increase in c-*fos* and c-*myc* expression (4); vasoconstriction (5); and chemoattraction of fibroblasts and smooth muscle cells (6). The effects of VEGF/VPF and PlGF seem to be primarily angiogenic. This stimulation of

Genetic Engineering, Vol. 17, Edited by J.K. Setlow
Plenum Press, New York, 1995

growth and proliferation of new blood vessels indicates a role for these factors in processes such as wound healing and embryogenesis. VEGF is also implicated in increased blood vessel permeability and is a potent mitogen for endothelial cells (7,8).

Structure of PDGF and Related Proteins

The PDGF family consists of a group of homologous disulfide-linked dimeric proteins that exhibit conservation of eight key Cys residues. The similarity in primary structure of these polypeptides implies that their tertiary structures are likely to be similar. The recently described crystal structure of PDGF-BB indicates that the peptide forms a highly-twisted structure with two sets of anti-parallel β-sheets (9). It also reveals the presence of a unique knot structure formed by intramolecular disulfide bonds (9). This "cysteine knot" motif is representative of a superfamily of growth factors, including TGF-β (transforming growth factor-β), NGF (nerve growth factor) and SIS (10). This structure, depicted in Figure 1, consists of three disulfide bonds that form between Cys16-Cys60, Cys49-Cys97 and Cys53-Cys99. There is also a fourth disulfide bond that connects the two subunits in an anti-parallel orientation, that forms between Cys43 of one molecule and Cys52 of the other, and vice-versa (9). The importance of these Cys residues will be discussed in more detail below.

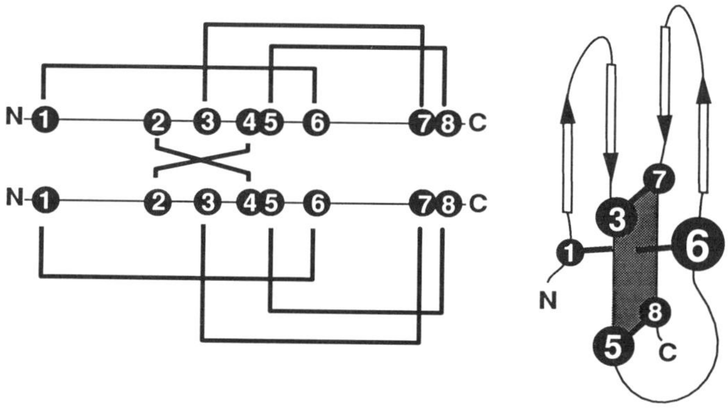

	1	2	3	4	5	6	7	8
SIS	127	154	160	163	164	171	208	210
PDGF-B	16	43	49	52	53	60	97	99
PDGF-A	10	37	43	46	47	54	91	93
VEGF	26	51	57	60	61	68	102	104
PIGF	52	77	83	86	87	94	128	130

Figure 1. Organization of disulfide bonds within the PDGF family. Shown in the upper left are the inter- and intra-molecular disulfide bonds that exist in the dimeric proteins of the PDGF family. The 8 Cys residues are indicated in order from the "N" terminus to the "C" terminus of each subunit. The connectivity of the disulfide bonds between Cys residues is indicated by the bold lines. The upper right panel demonstrates the organization of these disulfide bonds in the "cysteine knot" structure (9) within a single subunit. The table at the bottom shows the position of each conserved Cys residue within each of the PDGF family members.

PDGF was initially isolated from platelets and subsequently has been found to be secreted by activated macrophages (11,12), cultured arterial endothelial cells (13), stimulated fibroblasts (4,14), newborn rat arterial smooth muscle cells (15) and placental trophoblasts (16). PDGF is an approximately 30 kD protein that is cationic, heat-stable, and binds heparin. Both homodimeric forms and the AB heterodimeric form have been found to exist in nature (17,18).

The A and B chains of PDGF are encoded by genes that lie on separate chromosomes, suggesting independent regulatory mechanisms. The human PDGF-A gene is located on chromosome 7 (19,20) and human PDGF-B gene is located on chromosome 22 (21). Mature polypeptide chains resulting from the two genes share 56% identity at the amino acid level (19). The initial translation product of the PDGF-A gene is a 23 kD polypeptide, which in the absence of the B-chain undergoes homodimerization and processing at the N-terminus to yield the 31 kD secreted PDGF-AA (19,22). Isolation of the cDNA for the A-chain has shown that two unique splicing events occur which give rise to distinct forms, consisting of 196 and 211 amino acids, respectively (20,23). The longer, less-prevalent isolate includes a highly basic C-terminal addition of 18 amino acids, encoded by exon 6. These additional C-terminal residues can function as a cryptic nuclear targeting signal (24), but this motif has since been suggested to function as a compartmentalization signal to mediate association with heparin sulfate proteoglycans (25). This sequestration of PDGF may protect it from proteolytic degradation, and allows for localized retention, delivery and clearance of the growth factor.

The secreted form of PDGF-BB is a 30 kD protein. The initial translation product is a 28 kD polypeptide that, like the A-chain, is dimerized in the endoplasmic reticulum (ER) (26) and processed at both the N- and C-termini (27). Additionally, there exists an intracellular-retained 24 kD form of PDGF-BB resulting from additional N-terminal processing of the dimer in the Golgi complex, which directs the protein for lysosomal degradation (26). There is also a cryptic nuclear targeting signal embedded within the PDGF-B protein (28).

The transforming gene product of simian sarcoma virus (SSV), designated p28sis or SIS, (29), is derived by retroviral transduction from the PDGF-B chain (30). SIS is dimerized and processed at both the N- and C-terminus, similar to the PDGF-B polypeptide (31). A signal sequence, derived from the *env* gene of SSV, is required for secretion and biological activity of SIS (32). Processing at the N-terminal dibasic cleavage site, Lys110-Arg111, is not required, however, for the biological function of SIS (33). A consensus site for N-linked glycosylation exists in SIS at Asn93, but mutants that lack this site are still biologically active (34), suggesting that glycosylation is not necessary for biological activity. There are several possible sites for both N- and O-linked glycosylation within the A and B chains of PDGF as well. The functional relevance of this glycosylation is not known, since PDGF-BB expressed in *E. coli* retains complete biological activity (35). Thus, alternative dimeric forms, differential glycosylation and variable proteolysis likely account for all of the naturally-occurring variants of PDGF.

Characterization of the two newest members of the PDGF family, VEGF and PlGF, is not nearly as complete. VEGF was first purified from conditioned media of U937 lymphoma cells as a glycoprotein that stimulates permeability of blood vessels (36,37). Cloning of VEGF revealed homology to PDGF. There is only 15 to 18% identity of human VEGF with PDGF-A and B (as opposed to 44% identity between the A and B chains of PDGF); however, complete conservation of the eight Cys residues found in PDGF define this as a member of the PDGF family (38). VEGF is a 40 to 46 kD dimeric glycoprotein. The primary translation product of rat glioma-derived VEGF is a 190 amino acid polypeptide, which is dimerized and processed on the N-terminus to yield a mature homodimer of two 164 amino acid subunits (39). As with PDGF, there are multiple splice variants of VEGF. Four of these alternate splice products have been identified from human VEGF (40). These code for proteins of 121, 165, 189 and 206 amino acids. $VEGF_{121}$ and $VEGF_{165}$ are efficiently secreted from cells, while $VEGF_{189}$ and $VEGF_{206}$ remain cell-associated. This may regulate the function of these proteins.

PlGF is the most recently cloned PDGF family member. It was isolated from a placental cDNA library (41). The gene was later shown to be localized on chromosome 14 (42). The primary product of this gene is a 25 kD, 149 amino acid protein that is N-glycosylated and dimerized to a 50 kD form. PlGF is 71% homologous to the PDGF-like region of VEGF, and also retains the conserved spacing of eight Cys residues, placing it in the PDGF family of growth factors (40). A second form of PlGF has recently been identified. This form has a 21 amino acid, highly basic insertion near the C-terminus (42,43). Again, alternative splicing is involved in the generation of these variant forms of PlGF (42). Both forms of PlGF are efficiently secreted, and have similar mitogenic potency for cells, but the $PlGF_{170}$ form binds tightly to heparin (43).

Structure/Function Studies

Extensive structure and function studies have been performed on SIS, the oncogenic product of simian sarcoma virus (SSV). These studies have shed light on the function of PDGF and related growth factors, and have greatly increased understanding of the involvement of SIS in autocrine transformation. SIS was one of the first examples of a transforming viral gene product derived from a normal cellular growth factor or receptor (29). Thus, SIS has played a key role in highlighting the importance of research on viral oncogenes and their normal cellular counterparts. As mentioned previously, information gained from the study of SIS and PDGF structure and function should be relevant to all members of the PDGF family.

Complete reduction of the disulfide bonds of PDGF causes loss of its biological activity and reversion to a monomeric form. This observation highlights the importance of dimerization of PDGF and related factors for proper biological function (44). Progressive deletions of the C-terminal portion of SIS indicate that up to 57 amino acids can be deleted without significantly altering its biological activity. However, deletion of 67 or more C-terminal residues (yielding a protein containing 214 of the 271 amino acids of SIS) causes loss of transforming activity and loss of dimerization (45). The eight conserved Cys residues in PDGF and SIS participate in disulfide bonding necessary to dimerize these growth factors. Characterization of each individual Cys→Ser mutant indicated that Cys127, Cys160, Cys171 and Cys208 are required for transforming activity, whereas Cys154, Cys163, Cys164 and Cys210 are not required (46,47). Sauer and Donoghue (47) also demonstrated that the nontransforming single mutants of SIS were unable to dimerize. In addition, various double and triple mutants were characterized. Mutation of Cys163/Cys164 or Cys154/Cys163/Cys164 abolished transforming activity of SIS, and decreased dimer formation. Mutation of Cys154/Cys163 or Cys154/Cys164 had minimal effect on transforming activity, but significantly reduced the level of dimer formation (47). These results indicate that SIS may function as a monomer, and predicted that Cys154 and Cys163 participate in a non-essential disulfide bond. Mutations in analogous Cys residues of PDGF-A appear to confirm the existence of an intermolecular bond between corresponding residues (48), and demonstrate that an intramolecular disulfide bond forms between Cys10 and Cys91 of PDGF-A (Cys127 and Cys208 of SIS). Recent research has demonstrated that a PDGF-B protein with mutations at Cys43 and Cys52 (analogous to the Cys154/Cys163 double mutant of SIS described by Sauer and Donoghue) is mitogenically active and can form noncovalent dimers at pH 4-7 (49). Thus, dimerization is required for activity of these growth factors, but this dimerization does not necessarily require disulfide bond formation.

Given this recent realization, it is likely that other residues in PDGF and related proteins assist in stabilizing dimer formation. There are 13 non-cysteine residues in SIS that are conserved in VEGF. Maher et al. (50) used alanine-scanning mutagenesis to analyze the relative importance of these residues for both dimerization and transforming activity of *SIS*. They found that Pro152, Gly162, Arg159 and His205 were important in mediating dimerization. Four other residues -- Asn147, Phe148, Leu149 and Lys185 -- were not required for dimerization, but did yield proteins with decreased transforming activity if converted to alanine (50).

Mutational analysis of PDGF has also yielded information regarding the residues required for interaction with the PDGF receptors. LaRochelle et al. (51) created chimeric PDGF molecules by swapping different domains of PDGF-A and PDGF-B. They localized the ability to activate PDGFR-β to a region of SIS that lies between amino acids 105 and 144. This region was further mapped to within amino acids 136 and 148 of SIS (amino acids 25-37 of PDGF-B), by using an epitope-specific mAb to PDGF-B (52). Mutations in this region do not affect dimerization, but prevent receptor activation and hence biological activity; binding of the mAb (which recognizes an epitope within amino acids 136 to 139) also blocks receptor activation. Clements et al. (53) further localized the residues responsible for receptor activation to Arg27 and Ile30 of PDGF-B (53). Mutation of Arg27 led to a molecule that was wild type in its ability to stimulate DNA synthesis, but which exhibited decreased receptor binding and also decreased stimulation of PI (phosphatidylinositol) turnover. Mutation of Ile30 adversely affected activity of the molecule in all three assays (53). Similar studies have been undertaken with the A-chain. Jaumann et al. (54) exchanged various non-conserved residues in PDGF-A with the corresponding residues of PDGF-B in order to map residues that confer specificity of binding to the PDGFR-β. They showed that mutation of Pro26→Arg and Ser28→Asn in PDGF-A was sufficient to switch the binding specificity of the A-chain from the α-receptor to the β-receptor (54). It has also been demonstrated that a basic stretch of amino acids -- Arg159-Lys160-Lys161 -- which is present in both the A and B chains of PDGF, is critical for binding to the α-receptor (55).

Under some conditions, PDGF-BB is significantly more potent in inducing transformation than PDGF-AA (56). This may simply reflect the ability of the B-chain to activate both α- and β-receptors, or may be due to inherent differences in the amino acid sequences of these two proteins. SIS, a homologue of the B-chain, contains additional amino acids at both its N- and C-termini that are not present in PDGF-B. Deletion of these regions defined the minimal transforming region of this oncoprotein as residues 127-214 (34). All of these residues lie within the PDGF-B-homologous region. The studies of Aaronson and colleagues described above also demonstrated that amino acids 105 to 144 of the PDGF-B chain are responsible for the increased transforming activity of this PDGF isoform (51). More recent work by these researchers indicates that substitution of five amino acids of PDGF-A with the analogous residues from the B-chain can convert PDGF-A to a highly-transforming molecule. These B-chain residues are Arg109, Asn115 (or Arg109 with Leu110 and Arg113), Asn136, Arg137 and Arg142 (57). Perhaps these residues create tighter binding with PDGFR-α and/or PDGFR-β, allowing for more efficient and persistent activation of these receptors, and thus of signalling pathways that lead to transformation. Whatever the case, it is clear that specific amino acid residues in the PDGF family members play specialized roles in conferring specificity and activity on these molecules.

Given the knowledge that PDGF isoforms and PDGF receptors are often coexpressed in some tumor cells and cell lines, there has been much effort invested in the identification of dominant inhibitory mutants of PDGF. The hope is that such proteins would interfere with binding of PDGF to its receptors in these tumors, allowing for treatment of cancers that involve such autocrine loops. Mercola et al. (58) used site-directed mutagenesis to alter PDGF-A in an attempt to identify dominant negative mutants. One mutant, called 1308, has Cys129 changed to Ser, and can inhibit activity of both A- and B-chains of PDGF by forming dimers with the wild-type species (58). Another mutant, 1317, has the basic stretch Arg84-Arg85-Lys86-Arg87 changed to Arg-Ser-Asn--Gly. Interestingly, this mutation prevents processing of PDGF-A at the propeptide cleavage site, and renders homodimers of this mutant PDGF-A chain inactive. This contradicts results obtained with the PDGF B-chain, that show processing is not required for biological activity (33). However, the 1317 mutant forms heterodimers with PDGF-B, and these still function. Subsequently, these researchers demonstrated that these two mutants, 1308 and 1317, can revert the transformed phenotype of cells overexpressing PDGF-A (1308 and 1317) or PDGF-B (1308 only) (59). They also demonstrate that the 1308 mutant can revert the transformation of some astrocytoma cell lines, such as U343 and U87 (59).

Vassbotn et al. (60) designed a mutant of PDGF-A that was still able to form homodimers and heterodimers with wild-type PDGF-B, but that no longer bound PDGF receptors. This was accomplished by substituting a region of PDGF-A (amino acids 156-162) with the analogous region from VEGF. This region, at least in PDGF-B, has been implicated in mediating the binding of PDGF-B to the PDGFR-β (61). Expression of this mutant PDGF-A, called PDGF-0, in cells transformed by PDGF-B resulted in reversion of the transformed phenotype, indicating a dominant mechanism of interference of transformation by this mutant. Taken together, these results indicate that it may be possible to treat some tumors that exhibit an autocrine transformation loop with dominant interfering proteins.

RECEPTORS FOR THE PDGF FAMILY

Receptor tyrosine kinases have proven to be critical modulators of growth factor responses. These receptors function by ligand-induced dimerization and activation of their intracellular kinase domains. Both PDGF and VEGF bind to receptors that are tyrosine kinases. This section will describe the structure of these and other related tyrosine kinase receptors and summarize the current knowledge about what cellular proteins participate in downstream signal transduction.

"Split-kinase" Domain Characterizes PDGF and Related Receptors

There are two types of receptors for PDGF, designated PDGFR-α and PDGFR-β (62,63). These receptors form homodimers and heterodimers, and these dimeric receptors differ in their ability to bind the various PDGF isoforms. The PDGFR-α binds all three dimeric forms of PDGF (AA, AB and BB) with comparable affinities, while PDGFR-β binds the BB isoform with a much greater affinity, and binds AA poorly, if at all (64,65).

Other members of the PDGF family activate receptors that are structurally similar to PDGFR-α and -β. The initial characterization of receptors for VEGF was performed by Vaisman et al. (66) and suggested the presence of more that one receptor class. A VEGF receptor was subsequently cloned and identified as the *fms*-like tyrosine kinase Flt1 (67). The gene encoding Flt4, a related receptor, was then cloned (68) and localized to the q arm of chromosome 5 (69). Two isoforms of Flt4 were detected with different C-terminal domains as a result of alternative splicing (70). Birnbaum and colleagues demonstrated that *flt4* was a receptor tyrosine kinase that strongly resembled Flt1 in structure, and is thus likely a receptor for VEGF as well (71). *KDR/FLK1* has also been cloned, and appears to be a third member of the VEGF receptor family (72). Studies on the expression patterns of these three receptors have shown distinct but overlapping distribution in tissues, especially in vascular endothelial cells (73,74). These receptors are structurally related to the PDGF receptors (see Figure 2); however, it is not clear whether these receptors form only homodimers, or can heterodimerize with each other. The receptor or receptors for

PlGF have not yet been identified, but it has been demonstrated that PlGF can compete with $VEGF_{165}$ for endothelial cell receptor binding (43), indicating that PlGF and VEGF may exhibit overlapping patterns of receptor binding.

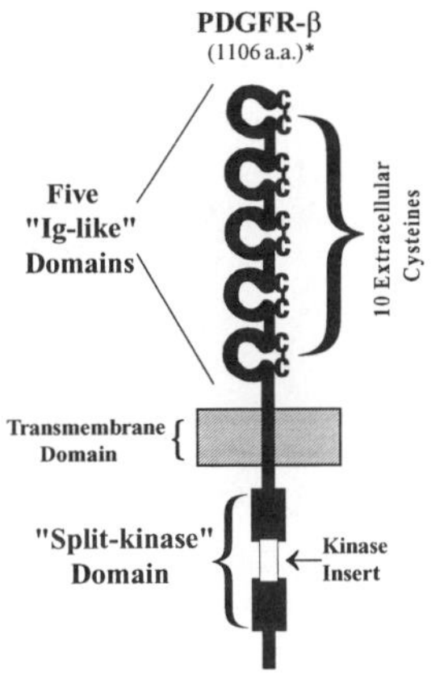

Related Receptors	amino acids (human)	References
PDGFR-α	1089	63
CSF-1R/c-*fms*	972	79
c-*kit*	976	76
*flt1***	1338	67
*flt4***	1298	71
KDL/*FLK1***	1356	72

*Reference for PDGFR-β (62).
**These receptors have seven immunoglobulin-like ("Ig-like") domains and 12 extracellular cysteines.

Figure 2. Structure of PDGFR-β and related receptors. The figure on the left shows the important structural features of PDGFR-β and related receptors, including the immunoglobulin-like ("Ig-like") domains, the single membrane spanning region, and the "split-kinase" domain with the kinase insert region. Listed in the table on the right are the related receptors and their amino acid lengths.

As shown in Figure 2, the basic structure of these receptors consists of an extracellular ligand-binding domain with either five (for PDGF receptors) or seven (for VEGF receptors) immunoglobulin-like domains, a single membrane-spanning domain, and an intracellular tyrosine kinase domain. There are 10 Cys residues in the extracellular domains of PDGFR-α and -β, and the spacing of these residues is absolutely conserved between the two receptors. The kinase domains of these receptors are unique in that they all have a "kinase-insert" domain of approximately 100 amino acids between the two segments of the kinase domain. The location but not the sequence of the "kinase-insert" domain is conserved between members of the receptor family. The CSF-1R, c-*fms*, and c-*kit* are also members of this family of "kinase-insert" receptors (75-79). The sequence similarity between PDGFR-α and -β is 44%, with the most related regions being the kinase domain and juxtamembrane regions (80). There is also sequence variation within the C-terminal domain of the receptors, but the relationship of this structural difference with function of individual receptor subtypes is not known.

Biosynthesis of the PDGF receptors has been studied with the use of metabolic labeling of cells and subsequent immunoprecipitation with antibodies to the receptor protein (81,82). In BALB/c 3T3 fibroblasts, the β-receptor is synthesized as a 160 kD precursor, which then gets converted to the mature, 180 kD form within 45 minutes (81). Similar results were obtained when receptor processing in human foreskin fibroblasts was studied (82). The 32 amino acid signal sequence is removed, and treatment of the receptor with glycosidases to remove N-linked oligosaccharides and sialic acid further reduces the molecular weight of the mature receptor to 120 kD (62,83). Thus, approximately 30 to 40 kD of the apparent molecular weight of the receptor when analyzed by SDS-PAGE is due to N-glycosylation and addition of sialic acid. This was confirmed by treatment of cells expressing PDGF receptors with endo-β-N-acetylglucosaminidase H (Endo H), which reduces the molecular weight of the immature, 160 kD form, but not the 180 kD form of the receptor, indicating the presence of complex forms of carbohydrate. Treatment with swainsonine, which inhibits the action of mannosidase II, leading to incomplete processing of "high-mannose" to "complex" types of carbohydrate, yielded 168 kD forms of the receptor (81). Interestingly, this partially processed form of the receptor was still responsive to PDGF, indicating that some post-translational modifications of PDGF receptors are not required for biological activity. Other modifications, such as O-linked glycosylation, ubiquitination and phosphorylation account for the remaining discrepancy in molecular weight. The half-life of the 180 kD, mature receptor is approximately 3 hours, but decreases to 45 minutes upon stimulation with PDGF (81). This is due to an increase in receptor downregulation or degradation, induced by ligand binding. This downregulation of the receptor may be important in regulating the activity of PDGF, but can also be exploited in order to study activation of cell surface versus intracellular receptor forms. Stimulation of cells expressing SIS or derivatives with PDGF-BB, followed by immunoprecipitation and assay of the receptors by *in vitro* kinase reaction, can be used to demonstrate the relative levels of available, activatable surface PDGF receptors and thus confirm expression of a construct at the cell surface (84-86). The mechanism responsible for downregulation of activated PDGF receptors is not known, and many of the studies described below involving mutant receptors show that most mutations do not affect this activity. However, it was shown that lack of PI3K (phosphatidylinositol-3-kinase) binding prevents internalization of PDGF receptors (87), consistent with the observed homology between the p110 subunit of PI3K and VPS34P, a yeast protein involved in intracellular trafficking (88).

Binding of PDGF ligand also induces dimerization of PDGF receptors (89). The stoichiometry of this process is one ligand molecule interacting with two receptor molecules. The species of dimer formed will vary, depending on the distribution of the two receptor subtypes on the particular cell of interest, and the ligand used to stimulate the cells (PDGF-AA, -AB or -BB). For example, $\beta\beta$-receptors can be activated only by PDGF-BB, whereas $\alpha\beta$-receptors can be activated by AA, AB or BB isoforms (65,89). Under most circumstances, dimerization of the receptor appears to be required for activation of its tyrosine kinase activity (89). The receptors, upon activation, undergo autophosphorylation at a number of specific tyrosine residues (see below). This generates binding sites on the receptor to which SH2-domain-containing proteins can bind (90-94). Specificity in binding of SH2-containing-proteins to tyrosine-phosphorylated proteins is conferred by the amino acid context of the phosphotyrosine residue (90,91,95). Thus, as one can imagine, there is an association of a variety of different proteins with the activated PDGFR.

A variety of mutagenesis approaches have been used in studying the PDGF receptors and associated proteins. Deletion studies of different domains or regions of the receptors have demonstrated the importance of such structures as the kinase-insert and C-terminal domains in the function of the receptors. Escobedo and Williams (96) utilized this approach to show that deletion of the kinase-insert from the receptor affected some, but not all, responses to PDGF (96). For example, downregulation of the receptor, PI hydrolysis, changes in cellular pH and calcium levels were unaffected by the deletion, but mitogenesis was severely impaired. The size of the deletion was also important, as Heidaran et al. showed by the effects of two different size deletions of the kinase-insert of PDGFR-α (97). These authors found that progressive deletions progressively altered binding of certain proteins to the receptor. In other experiments, Escobedo et al. (93) deleted the C-terminal domain of PDGFR-β, to determine what effect this would have on signalling. The strategy of deleting regions of a protein suffers, however, from the lack of knowledge of how the deletion will affect the overall structure of the protein. Some deletions may grossly alter tertiary structure, and cause a loss of activity or an effect that then may be unfairly attributed to that domain. Chimeric molecules have also been generated by swapping analogous regions between the α- and β-receptors, or between receptors within the same family. Substitution of the kinase-insert domain of PDGFR-α with that of c-*fms* provided insight into the function of the kinase-insert domains of these receptors (97). This approach was also exploited to create a chimeric receptor containing the ligand binding domain from PDGFR-α and the transmembrane domain and intracellular domain from PDGFR-β (98), allowing examination of the substrate specificities of the two receptors. Substitution of related domains is probably a better way to study the function of a protein than simply deleting a region of interest. This method preserves the tertiary structure of the protein (as long as the swapped regions are significantly homologous), and thus makes interpretation of results simpler. Point mutations have yielded the largest amount of information about the structure/function relationships within the PDGF receptors. To determine if the kinase activity of the receptor is required for any of its signalling functions, Williams and colleagues (99) mutated Lys602→Ala (K602A) in the putative ATP-binding site of PDGFR-β. Many other point mutations have been made in the receptor, primarily at sites of autophosphorylation. In the case of a mutation of Tyr857→Phe, it was discovered that this is a critical site for autophosphorylation and activation of the receptor itself (100). Such mutants have, in some cases, allowed for determination of specific binding sites for effector molecules. The knowledge that has been gained by mutational analysis of the PDGF receptors in relation to these effector molecules will be discussed below.

Mediators of the Signal Transduction Pathway of PDGF

PDGF stimulates a variety of cellular responses upon activation of its receptors, including increased DNA synthesis, PI turnover, increased mobilization of intracellular calcium, alterations in cellular pH, and increased phosphorylation of cellular proteins (101-103). As mentioned earlier, autophosphorylation of the receptor generates binding sites for effector molecules. To date, the following molecules are known to directly bind to the PDGF receptor complex: PI3-kinase (104,105), PLC-γ (106-109), rasGAP (110,111), Raf-1 (112), pp60*src*, pp62*yes* and pp59*fyn* (113), Syp (114,115) and GRB2 (116,117). In addition, other molecules have been detected in association with the activated receptor, but the identities and functions of these proteins are not known. Some of these molecules appear to overlap in their functions, but all bind to a distinct site on the activated, phosphorylated receptor. These effector molecules are discussed below and summarized in Table 1.

Phosphatidylinositol-3′-Kinase (PI3K)

PI3K consists of two subunits -- a 110 kD catalytic subunit and an 85 kD regulatory subunit. There appear to be alternative forms of the 110 kD protein, which may confer additional specificity. The enzyme catalyzes the addition of a phosphate group to the D3 position of the following inositol phosphates: PI, PI-4-P and PI-4,5-P_2 (118,119). A PI kinase activity co-immunoprecipitated with PDGF-stimulated phosphotyrosine-containing proteins (120,121). Some of these novel second messengers generated by PI3K may be important in mediating PDGF-induced effects.

To examine if PI kinase directly associates with PDGFR-β, Coughlin et al. (105) used PDGFR-β-specific antisera to co-immunoprecipitate PI kinase activity. In these studies, receptor mutants were tested to examine the structural requirements for association. A Y825F receptor (containing a mutation of a putative Tyr autophosphorylation site to Phe) (122) did not lack kinase activity, and behaved like the wild-type receptor in being able to associate with PI kinase upon stimulation with PDGF. Both of the receptor mutants, K602A (mutation of Lys602 at the ATP-binding site to Ala) and Δ-ct (deletion of a portion of the carboxyl-terminus), lacked kinase activity (99) and were unable to associate with the PI kinase activity when expressed in CHO cells (105). Additionally, a kinase-insert deletion mutant (Δ-ki) of PDGFR-β (96) retained autophosphorylation ability, but was also unable to associate with PDGF-stimulated PI kinase activity, indicating that both enzymatic activity of the receptor and an intact kinase-insert domain are important for retention of receptor-associated PI kinase activity (105). Deletion of the kinase-insert domain also abolished PDGF-stimulated mitogenic activity in CHO cells, but with no effect on PI turnover (96), implicating activation of a PI kinase in the generation of mitogenic signals from the PDGF receptor complex. Heidaran et al. (97) also analyzed the role of the kinase-insert domain in the homologous receptor PDGFR-α. Progressive deletion of the kinase-insert domain demonstrated that this domain of the PDGFR-α is also important for association of PI kinase activity. Exchanging the PDGFR-α kinase-insert domain with that of c-*fms* yielded an essentially wild-type phenotype (97), indicating that although this region is not closely related between the two receptors, it may serve a similar function, namely allowing association of a PI kinase with the receptor.

Since Tyr751 lies within the kinase-insert domain and is a major site of autophosphorylation in PDGFR-β (104), Kazlauskas and Cooper mutated Tyr751→Phe to examine the effect on associated PI kinase activity. They demonstrated that phosphorylation of Tyr751 is required for retention of a PI kinase. Analogous mutations were made in PDGFR-α by mutating Tyr731 or Tyr742 to Phe. PI3 kinase is no longer able to associate with these mutant receptors (123). Williams and colleagues used an *in vitro* system that tests the ability of specific peptides to block interaction of the receptor with associated proteins to verify these findings, and to determine that both phosphorylated Tyr751 and Tyr740 (Tyr719 and Tyr708 in the mouse β receptor) are able to bind PI kinase (93,94). They also demonstrated that binding of a 110 kD protein (the catalytic subunit of PI3K) was inhibited by the peptides, indicating that perhaps the 85 kD protein serves as an adaptor for association of the active subunit of the enzyme. The peptide containing phosphorylated Tyr719 also blocked binding of PI3K to c-*fms*, although the peptide is not an exact homologue of any phosphorylated region on c-*fms* (124), which again illustrates that the kinase-insert motif may play similar structural roles in different receptors. The binding sites for PI3K were further confirmed by binding studies performed with receptors containing point mutations at Tyr741, Tyr751 or both (all mutated to Phe) (125). It is likely that the two binding sites have differing affinities for PI3K, and that phosphorylation of both sites yields maximum binding.

PI3K has two SH2 and one SH3 (src-homology) domains. The C-terminal SH2 domain is responsible for the high affinity binding of PI3K to PDGF receptors (126,127). Although the binding of the N-terminal SH2 domain appears to be less specific, binding of this second SH2 domain contributes additional stability to the complex (126). Kavanaugh et al. (128) demonstrated that the p85 subunit of PI3K is modified by phosphorylation in or near the N-terminal SH2 domain upon stimulation of cells with PDGF. This modification inhibits the binding to the receptor, consistent with its decreased affinity. Perhaps such modifications allow a region of p85 to be available for other effector molecules or substrates to associate with the receptor signalling complex.

In TRMP cells (canine epithelial cells), lack of PI3K binding to the receptor (due to mutation of Tyr740 and Tyr751 to Phe in the receptor binding domains) does not drastically affect stimulation of DNA synthesis (125), from which one might conclude that PI3K activity is not important in this response. Yu et al. (123) also mutated the corresponding tyrosines in PDGFR-α (Tyr731 and Tyr742) and showed no correlation between reduced PI3K association and DNA synthesis response. However, as mentioned previously, Δ-ki receptors expressed in CHO cells also abolished mitogenesis in response to PDGF (96). Also, the mutational studies performed with the mouse PDGFR-β and BALB/c 3T3 cells, and the corresponding mutations abolishing PI3K association (Tyr708→Phe and Tyr719→Phe), also severely impaired PDGF-mediated DNA synthesis (94). Additionally, the F5 mutant receptor (with all five autophosphorylation sites -- Tyr740, Tyr751, Tyr771, Tyr1009 and Tyr1021 -- mutated to Phe) was impaired in its ability to mediate a DNA synthesis response; however, allowing association of just PI3K (by changing Phe740 and Phe751 back to Tyr) completely restored this response to PDGF (129). This apparent paradox may be due to cell type-specific differences in the mitogenic signalling pathways of PDGF. Alternatively, multiple mitogenic pathways may be activated upon PDGF-stimulation, so that lack of binding of one protein may be compensated for by other associated proteins. Thus, the "add-back" mutants described by Valius and Kazlauskas (129) may simplify the interpretation of structure/function relationships within the PDGF receptors, by eliminating potentially compensatory molecules from binding (although there still may be other as yet unidentified proteins associated with the receptor that play a role). The conclusion that PI3K plays a central role in DNA synthesis is strengthened by the observation that microinjection into cells of antibodies against the catalytic p110 subunit completely inhibits PDGF-induced mitogenesis (130).

Another response that might be attributed to PI3K is PI turnover. However, the putative product of PI3K activity, PI-3,4,5-P_3, has no described turnover pathway. Furthermore, studies with mutant receptors show no effect on PI turnover due to lack of PI3K binding to receptors (123,129) Another response to PDGF, chemotaxis, does seem to require activation and association of PI3K (97,131). Downregulation and trafficking of PDGFR-β receptors also require PI3K (87). Thus, PI3 kinase plays a part in a variety of the PDGF-induced signal transduction pathways, but further study is needed precisely to define its role.

Phospholipase-C-γ

PLC-γ is an approximately 150 kD phosphatidylinositol-4,5-bisphosphate-specific phosphodiesterase that catalyzes the breakdown of phosphoinositides from the membrane into inositol triphosphate (IP_3) and diacylglycerol (DAG), which increase intracellular calcium levels and activate protein kinase C, respectively (132,133). PLC-γ becomes phosphorylated on Tyr771, Tyr783 and Tyr1254 in response to PDGF (106,109), and indeed associates with activated PDGF receptors (108). The generation of the second messengers IP_3 and DAG may play a crucial role in stimulation of PDGF-related responses, either directly or by PKC activation. However, activation of PKC may occur by other pathways, as demonstrated by the observation that in the 32D murine myeloid cell line, PKC-δ becomes phosphorylated and translocates to the membrane in response to PDGF stimulation (134). This activation leads to differentiation of these monocytic cells.

Direct association between PLC-γ and PDGFR-β was shown to be PDGF-dependent, and requires phosphorylation of the receptor (107). Δ-ki or Y825F receptor mutants were unable to abolish association of PLC-γ; however, a kinase-defective receptor (K602A) did not bind the enzyme (107). Similar observations were made with a PDGFR-α Δ-ki mutant -- no effect was seen on phosphorylation of PLC-γ (97). Binding to the receptor leads to increased phosphorylation of PLC-γ on both serine and tyrosine. Immunoprecipitation of PLC-γ also brings down many other associated proteins, including PI3K, Raf-1, a 55 kD protein and a 120 kD protein. Presumably this is through a complex with the receptor, which is also co-immunoprecipitated. Perhaps one of these unidentified receptor-associated proteins, upon activation by the PDGF receptor, contributes to the increased serine phosphorylation of PLC-γ (Raf-1 is a serine/threonine kinase -- see more below). It was originally thought that PLC-γ and rasGAP bound to similar sites on phosphorylated PDGF receptors. This binding occurs through SH2 domains, and perhaps these domains are similar in structure and/or specificity between PLC-γ and GAP (92). However, Kazlauskas and colleagues demonstrated by peptide mapping of activated receptors that Tyr1021 was required for stable association of PLC-γ, but not GAP (114). This was confirmed by mutating Tyr1021→Phe and showing loss of PLC-γ binding. The N-terminal SH2 domain of the protein has the higher affinity for the receptor (92). A series of "add-back" mutants constructed by the same researchers further confirmed the requirement for phosphorylation of Tyr1021 in binding of PLC-γ to activated PDGFR-β (129). It has also been shown that the α-receptor has approximately the same affinity for PLC-γ as the β-receptor (98).

The activation of PLC-γ, as with PI3K, appears to play variable roles in DNA synthesis, depending on the cell type examined. In C3H10T1/2 cells and in porcine aortic endothelial cells, inhibition of PLC-γ binding did not affect PDGF-induced DNA synthesis (135,136). However, in TRMP cells and HepG2 cells expressing Y1021F mutant receptors, which no longer associate with PLC-γ, the DNA synthesis response was decreased to approximately one-third that of wild type (114). Intriguingly, if only PLC-γ is allowed to bind to receptors (in the context of the F5 mutant, lacking tyrosine phosphorylation sites), DNA synthesis is fully restored in these two cell types (129). Again, these results illustrate the complex and often redundant nature of the signalling pathways activated in response to PDGF.

Association of PLC-γ with the receptor correlates well with ability of PDGF to simulate PI hydrolysis (107). This is consistent with the function of PLC-γ, and also with the observation that Δ-ki and Y825F mutant receptors, while defective in some responses, can still bind PLC-γ and still initiate PI hydrolysis (96). However, receptors with point mutations in the ATP binding site (K602A) or deletions in the C-terminal domain are unable to stimulate this response (99), suggesting that kinase function of the receptor is important for this process, and possibly that association of the enzyme with the receptor requires residues within the C-terminal domain. Indeed, it was later shown that mutation of Tyr1021→Phe (the PLC-γ binding site) led to decreased ability of the receptor to mediate PI turnover (114), and restoration of PLC-γ binding restored the PI turnover response (129). Interestingly, Δ-ki PDGFR-α mutants were significantly impaired in PI turnover, even though this region is not involved in directly binding PLC-γ (97). Swapping the kinase-insert domain of the α-receptor with that of the c-*fms* receptor restored this response to PDGF. Thus, the kinase-insert may function differently in the α-receptor and participate in some way in the PI turnover response, whereas the β receptor does not require the kinase-insert to stimulate PI hydrolysis. PLC-γ has also been implicated in promotion of chemotaxis toward PDGF in CHO cells, TRMP cells and NIH 3T3 cells (131). Thus, it appears that PI3K and PLC-γ overlap significantly in their mediation of PDGF-induced responses. Perhaps this overlap provides a sort of "backup" system so that, if one pathway is not functioning, the other will overcome the deficiency. Otherwise, fine-tuning of the responses to PDGF may involve modulating both pathways.

ras-GTPase Activating Protein (rasGAP)

RasGAP is a 120 kD protein that regulates function of p21ras in vivo by catalyzing the conversion of bound GTP to GDP, thus inactivating p21*ras* (137). The level of p21*ras*-GTP in PDGF-stimulated cells increases 2 to 3-fold above the basal level (138), indicating activation of p21*ras* may be involved in PDGF signal transduction. PDGF stimulation of cells initiates a rapid phosphorylation of GAP on tyrosine (110,111,139). Since GAP is a negative regulator of p21*ras*, it is likely that this phosphorylation inhibits GAP function, allowing conversion of p21*ras* to its GTP-bound, active form. Alternatively, phosphorylation of GAP may allow it to perform other effector functions in PDGF-stimulated signal transduction.

The Δ-ki mutant PDGFR-β has decreased GAP binding (110). The precise binding site for GAP on the receptor was subsequently mapped to a residue within 20 amino acids of the Tyr719 binding site for PI3K (Tyr751 in human PDGFR-β) by testing the ability of synthetic tyrosine-containing peptides corresponding to receptor sequences to block GAP binding (94). This site, Tyr739 of the mouse PDGFR-β, was confirmed to be the GAP binding site in the human receptor (Tyr771) by Kazlauskas and colleagues (125). GAP also binds PDGFR-α at the analogous site (Tyr762), but the affinity of the α-receptors for GAP is about 10-fold lower than that of the β-receptor (98). In fact, the use of a chimeric receptor with an extracellular α-receptor ligand-binding domain, but an intracellular β-receptor kinase domain, demonstrated a correlation between the increased ability of PDGF-BB to transform cells (*versus* PDGF-AA) and the ability of the β-receptor to associate with GAP (98).

Structurally, rasGAP consists of two SH2 domains separated by an SH3 domain in the N-terminal half, and the GTPase activating domain in the C-terminal half of the protein. Cooper and Kashishian (126) used intricate swaps of the SH2 domains between PI3K and GAP to demonstrate that the C-terminal SH2 domain of GAP binds to Tyr771. The N-terminal SH2 domain is also required for stable binding, but the site to which it binds is unclear. Lack of GAP phosphorylation correlates with lack of receptor binding (140), but GAP does not need to be phosphorylated in order to bind the receptor (92,125).

RasGAP function in mediating DNA synthesis in response to PDGF is well documented. Lack of GAP binding to the PDGFR-β has no effect on DNA synthesis in PDGF-treated NMuMG cells (94). This is consistent with results obtained in TRMP cells (125), indicating that GAP binding to receptors is not a prerequisite for mitogenesis in response to PDGF. GAP does however appear to be important in suppressing chemotaxis of cells toward PDGF (131). Prevention of GAP binding to receptors does not affect the ability of PDGF to activate p21*ras* in CHO cells (141). MAP kinase activation also occurs normally in cells expressing receptors that lack GAP binding sites, indicating PDGF signalling may feed into the MAP kinase pathway *via* p21*ras*, but this does not involve GAP. It has recently been shown that p21*ras* can directly interact with the PI3K p110 subunit and enhance its enzymatic activity (142). Association of p85 with PDGF receptors is probably required to localize PI3K to the plasma membrane, where its activity can be modulated by p21*ras*. Thus, reduction of GAP activity (by its association with and phosphorylation by PDGF receptors) is not responsible for activation of p21*ras*, and GAP likely plays other effector roles, perhaps serving as an adaptor for other signalling molecules to localize to the PDGF receptor complex.

Raf-1

The 74 kD, ubiquitously-expressed proto-oncogene Raf-1 (143) is a serine/threonine-specific kinase. Initial evidence that Raf-1 was involved in signal transduction by receptor tyrosine kinases came from observations that Raf-1 becomes hyperphosphorylated in response to some growth factors, including PDGF (144). Subsequently, Raf-1 was shown to be directly activated by PDGF stimulation of cells expressing wild type PDGF-β receptors, but not in cells expressing mutant PDGF-receptors (112). This activation consists of phosphorylation on tyrosine and significant increase in its serine/threonine kinase activity. Raf-1 was also shown to bind directly to the activated PDGF receptor, although the specific binding site has not yet been elucidated.

The role of Raf-1 in PDGF-mediated signal transduction remains a mystery. Demonstration of a direct association between Raf-1 and p21*ras* (145,146) suggests that

Raf-1 may propagate a signal downstream of p21*ras*. Raf-1 has also been implicated as a player in the MAP kinase pathway (147), further placing Raf-1 downstream of p21*ras*. Raf-1 may simply be translocated to the plasma membrane by interaction with p21*ras*, bringing it into proximity with other PDGF receptor signalling molecules, where it can be acted on or activate other proteins (148,149). *In vitro*, Raf-1 is able to phosphorylate PLC-γ on serine, at sites that are phosphorylated *in vivo* in response to PDGF (107), indicating a possible role in modulating PLC-γ activation. Future studies on Raf-1 should shed additional light on its importance in PDGF-mediated signal transduction.

Src Family Kinases

Three members of the src-family of tyrosine kinases are ubiquitously expressed -- pp60*src*, pp59*fyn* and pp62*yes*, and are present at approximately equal levels in fibroblasts (113). Their localization at the plasma membrane of cells indicates they may be substrates for activated receptor tyrosine kinases. The activity of c-*src* increases transiently in PDGF-stimulated fibroblasts (150,151), possibly suggesting participation in this signal transduction pathway. Increased autophosphorylation of *src, fyn* and *yes* kinases was observed in response to PDGF-treatment of cells (113). These researchers also demonstrate association of the *src*-family proteins and PI3K, although receptor is also co-immunoprecipitated. Thus, the *src*-family kinases are likely members of the PDGF receptor signalling complex. Only 5 to 10% of the activated PDGF receptors associate with *src* kinases, but this low stoichiometry is also seen for PLC-γ (108) and Raf-1 (112). Thus, there may be only a threshold level of proteins that need to be activated in order to see a growth-factor mediated response.

Further examination of the requirements for association of *src, fyn* and *yes* kinases with the PDGF receptor led to discovery that this interaction requires no intermediate molecules, and can be mediated by the SH2 domain alone (in the case of *fyn*) (100). The association only occurs with phosphorylated receptors, and *src, fyn* and *yes* kinases appear to become phosphorylated as a result of receptor association. In the case of *fyn*, the phosphorylation occurs in the N-terminal region of the protein, and involves both tyrosine and serine/threonine residues. Subsequent research has shown that activated pp60*src* in PDGF-stimulated cells translocates from the plasma membrane to the cytosol, and that the serine/threonine phosphorylation of the N-terminus of pp60*src* is catalyzed by cAMP-dependent protein kinase (152). Phosphorylation of the human PDGFR-β at Tyr857 is required for efficient association of these kinases with the receptor. Lack of phosphorylation at Tyr751 prevents activation, but not association of these proteins, indicating that receptor association does not automatically lead to activation (153). The actual binding sites for these proteins, however, are in the juxtamembrane portion of the receptor. Specifically, mutation of Tyr579 and Tyr581 to Phe generates receptors with decreased ability to bind and activate *src, fyn* and *yes* kinases (154). The specificity of this binding was confirmed by binding of synthetic peptides corresponding to this region (154).

Inhibiting the binding of *src*-family kinases to either Tyr579 or Tyr581 on PDGF β-receptors did not affect PDGF-induced mitogenesis. However, a double mutation of both of these Tyr residues to Phe renders the receptor kinase-inactive; thus no firm conclusion can be drawn regarding mitogenesis and *src*-family kinases (154). However, the microinjection of dominant negative mutants of *src*, *fyn*, or *yes* leads to arrest of cells before S phase, and inhibition of PDGF-mediated signalling (155). Similarly, injection of antibodies to *src*-family kinases reduces signalling through PDGF receptors (156). Taken together, these results indicate that src-family kinases may mediate some of the mitogenic effects of PDGF.

Syp (SHPTP2)

A 64 kD protein is consistently found to be phosphorylated and associated with PDGF receptors in PDGF-stimulated cells. The level of phosphorylation of this protein, however, does not appear to be altered greatly due to PDGF treatment in cells overexpressing c-*ras* or v-*ras* (157). Analysis of PDGFR-β containing mutations of Tyr1009→Phe demonstrated that this residue is required for association of this 64 kD protein with the receptor (114). The identity of this protein was recently determined to be the protein tyrosine phosphatase Syp (also known as SH-PTP2, PTP-1D, or PTP2C) (115).

Syp is tyrosine-phosphorylated in response to PDGF, but association with the receptor is not required for this modification. The receptor does not appear to be a substrate for Syp, since the level of its phosphorylation does not change upon Syp activation (115). Receptor mutants that were unable to associate with Syp were also shown to be unable to activate p21*ras*, indicating a possible role for Syp in this pathway (129). More recent research has defined the major site of phosphorylation of Syp as Tyr542, whose surrounding sequence resembles the binding site for the GRB2 SH2 domain (158). It was also demonstrated that GRB2 and Syp co-immunoprecipitate in PDGF-stimulated cells, indicating that one function of Syp is to couple GRB2 to the PDGF receptor (134,158). Syp may also play other roles, such as promoting dephosphorylation of sites of negative regulation on other effector proteins, or modulating PDGF-stimulated activity by stimulating dephosphorylation of regulatory tyrosine residues.

GRB2

GRB2 is a widely-expressed protein of 25 kD that is composed of two SH3 domains separated by an SH2 domain (116). GRB2 was shown to associate with PDGF receptors in a PDGF-dependent manner, and this is mediated by the SH2 domain. This binding is independent of phosphorylation of GRB2 itself; in fact, GRB2 doesn't appear to be phosphorylated to any significant level *in vivo*. Co-injection of GRB2 and p21*ras* led to a significant increase in DNA synthesis in REF-52 cells, suggesting that GRB2 is able to activate p21*ras*. A number of recent reports indicate that GRB2 associates with a 55 kD protein called Sos, which acts as a nucleotide exchange factor for p21*ras* (159-161). As mentioned above, Syp binds to GRB2, linking GRB2 indirectly to the PDGF receptor. Arvidsson et al. (117) have very recently identified a site on the PDGF receptor, Tyr716, that, when phosphorylated, allows GRB2 to bind to the receptor directly. Mutation of this site to Phe, while abolishing direct receptor binding of GRB2, still allowed for PDGF-stimulated increase in DNA synthesis and activation of the MAP kinase pathway, providing further evidence that there are at least two ways for GRB2 to activate p21*ras* in a PDGF-dependent manner (117).

Unresolved Issues

A variety of proteins associate with the activated, tyrosine-phosphorylated PDGF receptors. This association involves recognition of a phosphotyrosine residue by an SH2 domain contained within the associated protein. The binding is sequence-specific, in that the residues within the SH2 domain and around the phosphotyrosine residue determine what proteins can bind to which sites. The known binding sites for effectors of PDGF-mediated signal transduction are summarized in Table 1. Other proteins that have recently been shown to associate with the PDGF receptor include Nck (162,163) and Shc (164), but their roles in PDGF-mediated signalling are not yet well characterized. There are probably other unidentified proteins that associate with the receptor, either at novel sites, or compete with these proteins for known autophosphorylation sites. It appears that the effector molecules required for mediating biological responses to PDGF may vary depending on the cell type in question. PI3K and PLC-γ appear to play a major role in mediating DNA synthesis in TRMP and HepG2 cells (114), although this is not consistent when 10T1/2 or porcine aortic endothelial cells are examined (135,136). There is evidence that supports a role for almost all of these proteins in the mitogenic response to PDGF, and possibly two or more proteins must cooperate in order to yield full mitogenic output. It is fairly clear that PI3K, but no other protein, is required for receptor downregulation. It is also apparent that GAP inhibits chemotaxis, while PI3K and PLC-γ stimulate it. However, the precise role of each associated protein has not yet been defined.

Table 1. Binding Sites for Effectors of PDGF-Mediated Signal Transduction

Effector	**Molecular Weight**	**Binding Site on PDGFR-β**	**References**
PI-3-kinase	85 kD (regulatory subunit)	Tyrosines 740 and 751	93,94,104
PLC-γ	150 kD	Tyrosine 1021	114,129
rasGAP	120 kD	Tyrosine 771	125
Raf-1	74 kD	unknown	112
c-*src*	60 kD	Tyrosines 579 and 581	154
c-*fyn*	59 kD	Tyrosines 579 and 581	154
c-*yes*	62 kD	Tryosines 579 and 581	154
Syp	64 kD	Tryosine 1009	114,115
GRB2	25 kD	Tyrosine 716	117
Nck	45 kD	Tyrosine 751	163
Shc	46 kD and 52 kD	multiple sites	164

PDGF AND RELATED FACTORS AND DISEASE

The importance of PDGF and related growth factors in maintenance of normal cellular processes is clear. PDGF is a key mediator of wound healing (165) and functions in conjunction with IGF-I to stimulate osteogenesis (166). The abnormal expression of PDGF has been linked to a variety of proliferative processes, including atherosclerosis (167) and fibrosis (168,169). VEGF has also been implicated in proliferative processes such as psoriasis (170), rheumatoid arthritis (171) and cancers such as kidney and bladder carcinoma (172). This section will concentrate on several specific examples of the abnormal function of PDGF-related growth factors.

PDGFR-α and the *Patch* Locus

In mice, the dominant white spotting (W) locus is present on chromosome 5, and lies in a region of the chromosome that contains a number of PDGFR-related genes, such as *Kit* and PDGFR-α. A mutation known as *Patch* causes a white spotting phenotype in heterozygous mice. Mice that are homozygous for this mutation (*Ph/Ph*) exhibit gross anatomical abnormalities and experience a high mortality rate midway through gestation. The *Patch* mutation has recently been mapped to the PDGFR-α locus on chromosome 5 in mice (173). Schatteman et al. (174) examined the developmental expression of PDGFR-α and its relationship to the *Patch* phenotype. They found that by 9.5 days, homozygotes display gross deficiencies in mesodermal structures, and that this correlates with the lack of PDGFR-α gene expression. Approximately one-half of these homozygotes die, and the other half survives until succumbing to a second round of defects that occur in developing connective tissue, dermis and other tissues of mesodermal and neural crest derivation. Again, PDGFR-α is not expressed in these tissues in the mutants, but is expressed in normal mice (174). Thus, PDGFR-α is differentially expressed in development, and some tissues absolutely require this gene in order to develop normally. The *Patch* mutation has also been shown to affect development of non-neuronal neural crest-derived cells and tissues, for example the thymus, cornea and teeth (175). The visceral endoderm and mesoderm derivatives also require expression of PDGFR-α (176). Mutations in the PDGFR-α/*Patch* locus also affect melanocytes, decreasing receptor tyrosine kinase activity and thus preventing proper development and migration to the skin of these pigment cells, leading to the "spotted" phenotype in heterozygotes (177).

PDGFR-β in Chronic Myelomonocytic Leukemia

The PDGFR-β gene has also been implicated in the development of disease. Specifically, it appears that a translocation that occurs in CMML (chronic myelomonocytic leukemia) results in a fusion of a portion of PDGFR-β and *tel*, an *ets*-like gene (178). CMML is a subtype of MDS (myelodysplastic syndrome), which can be thought of as the transition between normal blood cell development and maturation and the development of AML (acute myeloid leukemia). The t(5;12)(q33;p13) translocation occurs in a number of CMML cases, and was shown to result in the production of a fusion transcript consisting of the HLH (helix-loop-helix) domain of *tel* in the 5′ end and the transmembrane and tyrosine kinase domains of the PDGFR-β in the 3′ end (178). The reciprocal transcript is not expressed, and the specific consequences of expression of the fusion transcript are not yet known. However, one can postulate a variety of mechanisms by which this transcript and the resulting fusion protein can lead to transformation and leukemia. *tel* is a member of the *ets* family of transcription factors, and normally has the HLH transactivating domain and a DNA binding domain. It is possible that the fusion protein resulting from the leukemia translocation leads to an inhibition or alteration of normal *tel* function by the HLH domain expressed in a heterologous environment. Since HLH domains can mediate protein-protein interactions, it is also feasible that the fusion protein undergoes ligand-independent dimerization, which results in constitutive tyrosine kinase activity of the PDGFR-β domain. Alternatively, the *tel*-derived domain may serve no real function in the protein, but instead provides a strong promoter that drives expression of the truncated and hence unregulated tyrosine kinase domain. Whatever the case, this and the above example indicate that the PDGF receptors and other family members play important roles in development and differentiation.

VEGF and Ocular Angiogenesis

As a major mediator of neovascularization, or the growth and proliferation of new capillaries, VEGF is critical for general growth and survival. The observations that ischemia (deficiency of blood and oxygen) is often associated with ocular neovascularization, and that hypoxia has been shown to upregulate expression of VEGF *in vitro* (179), suggest that VEGF may play a role in this pathology. Miller et al. (180), using a monkey model of iris neovascularization, have demonstrated a clear correlation between increased vascularization and increased secretion of VEGF. Laser vein occlusion was used to induce hypoxia in the retina. Levels of VEGF were undetectable in the aqueous of the anterior chamber before occlusion, but rose within 2 days after the injury. The rise in VEGF levels was followed by increased vascularization of the iris. VEGF is secreted by cells in the retina, and diffuses through the vitreous humor to the anterior chamber, where it induces blood vessel growth in the iris. The VEGF-producing cells were not definitively determined, but could either be Mueller cells or the smooth muscle cells (pericytes) of the vasculature in the retina, which are known producers of VEGF (181). Considering that diabetic retinopathy is a major cause of blindness in patients with diabetes, and involves a neoproliferative process, this finding is highly significant, and perhaps will help the development of a treatment for these and other disorders.

SIS and Autocrine Transformation

The coexpression of both PDGF ligands and PDGF receptors has been detected in a large number of tumors and transformed cell lines, including astrocytomas (182,183), meningiomas (184), the SMS-SB and NALM-6 pre-B-cell acute lymphocytic leukemia cell lines (185), and various fibro-, rhabdo-, leiomyo- and osteogenic-sarcoma cell lines (183,186). Coexpression of ligand and receptor can allow a cell to stimulate its own growth in an unregulated manner. This implies that an autocrine mechanism of growth stimulation may be the key to the development of some forms of cancer (1). Autocrine transformation, especially by SIS has been a topic of considerable controversy (see Table 2). A copious amount of evidence exists both in support of the notion that intracellular interactions between SIS and its receptors can lead to transformation, and in contradiction to such interactions being functionally relevant (19,84-86,187-194). Despite extensive efforts, the precise mechanism of autocrine transformation has not yet been defined.

Depicted schematically in Figure 3 are the possible sites of interaction between SIS and PDGF receptors in a cell that expresses both proteins. Since both are synthesized and processed by the same biosynthetic and secretory pathways, the possibility that these proteins can interact exists along the entire route, including the ER, Golgi and TGN (*trans*-Golgi network). Once the proteins reach the cell surface, they can of course interact as normal growth factors and receptors do. The question remains -- do the two proteins have to reach a cell-surface location in order to trigger the pathways that lead to transformation, or can these signalling pathways be activated from inside the cell by interaction between SIS and PDGF receptors?

Recent work has focused on directly assessing whether SIS can interact within the cell with PDGF receptors to elicit the transformed phenotype. Hannink and Donoghue (195) first tested the ability of SIS to function as a membrane-anchored growth factor. They made a fusion protein of SIS and the transmembrane domain of the G protein from vesicular stomatitis virus to form a Type I integral membrane protein. The resulting protein dimerized normally, and was proteolytically processed and glycosylated during transport to the cell surface. Most important, however, was the observation that the construct was still able to induce focus formation in NIH 3T3 cells, indicating that SIS can activate PDGF receptors even if membrane-anchored (195). More recently, Xu et al. demonstrated that SIS can retain biological activity if anchored to the membrane in a Type II orientation instead (196).

Lee and Donoghue (85) then examined whether SIS could transform NIH 3T3 cells if anchored within the cell using an ER retention signal, containing the dibasic motif DEKKMP, derived from an adenoviral transmembrane protein. Targeting to the ER prevented secretion of SIS, and also abolished its ability to cause transformation. Thus,

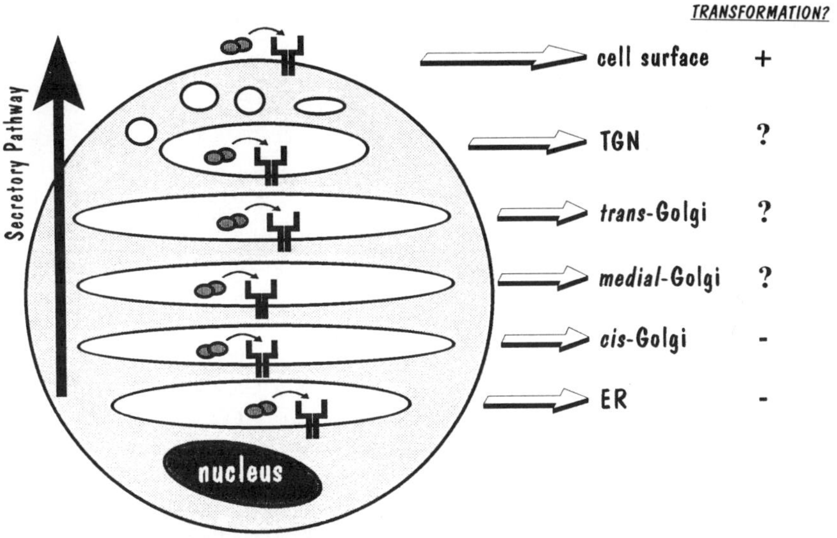

Figure 3. Localization of autocrine and "intracrine" interactions between SIS and PDGF receptors. As indicated, interactions between the ligand and receptor leading to transformation do not occur in the ER or *cis*-Golgi (85,86). It is not yet clear if these interactions can occur in later Golgi compartments, but it is very clear that the cell surface is a site for autocrine interactions.

autocrine transformation cannot occur from within the ER (85). Hart et al. (86) then examined whether the *cis*-Golgi compartment allows for transforming interactions between SIS and PDGF receptors. Attachment of the well-characterized *cis*-Golgi retention signal from the first transmembrane domain of the E1 glycoprotein of IBV (infectious bronchitis virus) to SIS caused highly efficient retention of the protein within the early Golgi, as determined by lack of proteolytic processing, lack of surface expression, and co-localization with known Golgi markers by immunofluorescence. This fusion protein was also unable to cause transformation of NIH 3T3 cells, suggesting that this early Golgi compartment also does not allow for autocrine interactions to occur (86). In progress are studies aimed at assessing the ability of autocrine interactions to occur in later Golgi regions, such as the *medial-* and *trans*-Golgi cisternae and TGN, with the use of retention signals derived from glycosyltransferases, which are resident Golgi proteins. The results from these and other studies should finally allow for a conclusive determination to be made regarding the mechanism of autocrine transformation by SIS.

Table 2. Evidence For and Against Intracellular Autocrine Transformation

Evidence For Intracellular Autocrine Interactions

Huang et. al (1984) (187):
- some v-*sis*-transformed cells do not secrete v-*sis* protein
- in SSV-NP1 cells, α-PDGF antibodies did not affect transformation
- transformation, as opposed to cell growth, does NOT require secretion of v-*sis*

Huang and Huang, 1988 (84):
- Protamine (a competitive inhibitor of v-*sis*/PDGF binding to surface receptors) did NOT affect rapid turnover of PDGF receptors in v-*sis*-transformed cells
- processing of PDGF receptors is altered in v-*sis*-transformed cells
  - increased amount of 160 kD, immature receptor form
  - autophosphorylation of immature receptor in v-*sis* transformed cells

Keating and Williams, 1988 (188):
- the majority of receptor activation in v-*sis*-transformed NRK cells occurs in intracellular compartments

Bejecek et al., 1989 (189):
- Attachment of an ER-retention signal, SE kDEL, to v-*sis* did not abolish its transforming ability
- this protein was not detectably secreted

Lokeshwar et al., 1990 (190)
- intracellular forms of v-*sis* in SSV-transformed cell lines exhibit rapid turnover
- rate of v-*sis* intracellular turnover correlates with tumorigenicity of cell lines

Structural Requirements for Transformation:

Signal sequence?	Yes
Processing at dibasic cleavage site?	No
N-linked glycosylation?	No
O-linked glycosylation?	No

Evidence Supporting Surface Interactions

Huang et al., 1984 (187):
- mitogenesis can be blocked in v-*sis*-transformed cells by α-PDGF antibodies
- mitogenic activity from these cells competes with [^{125}]-PDGF binding, indicating a secreted protein

Garrett et. al., 1984 (191):
- Suramin treatment of SSV-transformed NRK cells leads to up-regulation of PDGF-stimulatable receptors on surface

Betsholtz et al., 1986 (19)
- Suramin reverts morphology of SSV-transformed human and rat fibroblasts

Huang and Huang, 1988 (84)
- Suramin inhibits binding of v-*sis* and PDGF to receptors and reverses the rapid receptor turnover in v-*sis*-transformed cells

Hannink and Donoghue, 1988 (192)
- Monensin treatment of cells expressing an inducible v-*sis* construct prevents autophosphorylation of mature PDGF receptors, prevents induction of c-*fos*

Fleming et al., 1989 (193):
- Suramin reverts transformation of v-*sis*-expressing cells
- only the 180 kD, mature form of PDGFR-β shows decreased phosphorylation in response to suramin
- supplemented cells with PDGF-BB (3ng/ml) each day, and observed transformed phenotype indistinguishable from v-*sis* transformation

Fleming et al., 1992 (194):
- Suramin reversion of increased DNA synthesis observed in some human tumor cell lines that co-express PDGF and PDGF receptors

Lee and Donoghue, 1992 (85):
- use of ER-retention signal (DEKKMP) to anchor v-*sis* in the ER led to complete abrogation of transforming ability

Hart et al., 1994 (in press) (86)
- retention of v-*sis* in the *cis*-Golgi by means of the TMD of E1 protein abolishes transforming ability

REFERENCES

1 Sporn, M.B. and Todaro G.J. (1980) N. Engl. J. Med. 303, 878-880.
2 Chait, A., Ross, R., Albers, J.J. and Bierman, E.L. (1980) Proc. Nat. Acad. Sci. U.S.A. 77, 4084-4088.
3 Owen, A.J. 3d., Geyer, R.P. and Antoniades, H.N. (1982) Proc. Nat. Acad. Sci. U.S.A. 79, 3203-3207.
4 Paulsson, Y., Bywater, M., Heldin, C.H. and Westermark, B. (1987) Exp. Cell Res. 171, 186-194.
5 Berk, B.C., Alexander, R.W., Brock, T.A., Gimbrone, M.A. Jr. and Webb, R.C. (1986) Science 232, 87-90.
6 Grotendorst, G.R., Chang, T., Seppa, H.E., Kleinman, H.K. and Martin, G.R. (1982) J. Cell. Physiol. 113, 261-266.
7 Tischer, E., Gospodarowicz, D., Mitchell, R., Silva, M., Schilling, J., Lau, K., Crisp, T., Fiddes, J.C. and Abraham, J.A. (1989) Biochem. Biophys. Res. Commun. 165, 1198-1206.
8 Senger, D.R., Perruzzi, C.A., Feder, J. and Dvorak, H.F. (1986) Cancer Res. 46, 5629-5632.
9 Oefner, C., D'Arcy, A., Winkler, F.K., Eggimann, B. and Hosang, M. (1992) EMBO J. 11, 3921-3926.
10 McDonald, N.Q. and Hendrickson, W.A. (1993) Cell 73, 421-424.
11 Shimokado, K., Raines, E.W., Madtes, D.K., Barrett, T.B., Benditt, E.P. and Ross, R. (1985) Cell 43, 277-286.
12 Martinet, Y., Bitterman, P.B., Mornex, J.F., Grotendorst, G.R., Martin, G.R and Crystal, R.G. (1986) Nature 319, 158-160.
13 DiCorleto, P.E. and Bowen-Pope, D.F. (1983) Proc. Nat. Acad. Sci. U.S.A. 80, 1919-1923.
14 Raines, E.W., Dower, S.K. and Ross, R. (1989) Science 243, 393-396.
15 Seifert, R.A., Schwartz, S.M. and Bowen-Pope, D.F. (1984) Nature 311, 669-671.
16 Goustin, A.S., Betsholtz, C., Pfeifer-Ohlsson, S., Persson, H., Rydnert, J., Bywater, M., Holmgren, G., Heldin, C.H., Westermark, B. and Ohlsson R. (1985) Cell 41, 301-312.
17 Hammacher, A., Nister, M., Westermark, B. and Heldin, C.H. (1988) Eur. J. Biochem. 176, 179-186.
18 Heldin, C.H., Johnsson, A., Wennergren, S., Wernstedt, C., Betsholtz, C. and Westermark, B. (1986) Nature 319, 511-514.
19 Betsholtz, C., Johnsson, A., Heldin, C.H., Westermark, B., Lind, P., Urdea, M.S., Eddy, R., Shows, T.B., Philpott, K., Mellor, A.L., Knott, T.J. and Scott, J. (1986) Nature 320, 695-699.
20 Bonthron, D.T., Morton, C.C., Orkin, S.H. and Collins, T. (1988) Proc. Nat. Acad. Sci. U.S.A. 85, 1492-1496.
21 Swan, D.C., McBride, O.W., Robbins, K.C., Keithley, D.A., Reddy, E.P and Aaronson, S.A. (1982) Proc. Nat. Acad. Sci. U.S.A. 79, 4691-4695.
22 Ostman, A., Rall, L., Hammacher, A., Wormstead, M.A., Coit, D., Valenzuela, P., Betsholtz, C., Westermark, B. and Heldin, C.H. (1988) J. Biol. Chem. 263, 16202-16208.
23 Rorsman, F., Bywater, M., Knott, T.J., Scott, J. and Betsholtz, C. (1988) Mol. Cell. Biol. 8, 571-577.
24 Maher, D.W., Lee, B.A. and Donoghue, D.J. (1989) Mol. Cell. Biol. 9, 2251-2253.
25 Raines, E.W. and Ross, R.J. (1992) Cell. Biol. 116, 533-543.

26 Ostman, A., Thyberg, J., Westermark, B. and Heldin, C.H. (1992) J. Cell Biol. 118, 509-519.
27 Westermark, B. and Heldin, C.H. (1993) Acta Oncologica 32, 101-105.
28 Lee, B.A., Maher, D.W., Hannink, M. and Donoghue, D.J. (1987) Mol. Cell. Biol. 7, 3527-3537.
29 Devare, S.G., Reddy, E.P., Law, J.D., Robbins, K.C. and Aaronson, S.A. (1983) Proc. Nat. Acad. Sci. U.S.A., 80, 731-735.
30 Waterfield, M.D., Scrace, G.T., Whittle, N., Stroobant, P., Johnsson, A., Wasteson, A., Westermark, B., Heldin, C.H., Huang, J.S. and Deuel, T.F. (1983) Nature 304, 35-39.
31 Robbins, K.C., Leal, F., Pierce, J.H. and Aaronson, S.A. (1985) EMBO J. 4, 1783-1792.
32 Hannink, M. and Donoghue, D.J. (1984) Science 226, 1197-1199.
33 Hannink, M. and Donoghue, D.J. (1986) Mol. Cell. Biol. 6, 1343-1348.
34 Sauer, M.K., Hannink, M. and Donoghue, D.J. (1986) J. Virol. 59, 292-300.
35 Hoppe, J., Weich, H.A. and Eichner, W. (1989) Biochemistry 28, 2956-2960.
36 Connolly, D.T., Olander, J.V., Heuvelman, D., Nelson, R., Monsell, R., Siegel, N., Haymore, B.L., Leimgruber, R. and Feder, J. (1989) J. Biol. Chem. 264, 20017-20-024.
37 Connolly, D.T. (1991) J. Cell. Biochem. 47, 219-223.
38 Keck, P.J., Hauser, S.D., Krivi, G., Sanzo, K., Warren, T., Feder, J. and Connolly, D.T. (1989) Science 246, 1309-1312.
39 Conn, G., Bayne, M.L., Soderman, D.D., Kwok, P.W., Sullivan, K.A., Palisi, T.M., Hope, D.A. and Thomas, K.A. (1990) Proc. Nat. Acad. Sci. U.S.A. 87, 2628-2632.
40 Houck, K.A., Ferrara, N., Winer, J., Cachianes, G., Li, B. and Leung, D.W. (1991) Mol. Endocrinol. 5, 1806-1814.
41 Maglione, D., Guerriero, V., Viglietto, G., Delli-Bovi, P. and Persico, M.G. (1991) Proc. Nat. Acad. Sci. U.S.A. 88, 9267-9271.
42 Maglione, D., Guerriero, V., Viglietto, G., Ferraro, M.G., Aprelikova, O., Alitalo, K., Del Vecchio, S., Lei, K.J., Chou, J.Y. and Persico, M.G. (1993) Oncogene 8, 925-931.
43 Hauser, S. and Weich, H.A. (1993) Growth Factors 9, 259-268.
44 Robbins, K.C., Antoniades, H.N., Devare, S.G., Hunkapiller, M.W. and Aaronson, S.A. (1983) Nature 305, 605-608.
45 Hannink, M., Sauer, M.K. and Donoghue, D.J. (1986) Mol. Cell. Biol. 6, 1304-1314.
46 Giese, N.A., Robbins, K.C. and Aaronson, S.A. (1987) Science 236, 1315-1318.
47 Sauer, M.K. and Donoghue, D.J. (1988) Mol. Cell. Biol. 8, 1011-1018.
48 Jaumann, M., Hoppe, V., Tatje, D., Eichner, W. and Hoppe, J. (1991) Biochemistry 30, 3303-3309.
49 Kenney, W.C., Haniu, M., Herman, A.C., Arakawa, T. Costigan, V.J., Lary, J., Yphantis, D.A. and Thomason, A.R. (1994) J. Biol. Chem. 269, 12351-12359.
50 Maher, D.W., Strawn, L.M. and Donoghue, D.J. (1993) Oncogene 8, 533-541.
51 LaRochelle, W.J., Giese, N., May-Siroff, M., Robbins, K.C. and Aaronson, S.A. (1990) Science 248, 1541-1544.
52 Giese, N., LaRochelle, W.J., May-Siroff, M., Robbins, K.C. and Aaronson, S.A. (1990) Mol. Cell. Biol. 10, 5496-5501.
53 Clements, J.M., Bawden, L.J., Bloxidge, R.E., Catlin, G., Cook, A.L., Craig, S., Drummond, A.H., Edwards, R.M., Fallon, A., Green, D.R., Hellewell, P.G.,

Kirwin, P.M., Nayee, P.D., Richardson, S.J., Brown, D., Chahwala, S.B., Snarey, M. and Winslow, D. (1991) EMBO J. 10, 4113-4120.
54 Jaumann, M., Tatje, D. and Hoppe, J. (1992) FEBS Lett. 302, 265-268.
55 Fenstermaker, R.A., Poptic, E., Bonfield, T.L., Knauss, T.C., Corsillo, L., Piskurich, J.F., Kaetzel, C.S., Jentoft, J.E., Gelfand, C., DiCorleto, P.E., and Kaetzel, D.M. Jr. (1993) J. Biol. Chem. 268, 10482-10489.
56 Beckmann, M.P., Betsholtz, C., Heldin, C.H., Westermark, B., Di Marco, E., Di Fiore, P.P., Robbins, K.C. and Aaronson, S.A. (1988) Science, 241, 1346-1349.
57 LaRochelle, W.J., Pierce, J.H., May-Siroff, M., Giese, N. and Aaronson, S.A. (1992) J. Biol. Chem. 267, 17074-17077.
58 Mercola, M., Deininger, P.L., Shamah, S.M., Porter, J., Wang, C.Y. and Stiles, C.D. (1990) Genes and Devel. 4, 2333-2341.
59 Shamah, S.M., Stiles, C.D. and Guha, A. (1993) Mol. Cell. Biol. 13, 7203-7212.
60 Vassbotn, F.S., Andersson, M., Westermark, B., Heldin, C.H. and Ostman, A. (1993) Mol. Cell. Biol. 13, 4066-4076.
61 Ostman, A., Andersson, M., Hellman, U. and Heldin, C.H. (1991) J. Biol. Chem. 266, 10073-10077.
62 Yarden, Y., Escobedo, J.A., Kuang, W.-J., Yang-Feng, T.L., Daniel, T.O., Tremble, P.M., Chen, E.Y., Ando, M.E., Harkins, R.N., Francke, U., Fried, V.A., Ullrich, A. and Williams, L.T. (1986) Nature 323, 226-232.
63 Matsui, T., Heidaran, M., Miki, T., Popescu, N., La Rochelle, W., Kraus, M., Pierce, J. and Aaronson, S. (1989) Science 243, 800-804.
64 Seifert, R.A., Hart, C.E., Phillips, P.E., Forstrom, J.W., Ross, R., Murray, M.J. and Bowen-Pope, D.F. (1989) J. Biol. Chem. 264, 8771-8778.
65 Hart, C.E., Forstrom, J.W., Kelly, J.D., Seifert, R.A., Smith, R.A., Ross, R., Murray, M.J. and Bowen-Pope, D.F. (1988) Science 240, 1529-1531.
66 Vaisman, N., Gospodarowicz, D. and Neufeld, G. (1990) J. Biol. Chem. 265, 19461-19466.
67 de Vries, C., Escobedo, J.A., Ueno, H., Houck, K., Ferrara, N. and Williams, L.T. (1992) Science 255, 989-991.
68 Pajusola, K., Aprelikova, O., Korhonen, J., Kaipainen, A., Pertovaara, L., Alitalo, R. and Alitalo, K. (1992) Cancer Res. 52, 5738-5743.
69 Galland, F., Karamysheva, A., Mattei, M.G., Rosnet, O., Marchetto, S. and Birnbaum, D. (1992) Genomics 13, 475-478.
70 Pajusola, K., Aprelikova, O., Armstrong, E., Morris, S. and Alitalo, K. (1993) Oncogene 8, 2931-2937.
71 Galland, F., Karamysheva, A., Pebusque, M.J., Borg, J.P., Rottapel, R., Dubreuil, P., Rosnet, O. and Birnbaum, D. (1993) Oncogene 8, 1233-1240.
72 Terman, B.I., Dougher-Vermazen, M., Carrion, M.E., Dimitrov, D., Armellino, D.C., Gospodarowicz, D. and Bohlen, P. (1992) Biochem. Biophys. Res. Commun. 187, 1579-1586.
73 Kaipainen, A., Korhonen, J., Pajusola, K., Aprelikova, O., Persico, M.G., Terman, B.I. and Alitalo, K. (1993) J. Exp. Med. 178, 2077-2088.
74 Barleon, B., Hauser, S., Schollmann, C., Weindel, K., Marme, D., Yayon, A. and Weich, H.A. (1994) J. Cell. Biochem. 54, 56-66.
75 Sherr, C.J., Rettenmier, C.W., Sacca, R., Roussel, M.F., Look, A.T. and Stanley, E.R. (1985) Cell 41, 665-676.
76 Yarden, Y., Kuang, W.J., Yang-Feng, T., Coussens, L., Munemitsu, S., Dull, T.J., Chen, E., Schlessinger, J., Francke, U. and Ullrich, A. (1987) EMBO J. 6, 3341-3351.

77 Qiu, F.H., Ray, P., Brown, K., Barker, P.E., Jhanwar, S., Ruddle, F.H. and Besmer, P. (1988) EMBO J. 7, 1003-1011.
78 Vandenbark, G.R., deCastro, C.M., Taylor, H., Dew-Knight, S. and Kaufman, R.E. (1992) Oncogene 7, 1259-1266.
79 Coussens, L., VanBeveren, C., Smith, D., Chen, E., Mitchell, R.L., Isacke, C.M., Verma, I.M. and Ullrich, A. (1986) Nature 320, 277-280.
80 Westermark, B., Claesson-Welsh, L., and Heldin, C.H. (1990) Ciba Foundation Symp. 150, 6-14., discussion 14-22.
81 Keating, M.T. and Williams, L.T. (1987) J. Biol. Chem. 262, 7932-7937.
82 Claesson-Welsh, L., Ronnstrand, L. and Heldin, C.H. (1987) Proc. Nat. Acad. Sci. U.S.A. 84, 8796-8800.
83 Daniel, T.O., Milfay, D.F., Escobedo, J. and Williams, L.T. (1987) J. Biol. Chem. 262, 9778-9784.
84 Huang, S.S. and Huang, J.S. (1988) J. Biol. Chem. 263, 12608-12618.
85 Lee, B.A. and Donoghue, D.J. (1992) J. Cell Biol. 118, 1057-1070.
86 Hart, K.C., Xu, Y.F., Meyer, A.N., Lee, B.A. and Donoghue, D.J (1994) J. Cell Biol. (in press).
87 Joly, M., Kazlauskas, A., Fay, F.S. and Corvera, S. (1994) Science 263, 684-687.
88 Klionsky, D.J., Herman, P.K. and Emr, S.D. (1990) Microbiol. Rev. 54, 266-292.
89 Heldin, C.H., Ernlund, A., Rorsman, C. and Ronnstrand, L. (1989) J. Biol. Chem. 264, 8905-8912.
90 Koch, C.A., Anderson, D., Moran, M.F., Ellis, C. and Pawson, T. (1991) Science 252, 668-674.
91 Heldin, C.H. (1991) Trends Biochem. Sci. 16, 450-452.
92 Anderson, D., Koch, C.A., Grey, L., Ellis, C., Moran, M.F. and Pawson, T. (1990) Science 250, 979-982.
93 Escobedo, J.A., Kaplan, D.R., Kavanaugh, W.M., Turck, C.W. and Williams, L.T. (1991) Mol. Cell. Biol. 11, 1125-1132.
94 Fantl, W.J., Escobedo, J.A., Martin, G.A., Turck, C.W., del Rosario M., McCormick, F. and Williams, L.T. (1992) Cell 69, 413-423.
95 Cantley, L.C., Auger, K.R., Carpenter, C., Duckworth, B., Graziani, A., Kapeller, R. and Soltoff, S. (1991) Cell 64, 281-302.
96 Escobedo, J.A. and Williams, L.T. (1988) Nature 335, 85-87.
97 Heidaran, M.A., Pierce, J.H., Lombardi, D., Ruggiero, M., Gutkind, J.S., Matsui, T. and Aaronson, S.A. (1991) Mol. Cell. Biol. 11, 134-142.
98 Heidaran, M.A., Beeler, J.F., Yu, J.C., Ishibashi, T., LaRochelle, W.J., Pierce, J.H. and Aaronson, S.A. (1993) J. Biol. Chem. 268, 9287-9295.
99 Escobedo, J.A., Barr, P.J. and Williams, L.T. (1988) Mol. Cell. Biol. 8, 5126-5131.
100 Twamley, G.M., Kypta, R.M., Hall, B. and Courtneidge, S.A. (1992) Oncogene 7, 1893-1901.
101 Berridge, M.J. (1983) Biochem. J. 212, 849-858.
102 Coughlin, S.R., Lee, W.M., Williams, P.W., Giels, G.M. and Williams, L.T. (1985) Cell 43, 243-251.
103 Ives, H.E. and Daniel, T.O. (1987) Proc. Nat. Acad. Sci. U.S.A. 84, 1950-1954.
104 Kazlauskas, A. and Cooper, J.A. (1989) Cell 58, 1121-1133.
105 Coughlin, S.R., Escobedo, J.A. and Williams, L.T. (1989) Science 243, 1191-1194.
106 Meisenhelder, J., Suh, P.G., Rhee, S.G. and Hunter, T. (1989) Cell 57, 1109-1122.
107 Morrison, D.K., Kaplan, D.R., Rhee, S.G. and Williams, L.T. (1990) Mol. Cell. Biol. 10, 2359-2366.
108 Kumjian, D.A., Wahl, M.I., Rhee, S.G. and Daniel, T.O. (1989) Proc. Nat. Acad. Sci. U.S.A. 86, 8232-8236.

109 Wahl, M.I., Olashaw, N.E., Nishibe, S., Rhee, S.G., Pledger, W.J. and Carpenter, G. (1989) Mol. Cell. Biol. 9, 2934-2943.
110 Kaplan, D.R., Morrison, D.K., Wong, G., McCormick, F. and Williams, L.T. (1990) Cell 61, 125-133.
111 Kazlauskas, A., Ellis, C., Pawson, T. and Cooper, J.A. (1990) Science 247, 1578-1581.
112 Morrison, D.K., Kaplan, D.R., Escobedo, J.A., Rapp, U.R., Roberts, T.M. and Williams, L.T. (1989) Cell 58, 649-657.
113 Kypta, R.M., Goldberg, Y., Ulug, E.T. and Courtneidge, S.A. (1990) Cell 62, 481-492.
114 Valius, M., Bazenet, C. and Kazlauskas, A. (1993) Mol. Cell. Biol. 13, 133-143.
115 Kazlauskas, A., Feng, G.S., Pawson, T. and Valius, M. (1993) Proc. Nat. Acad. Sci. U.S.A. 90, 6939-6943.
116 Lowenstein, E.J., Daly, R.J., Batzer, A.G., Li, W., Margolis, B., Lammers, R., Ullrich, A., Skolnik, E.Y., Bar-Sagi, D. and Schlessinger, J. (1992) Cell 70, 431-442.
117 Arvidsson, A.K., Rupp, E., Nanberg, E., Downward, J., Ronnstrand, L., Wennstrom, S., Schlessinger, J., Heldin, C.H. and Claesson-Welsh, L. (1994) Mol. Cell. Biol. 14, 6715-6726.
118 Whitman, M., Downes, C.P., Keeler, M., Keller, T. and Cantley, L. (1988) Nature 332, 644-646.
119 Yamamoto, K., Graziani, A., Carpenter, C., Cantley, L.C. and Lapetina, E.G. (1990) J. Biol. Chem. 265, 22086-22089.
120 Kaplan, D.R., Whitman, M., Schaffhausen, B., Pallas, D.C., White, M., Cantley, L. and Roberts, T.M. (1987) Cell 50, 1021-1029.
121 Whitman, M., Kaplan, D., Roberts, T. and Cantley, L. (1987) Biochem. J. 247, 165-174.
122 Fantl, W.J., Escobedo, J.A. and Williams, L.T. (1989) Mol. Cell. Biol. 9, 4473-4478.
123 Yu, J.C., Heidaran, M.A., Pierce, J.H., Gutkind, J.S., Lombardi, D., Ruggiero, M. and Aaronson, S.A. (1991) Mol. Cell. Biol. 11, 3780-3785.
124 Tapley, P., Kazlauskas, A., Cooper, J.A. and Rohrschneider, L.R. (1990) Mol. Cell. Biol. 10, 2528-2538.
125 Kazlauskas, A., Kashishian, A., Cooper, J.A. and Valius, M. (1992) Mol. Cell. Biol. 12, 2534-2544.
126 Cooper, J.A. and Kashishian, A. (1993) Mol. Cell. Biol. 13, 1737-1745.
127 Klippel, A., Escobedo, J.A., Fantl, W.J. and Williams, L.T. (1992) Mol. Cell. Biol. 12, 1451-1459.
128 Kavanaugh, W.M., Klippel, A., Escobedo, J.A. and Williams, L.T. (1992) Mol. Cell. Biol. 12, 3415-3424.
129 Valius, M. and Kazlauskas, A. (1993) Cell 73, 321-334.
130 Roche, S., Koegl, M. and Courtneidge, S.A. (1994) Proc. Nat. Acad. Sci. U.S.A. 91, 9185-9189.
131 Kundra, V., Escobedo, J.A., Kazlauskas, A., Kim, H.K., Rhee, S.G., Williams, L.T. and Zetter, B.R. (1994) Nature 367, 474-476.
132 Berridge, M.J. (1987) Annu. Rev. Biochem. 56, 159-193.
133 Rhee, S.G., Suh, P.G., Ryu, S.H. and Lee, S.Y. (1989) Science 244, 546-550.
134 Li, W., Yu, J.C., Michieli, P., Beeler, J.F., Ellmore, N., Heidaran, M.A. and Pierce, J.H. (1994) Mol. Cell. Biol. 14, 6727-6735.
135 Ronnstrand, L., Mori, S., Arridsson, A.K., Eriksson, A., Wernstedt, C., Hellman, U., Claesson-Welsh, L. and Heldin, C.H. (1992) EMBO J. 11, 3911-3919.

136 Hill, T.D., Dean, N.M., Mordan, L.J., Lau, A.F., Kanemitsu, M.Y. and Boynton, A.L. (1990) Science 248, 1660-1663.
137 Trahey, M. and McCormick, F. (1987) Science 238, 542-545.
138 Satoh, T., Endo, M., Nakafuku, M., Nakamura, S. and Kaziro, Y. (1990) Proc. Nat. Acad. Sci. U.S.A. 87, 5993-5997.
139 Molloy, C.J., Bottaro, D.P., Fleming, T.P., Marshall, M.S., Gibbs, J.B. and Aaronson, S.A. (1989) Nature 342, 711-714.
140 Kashishian, A., Kazlauskas, A. and Cooper, J.A. (1992) EMBO J. 11, 1373-1382.
141 Satoh, T., Fantl, W.J., Escobedo, J.A., Williams, L.T. and Kaziro, Y. (1993) Mol. Cell. Biol. 13, 3706-3713.
142 Rodriguez-Viciana, P., Warne, P.H., Dhand, R., Vanhaesebroeck, B., Gout, I., Fry, M.J., Waterfield, M.D. and Downward, J. (1994) Nature 370, 527-532.
143 Williams, N.G. and Roberts, T.M. (1994) Cancer Metastasis Rev. 13, 105-116.
144 Morrison, D.K., Kaplan, D.R., Rapp, U. and Roberts, T.M. (1988) Proc. Nat. Acad. Sci. U.S.A. 85, 8855-8859.
145 Warne, P.H., Viciana, P.R. and Downward, J. (1993) Nature 364, 352-355.
146 Vojtek, A.B., Hollenberg, S.M. and Cooper, J.A. (1993) Cell 74, 205-214.
147 Howe, L.R., Leevers, S.J., Gomez, N., Nakielny, S., Cohen, P. and Marshall, C.J. (1992) Cell 71, 335-342.
148 Leevers, S.J., Paterson, H.F. and Marshall, C.J. (1994) Nature 369, 411-414.
149 Stokoe, D., Macdonald, S.G., Cadwallader, K., Symons, M. and Hancock, J.F. (1994) Science 264, 1463-1467.
150 Ralston, R. and Bishop, J.M. (1985) Proc. Nat. Acad. Sci. U.S.A. 82, 7845-7849.
151 Gould, K.L. and Hunter, T. (1988) Mol. Cell. Biol. 8, 3345-3356.
152 Walker, F., deBlaquiere, J. and Burgess, A.W. (1993) J. Biol. Chem. 268, 19552-19558.
153 Courtneidge, S.A., Kypta, R.M., Cooper, J.A. and Kazlauskas, A. (1991) Cell Growth Diff. 2, 483-486.
154 Mori, S., Ronnstrand, L., Yokote, K., Engstrom, A., Courtneidge, S.A., Claesson-Welsh, L. and Heldin, C.H. (1993) EMBO J. 12, 2257-2264.
155 Courtneidge, S.A., Fumagalli, S., Koegl, M., Superti-Furga, G. and Twamley-Stein, G.M. (1993) Development. Supplement 57-64.
156 Twamley-Stein, G.M., Pepperkok, R., Ansorge, W. and Courtneidge, S.A. (1993) Proc. Nat. Acad. Sci. U.S.A. 90, 7696-7700.
157 Molloy, C.J., Fleming, T.P., Bottaro, D.P., Cuadrado, A. and Aaronson, S.A. (1992) Mol. Cell. Biol. 12, 3903-3909.
158 Bennett, A.M., Tang, T.L., Sugimoto, S., Walsh, C.T. and Neel, B.G. (1994) Proc. Nat. Acad. Sci. U.S.A. 91, 7335-7339.
159 Buday, L. and Downward, J. (1993) Cell 73, 611-620.
160 Egan, S.E., Giddings, B.W., Brooks, M.W., Buday, L., Sizeland, A.M. and Weinberg, R.A. (1993) Nature 363, 45-51.
161 Li, N., Batzer, A., Daly, R., Yajnik, V., Skolnik, E., Chardin, P., Bar-Sagi, D., Margolis, B. and Schlessinger, J. (1993) Nature 363, 85-88.
162 Li, W., Hu, P., Skolnik, E.Y., Ullrich, A. and Schlessinger, J. (1992) Mol. Cell. Biol. 12, 5824-5833.
163 Nishimura, R., Li, W., Kashishian, A., Mondino, A., Zhou, M., Cooper, J. and Schlessinger, J. (1993) Mol. Cell. Biol. 13, 6889-6896.
164 Yokote, K., Mori, S., Hansen, K., McGlade, J., Pawson, T., Heldin, C.H. and Claesson-Welsh, L. (1994) J. Biol. Chem. 269, 15337-15343.

165 Deuel, T.F., Kawahara, R.S., Mustoe, T.A. and Pierce, A.F. (1991) Annu. Rev. Med. 42, 567-584.
166 Lynch, S.E., Colvin, R.B. and Antoniades, H.N. (1989) J. Clin. Invest. 84, 640-646.
167 Tennant, M. and McGeachie, J.K. (1991) Australian and New Zealand J. Surg. 61, 482-488.
168 Martinet, Y., Rom, W.N., Grotendorst, G.R., Martin, G.R. and Crystal, R.G. (1987) N. Engl. J. Med. 317, 202-209.
169 Antoniades, H.N., Bravo, M.A., Avila, R.E., Galanopoulos, T., Neville-Golden, J., Maxwell, M. and Selman, M. (1990) J. Clin. Invest. 86, 1055-1064.
170 Detmar, M., Brown, L.F., Claffey, K.P., Yeo, K.T., Kocher, O., Jackman, R.W., Berse, B. and Dvorak, H.F. (1994) J. Exp. Med. 180, 1141-1146.
171 Koch, A.E., Harlow, L.A., Haines, G.K., Amento, E.P., Unemori, E.N., Wong, W.L., Pope, R.M. and Ferrara, N. (1994) J. Immunol. 152, 4149-4156.
172 Brown, L.F., Berse, B., Jackman, R.W., Tognazzi, K., Manseau, E.J., Dvorak, H.F. and Senger, D.R. (1993) J. Pathol. 143, 1255-1262.
173 Stephenson, D.A., Mercola, M., Anderson, E., Wang, C.Y., Stiles, C.D., Bowen-Pope, D.F. and Chapman, V.M. (1991) Proc. Nat. Acad. Sci. U.S.A. 88, 6-10.
174 Schatteman, G.C., Morrison-Graham, K., van Koppen, A., Weston, J.A. and Bowen-Pope, D.F. (1992) Development 115, 123-131.
175 Morrison-Graham, K., Schatteman, G.C., Bork, T., Bowen-Pope, D.F. and Weston, J.A. (1992) Development 115, 133-142.
176 Orr-Urtreger, A., Bedford, M.T., Do, M.S., Eisenbach, L. and Lonai, P. (1992) Development 115, 289-303.
177 Halaban, R. and Moellmann, G. (1993) J. Invest. Dermatol. 100, 176S-185S.
178 Golub, T.R., Barker, G.F., Lovett, M. and Gilliland, D.G. (1994) Cell 77, 307-316.
179 Shweiki, D., Itin, A., Soffer, D. and Keshet, E. (1992) Nature 359, 843-845.
180 Miller, J.W., Adamis, A.P., Shima, D.T., D'Amore, P.A., Moulton, R.S., O'Reilly, M.S., Folkman, J., Dvorak, H.F., Brown, L.F., Berse, B., Yeo, T.K., and Yeo, K.T. (1994) J. Pathol. 145, 574-584.
181 Ferrara, N., Houck, K.A., Jakeman, L.B., Winer, J. and Leung, D.W. (1991) J. Cell. Biochem. 47, 211-218.
182 Guha, A. (1992) Pediatric Neurosurg. 17, 14-20.
183 Fleming, T.P., Matsui, T. and Aaronson, S.A. (1992) Exp. Gerontol. 27, 523-532.
184 Black, P.M., Carroll, R., Glowacka, D., Riley, K. and Dashner, K.J. (1994) Neurosurg. 81, 388-393.
185 Tsai, L.H., White, L., Raines, E., Ross, R., Smith, R.G., Cushley, W. and Ozanne, B. (1994) Blood 83, 51-55.
186 Silver, B.J. (1992) Biofactors 3, 217-227.
187 Huang, J.S., Huang, S.S. and Deuel, T.F. (1984) Cell 39, 79-87.
188 Keating, M.T. and Williams, L.T. (1988) Science 239, 914-916.
189 Bejcek, B.E., Li, D.Y. and Deuel, T.F. (1989) Science 245, 1496-1499.
190 Lokeshwar, V.B., Huang, S.S. and Huang, J.S. (1990) J. Biol. Chem. 265, 1665-1675.
191 Garrett, J.S., Coughlin, S.R., Niman, H.L., Tremble, P.M., Giels, G.M. and Williams, L.T. (1984) Proc. Nat. Acad. Sci. U.S.A. 81, 7466-7470.
192 Hannink, M. and Donoghue, D.J. (1988) J. Cell Biol. 107, 287-298.
193 Fleming, T.P., Matsui, T., Molloy, C.J., Robbins, K.C. and Aaronson, S.A. (1989) Proc. Nat. Acad. Sci. U.S.A. 86, 8063-8067.

194 Fleming, T.P., Matsui, T., Heidaran, M.A., Molloy, C.J., Artrip, J. and Aronson, S.A. (1992) Oncogene 7, 1355-1359.
195 Hannink, M. and Donoghue, D.J. (1986) J. Cell Biol. 103, 2311-2322.
196 Xu, Y.F., Meyer, A.N., Webster, M.K., Lee, B.A. and Donoghue, D.J. (1993) J. Cell Biol. 123, 549-560.

RECOMBINATION BETWEEN PROKARYOTIC AND EUKARYOTIC DNA: INTEGRATION OF *AGROBACTERIUM TUMEFACIENS* T-DNA INTO THE PLANT GENOME

Bruno Tinland and Barbara Hohn

Friedrich Miescher-Institut
P.O.Box 2543
CH-4002 Basel, Switzerland

INTRODUCTION: MECHANISM OF T-DNA TRANSFER

The gram-negative bacterium *Agrobacterium tumefaciens* is a fascinating organism, from many points of view: the mechanism of gene transfer, a unique means of transport of DNA from a prokaryotic to a eukaryotic organism; the mechanism of tumorigenicity; and finally the bacterium's usefulness for the generation of transgenic plants for molecular plant breeding.

Agrobacterium tumefaciens (and its relative *Agrobacterium rhizogenes*) is uniquely equipped to fulfill its role as sophisticated plant parasite. It contains a plasmid called Ti plasmid (tumor inducing) which encodes most of the major functions required for virulence. One segment of the plasmid, the T-DNA (transferred DNA), is designed to be transported to the plant, to be integrated into the plant nuclear DNA and to express its genes only in the eukaryotic cellular environment. These genes code for enzymes required for synthesis of new compounds termed opines and for enzymes involved in phytohormone production (and possibly phytohormone perception in the case of *A. rhizogenes*). The opines serve as food source for the bacterium whereas the hormones lead to the neoplastic phenotype of tumors (or proliferative root development for *A. rhizogenes*). The fact that none of the T-DNA genes is required for the T-DNA transfer step as such led to the development and wide use of *Agrobacterium*-based plant transformation technology. T-DNA in the bacterium is delimited and thereby defined by two 25 bp direct imperfect repeat-sequences called borders. Most functions necessary for transfer of T-DNA are encoded in the virulence region located on the Ti plasmid; some are located on the chromosome.

Only when a wounded plant is in "useful" distance from a bacterium are the virulence genes induced. Chemical signals emitted by a healing plant wound are, *via* a receptor/activator system (proteins VirA/VirG) converted into transcriptional activation of

Genetic Engineering, Vol. 17, Edited by J.K. Setlow
Plenum Press, New York, 1995

the virulence genes. Virulence proteins D1 and D2 introduce a specific cut into one strand of the border sequences. The C-proteins have ancillary functions in this process. Single-stranded molecules of T-DNA (T-strands), covalently bound to VirD2 and possibly coated by the single-stranded DNA binding protein VirE2, are transported through bacterial pore structures probably composed of VirB proteins into the plant. By a largely unexplored mechanism the T-complex reaches the plant cells, is actively targeted through nuclear pores to the nucleus and by recombination becomes part of the nuclear DNA.

In this review we focus on the integration step. Recombination between prokaryotic and eukaryotic DNA raises the interesting question of how potential barriers between these very different chromosomes are overcome and whether/how prokaryotic and eukaryotic enzymes collaborate to bring about integration. An especially interesting question is related to target site selection. This issue is of importance for the use of *A. tumefaciens* for transformation and gene isolation.

For general reviews on the mechanism of T-DNA transfer the interested reader is referred to refs. 1-8.

T-DNA INTEGRATION AT THE CHROMOSOME LEVEL

On the level of the plant chromosomes T-DNA insertion seems to be random. In transformation experiments with *A. rhizogenes* in the plant *Crepis capillaris* T-DNA inserts were found by *in situ* hybridization to be distributed to all 3 chromosomes (9). Similarly, restriction fragment length polymorphism (RFLP) analysis of tomato transformants revealed distributions of T-DNA insertions to different chromosomes (10); with the use of a genetically marked *Petunia hybrida* line T-DNA integration loci could be mapped to different chromosomes (11).

Tomato and *Arabidopsis thaliana* with their dense genetic maps serve as excellent plant systems to study the intrachromosomal distribution of T-DNA inserts. RFLP mapping of 37 T-DNA transformants of tomato revealed a distribution to 11 of the 12 chromosomes. The distribution of T-DNAs in different chromosomes deviated from the distribution expected based on euchromatin content of different chromosomes (12). However, the size of the sample may not have been large enough to allow a definite conclusion about randomness. This issue is important for gene tagging with transposons which preferentially move to chromosomal locations close to the original point of insertion. This issue is also important for tagging of genes directly with T-DNA (see below). Analysis of 51 T-DNA integration sites in *Arabidopsis thaliana* revealed a fairly even distribution with about one insert every 10 cM (K.A. Feldmann, personal communication). These latter data have been compiled for insertions causing mutations.

DNA of eukaryotic genomes is organized in chromosomal loops anchored on proteinaceous scaffolds (13). The scaffold attachment points are proposed to function as boundaries of chromosomal domains and their gene activity. DNA sequences called scaffold attachment regions (SARs) bind to the nuclear scaffold *in vitro* and possess an AT-rich sequence with domains conserved between different eukaryotic organisms. Although the exact *in vivo* function of plant SARs is unclear, in transformation experiments expression from a transgene flanked by two SARs showed less variability and in some cases a reduction of the mean expression level (14-16). A possible neighbourhood of integrated T-DNA and SARs has in most cases not been analyzed. The detection of a SAR active *in vitro* and *in vivo* in the immediate vicinity of a right T-DNA junction in *Petunia* (17) raises the question of the generality of the finding. Possibly the deregulation of transgene activity of the particular analyzed insert is related to the proximity of the SAR sequence.

T-DNA INTEGRATION AT THE LEVEL OF THE GENE

Already early studies revealed that T-DNA can integrate into unique as well as repeated plant DNA segments (e.g., 18,19). However, several lines of evidence allow the conclusion that T-DNA is preferentially integrated into potentially transcriptionally active, DNase I-sensitive domains of chromatin. Using T-DNAs containing a selectable gene, as well as a second but promoterless screenable or selectable gene, transcriptional fusions to plant promoters were recovered with a frequency of 30% (20-22). Since similar fusion frequencies were found for tobacco and *Arabidopsis*, species differing strongly in the genome complexity, transcriptionally active chromosomal regions must have been targeted. However, preselection for activity of the first marker gene may have predetermined the use of a target with an accessible chromatin configuration (see below). Integrated T-DNA in crown gall tumors was found to be more accessible to DNase I digestion than the bulk of the chromatin, but the parallel analysis of the respective preinsertion sites is missing (23). Integration into regions of highly repetitive DNA has been reported to lead to expressional silencing of the transgenes (24). As will be discussed later, plant genomic DNA has to be accessible or made accessible for T-DNA incorporation; processes such as transcription or replication may be helpful if not required for this step.

The wealth of T-DNA tagged mutants of *Arabidopsis thaliana* has allowed the mapping of T-DNA inserts within the different parts of a "typical" *Arabidopsis* gene. The insertions were randomly distributed between untranslated and translated regions, and their distribution did not show an apparent bias with respect to direction of transcription (K.A. Feldmann, personal communication).

Since no other element is known to invade the plant from the outside world and to integrate into nuclear DNA, comparisons of T-DNA integration remain restricted to systems in other kingdoms. Retroviruses require the viral integrase protein for insertion of viral DNA. This specialized recombination event is site-specific on the viral but not the host DNA part (see ref. 25 for review). The retroviral integration complex seems to prefer actively transcribing regions for insertion (26, 27), an important aspect for the use of retroviruses as insertion mutagens. However, PCR analysis of a multitude of unselected recombination events has revealed a large number of integration sites widely distributed in the analyzed genome (28). Although differences in the analyzed retroviruses and cell lines may contribute, an important difference between these results may be the presence or absence of selection for activity of the gene carried by the mobile element. In this context it would be of interest to study T-DNA integration in the absence of selection for expression of the transforming T-DNA.

Not much is known about the influence chromatin structure may have on selection of target sites by T-DNA. *In vitro* integration studies involving purified retroviral integrase and free or nucleosomal DNA as targets showed a preference for the latter. This was explained as preference for bent DNA (29). However, the involvement of bends in the natural integration event remains to be tested.

NUMBER AND CONFORMATION OF INTEGRATED T-DNA UNITS

Two or more T-DNA units can be transferred to the same plant cell by the same or different *Agrobacterium* cell(s) (e.g., 30,97). T-DNAs stemming from such inoculations can be localized on different or on the same chromosome(s) and can be separate or linked on the same chromosome. Oligomeric T-DNA at the same locus can be found in inverted and/or direct repeats (e.g.,31-37). Depending on the report, different arrangements have been found in various proportions and various combinations (see also section COMPLEX

T-DNA ARRANGEMENTS below). Several parameters have been proposed to be responsible for these differences: the *Agrobacterium* strain used, the transformation protocol, the plant recipient and different tissues thereof as well as the selection pressure exerted.

The plant material used for transformation is certainly an important parameter. It is difficult, however, to generalize on this aspect, as in different transformation experiments different combinations of the variables listed above have been used. In a recent study different tissues of *Arabidopsis* have been compared. Single-copy T-DNA insertions have been found predominantly in transformed root material whereas multiple insertions are found in leaf disk transformants (38). Since the mechanism of multimer formation is not at all understood it is difficult to speculate (see below). However, since the same agrobacterial strains were used for the different tissues some aspect in the plant cells probably accounts for the observed phenomena, at least in this case. Also relevant for this discussion is the finding that high levels of expression of transgenes were reproducibly correlated with single T-DNA insertions whereas inverted repeat insertions tended to give lower expression levels (39,40). Thus the exerted selection for establishment of transformed plants may be an important factor determining the structure of a transgene locus.

For practical purposes such as T-DNA insertional mutagenesis, T-DNA/plant DNA gene fusions and transgene expression, single-copy inserts are advantageous. This may be achieved with the use of *Agrobacterium* strains which are not too virulent on the host to be transformed, by reducing the concentration of the agrobacterial suspension used and/or by reducing the concentration of the selective agent. Analysis of a large number of transgenic potatoes indeed revealed a negative correlation between concentration of selective compound and the number of copies of T-DNA inserts (K. Blok, personal communication); similar correlations were found for tobacco transformation (B. Tinland, unpublished data).

SEQUENCE ANALYSIS OF T-DNA INTEGRATION

Early analysis of integrated DNA was performed on tumor lines. In a series of pioneering reports the border sequences of the Ti plasmid have been defined, junctions of T-DNA and plant DNA as well as T-DNA/DNA were isolated and sequenced and adjacent regions of plant DNA were characterized (18, 32, 41-44). The right T-DNA/plant DNA junction was found to be better conserved (on the T-DNA side) than the left one. Frequently internal junctions were observed (see above). Plant DNA sequences adjacent to T-DNA insertions were found to be AT-rich, different for each tumor line and seeded with repeated sequences close to the junctions. It it not clear whether these repeats were a consequence of T-DNA integration, as preinsertion sites were not analyzed. T-DNA in tumors was not always found to be colinear with bacterial T-DNA; major rearrangements of different parts of T-DNA have been observed as well (45-47). It is not clear whether a correlation exists between complex or truncated T-DNA arrangements and the tumorous phenotype of the transformed cell.

T-DNAs carrying large internal deletions were found transmitted to the offspring (48-50). A major breakthrough in the field was the establishment of plant transformation yielding normal, fertile plants transgenic for a chosen trait (51-53). In most studied cases Southern analysis revealed the T-DNA to be intact. However, one extensive screen uncovered that almost half of the transformants had lost one or both ends of the T-DNA (54).

Production of T-strands in the bacteria, and possibly movement of the T-DNA complex through the different pro- and eukaryotic membrane, seems to be a polar process

(55). Occasional overriding of the left border signal leads to the establishment of transgenic plants in which sequences from beyond the left border are incorporated into transgenic plants (56). This is clearly an unwanted characteristic for well-defined engineered crops or T-DNA tagged plant mutants. Mutations in the interacting partners, the left border sequence (57) or the VirD2 protein (58), indeed lead to "left border readthrough" and resulting establishment in the plant recipient of beyond-left-border sequences. T-DNA processing in bacteria under suboptimal conditions apparently can lead to the same phenomenon. Screening of transformants by Southern analysis or employment of a negative selectable marker located "left" of the left border on the binary plasmid should obviate the problem.

In order to be able to make conclusions about the mechanism of T-DNA integration, T-DNA/plant DNA junction sequences have to be compared with the particular target sequence used by a particular T-DNA. The major conclusion from such experiments was that extended regions of homology between T-DNA and plant DNA are not the rule. Although T-DNA in principle is able to recognize homologies in plant DNA, such events are rare and have to be selected for (59-61).

Analyses of several T-DNA inserts into the tobacco genome have yielded a complex picture. Figure 1A, B and C give a simplified representation of three integration events into tobacco tumorlines (A, 62; B, 63) and in a transgenic tobacco plant (C, 64). Invariably, as has been also found for the less complex situations (see below), a short stretch of target sequence is deleted. A high degree of sequence repetition is visible adjacent to the plant DNA/T-DNA junctions. The length of the repeated sequences varies between about 10 and 158 base pairs. A model for the generation of large repeats of target sequences, involving integration of T-DNA into a DNA replication bubble, has been proposed (65). As shown in the schematic representation of Figure 1, the right border is almost complete in A (a complete border junction includes the third base of the border, the position of the nick that is introduced in the *Agrobacterium* cell) and complete in C. The insert shown in B is highly truncated at both ends; this particular event may be comparable to integration of naked DNA rather than to T-DNA specific insertion. It is not clear whether the DNA rearrangements seen in events A to C are a consequence of T-DNA integration or whether T-DNA was specifically targeted to genomic regions in the very process of rearrangement. Such a scenario would have the advantage that the plant's enzymatic machinery involved in this rearrangement is already at hand.

The tobacco DNA into which the T-DNA units have inserted are at a junction between repeated and single-copy DNA for A, in unique DNA with short repeated elements in B and in repeated DNA for C; thus the presence or absence of repeated sequences does not seem to have a major influence on the degree of complexity of the insert. Also the tumorous state of a transgenic line may not have a strong effect.

We do not know whether the species of a plant or a particular transformation procedure influences the integration pattern, but many T-DNA inserts into *Arabidopsis thaliana* look less complex. Perhaps the small genome of this plant is less flexible and does not tolerate rearrangements as easily as the large one of the allotetraploid plant *Nicotiana tabacum*.

A total of 13 transgenic *Arabidopsis* plants were analyzed and sequences of the T-DNA/plant DNA junctions were compared to those of the respective preinsertions sites (65, 66). This analysis allowed the establishment of the following rules: 1) T-DNA integration is not site specific and does not occur by means of long stretches of homology; 2) integration of T-DNA leads to small (13 to 73 bp) deletions of the plant DNA at the sites of insertion; 3) the left end of the T-DNA is poorly conserved, i.e., in the vast majority of the studied cases it is truncated from three up to 100 nucleotides (Figure 2), and, in addition, it shows homologies with the preinsertion sites of at least 5 mostly

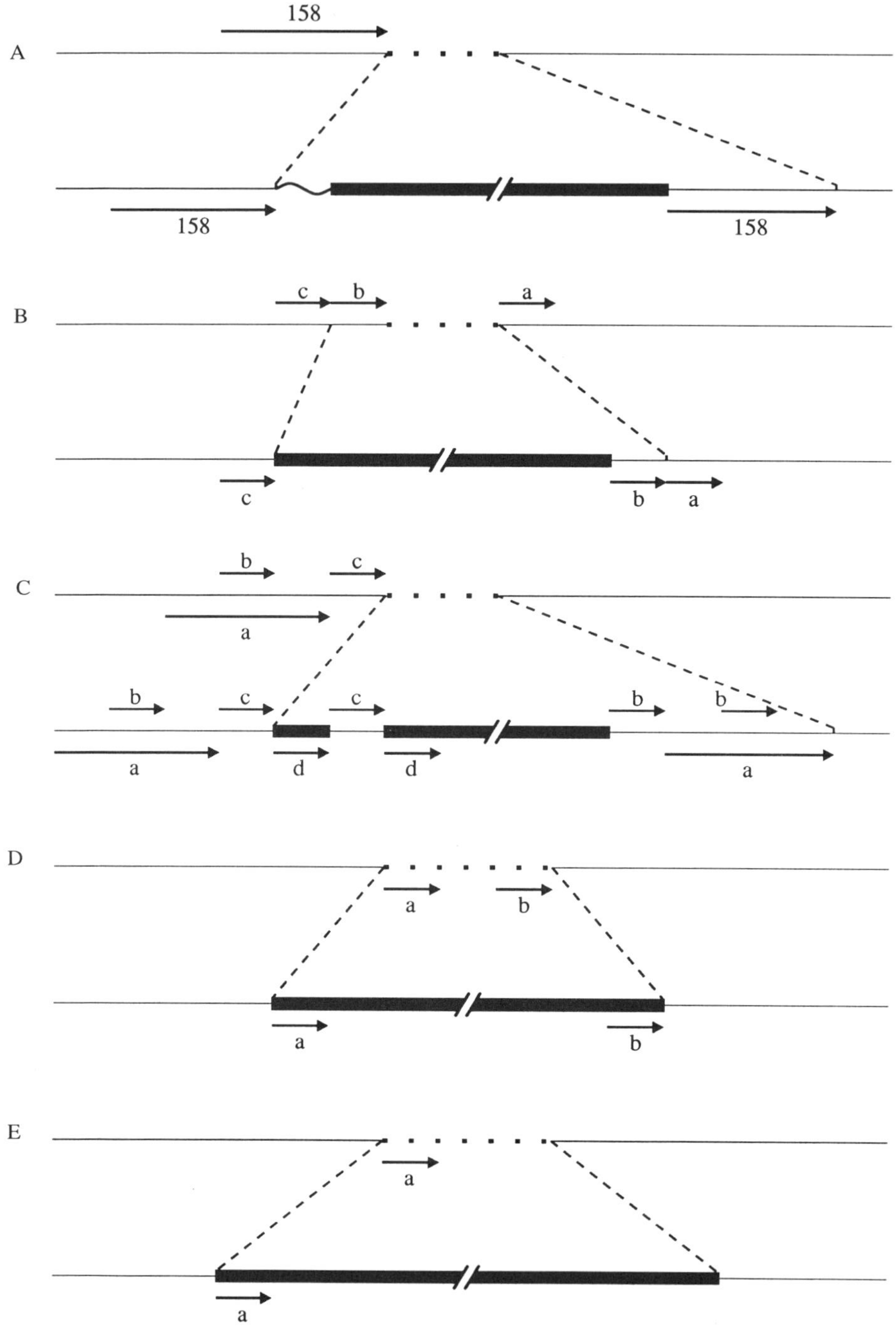

Figure 1. Schematic representations of integrated T-DNAs and their chromosomal environments. Black bars, T-DNA; thin lines, plant DNA; arrows, repeated sequences; dotted line, target site deletion, wavy line, filler DNA. A, from ref. 62; filler DNA 33 bp, target deletion 27 bp, right border junction 2 bp from nick site, (slightly simplified version). B, from ref. 63; target deletion 23 bp; a, 9 bp; b, 10 and 13 bp; c, 7 bp. b and c have partial homology to the adjacent but deleted part of T-DNA; both T-DNA ends truncated. C, from ref. 64; target deletion 16 bp; a = 74, b = 16, c = 17, d = 25 bp; right border junction at the nick site,

contiguous nucleotides; 4) the right end of the integrated T-DNA is frequently conserved up to the nucleotide which is cleaved in the bacterium (Figures 1E and 2). In such cases homology to the preinsertion site is usually restricted to one or very few bases. However, in transformants in which the right end of the T-DNA is not complete (such as in Figure 1D) short stretches of homology between T-DNA and plant DNA are seen. The mode of integration thus seems to be a chimeric one: some features, such as the involvement of small sequence homologies, are reminiscent of illegitimate (nonhomologous) recombination. Integration of plasmid DNA into animal cells has been reported following this mode of illegitimate recombination (67,68). Assimilation of the right T-DNA end, on the other hand, seems to invoke a special, T-DNA specific mechanism, probably involving virulence protein(s) (see below).

In a completely different approach T-DNA molecules were recovered from plants as replicating viral units (57). In this system T-DNA had to circularize or to "integrate" into T-DNA in order to become a viable viral DNA. Analysis of the junction sequences revealed a picture which is highly reminiscent of *bone fide* T-DNA integration: up to three bases from the right border sequence were found conserved whereas the sequences remaining at the left end were more variable and even included bases from beyond the left border in T-DNA molecules originating from a strain with a left border mutation. In addition, many rescued T-DNA molecules carried small insertions at their junctions. Their origin could in many cases be traced to a location close by on the viral genome. Interestingly, such "filler" sequences were also detected in *Arabidopsis* and tobacco T-DNA inserts (65) as well as in nonhomologous recombination in mammalian cells (68).

MODEL FOR T-DNA INTEGRATION

The data summarized so far have been used to suggest models for T-DNA integration (65,66,69). These models take into account the general notion that T-DNA integration is an efficient process, although the efficiency of the integration process *per se* has never been assessed. If there is indeed an enhancement in the efficiency of T-DNA integration over that of naked DNA it must be mediated by proteins produced in the bacterium and transferred to the plant cell. Thus the proteins contained in the T-complex (see INTRODUCTION) are the best candidates to assist in integration. The bacterial VirD2 protein which is attached to the right end of the T-strand and which targets the T-DNA to the nucleus (70-76) during the transfer process was recognized as an attractive candidate for a function in integration.

Our updated version of the T-DNA integration model includes the following sets of new data. a) Purified VirD2 protein has been shown to cleave single-stranded DNA containing a border sequence at the correct position (77, 78). VirD2 was found covalently attached to the 5′ terminus of cleaved DNA, through a tyrosine residue at position 29 (77). In addition, specific joining of cleavage products restored the original border sequence (77). Whether VirD2 actually actively performs a ligation step in the plant is not so clear, especially since this step in plants does not involve sequence specificity (see below). b) Sequence comparison between the right ends of intact T-DNAs and their respective preinsertion sequences revealed microhomologies (58). c) A mutant VirD2 protein leads to a changed integration pattern at the right junction with plant DNA (58). This implies that

left T-DNA end truncated. D, transformant 621 x 34, from ref. 66; target deletion 57 bp; a = 5 or 6 bp, b = 5 bp; right T-DNA end slightly truncated, left junction within the border. E, transformant N6H-Cs4, from ref. 66; target deletion 34 bp; a = 6 bp; right border junction at the nick site, left T-DNA end at the nick site.

the wild-type version of VirD2 is involved in the integration step. d) For this same mutant the integration step *per se*, estimated as fraction of transformants out of all T-DNA molecules that entered the plant cell nucleus, was unaltered (58). This can be interpreted to mean that the VirD2 mediated integration step is not the rate-limiting one and that synapsis between plant DNA and T-DNA is initiated at the left T-DNA end. e) T-DNA reaches the plant cell in a single-stranded form (79) and can be detected in the plant nucleus in this form (79a). This does not *a priori* imply that single-stranded T-DNA integrates but it is not obvious why a double-strand form should be invoked if the reactive VirD2-containing and homology-searching ends are on the same strand. In some cases, however, a direct involvement of single-stranded T-DNA in integration is less obvious (see later).

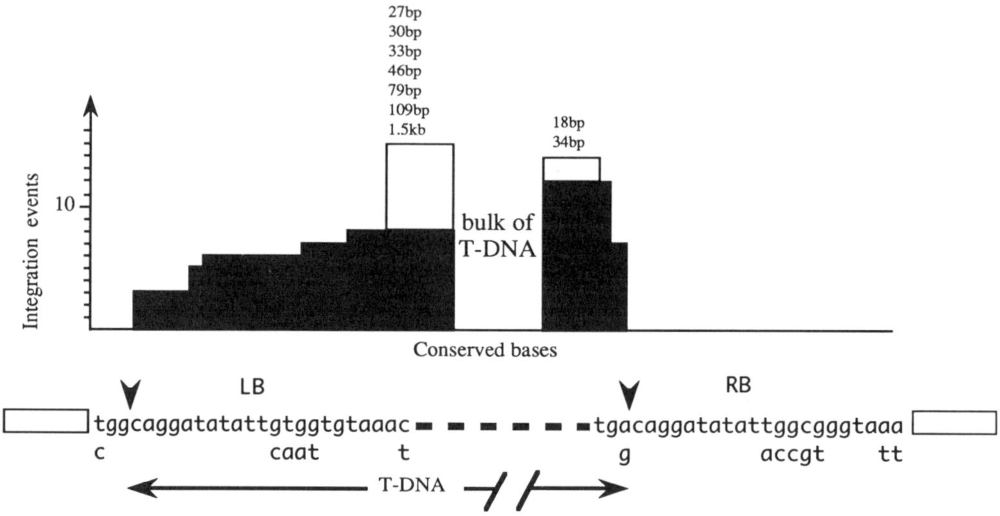

Figure 2. T-DNA integrated into plant DNA. Analysis of 15 independent integration events into the genome of *Arabidopsis thaliana* or tobacco (reproduced from ref. 126; data from refs. 65 and 66). The conservation of individual bases of the T-DNA borders is shown. Filled boxes represent integration events that preserved border bases to the extent shown, in comparison to the border sequences given below (the border sequences are given for nopaline strains, with the bases deviating from it for octopine strains; arrows indicate the position of T-DNA nicking, which however takes place in the complementary strand). Open boxes represent integration events at T-DNA positions which are located "inside" the T-DNA, removed from the nicking site by the indicated distances. The bulk of the T-DNA is to be imagined between the right and the left border or the left and the right border junctions, respectively.

Figure 3 depicts the model. In a) the 3′ end or adjacent sequences on the T-strand find homologies to the plant DNA. Single-stranded regions of the target are exposed due to a DNA-metabolic event such as replication or transcription (69). As a consequence, the displaced plant DNA strand is subjected to endonucleolytic digestion, at (1) and (2); in b) the overhanging 3′ end (if there is any) of the T-strand is digested away, either exonucleolytically or by an endonuclease cleaving at (3). The VirD2-complexed 5′ end of the T-strand screens for microhomologies on the complementary target strand. In c) the annealing of the nucleotide attached to VirD2 with the plant DNA strand (which either is of transient existence or is stabilized by plant proteins) exposes the phosphotyrosine bond to a 3' OH end of the lower plant DNA strand, which had been created upon digestion at (2) or which had been extended upon repair synthesis. Meanwhile the 3′ end of the annealed T-strand primes the synthesis of plant DNA which has become ligated at d). The upper strand of the plant DNA, nicked at (4) in c), is extended by repair replication to copy the T-DNA whereas the nucleotides up to the microhomology are degraded. This leads to deletion of a small stretch in the target DNA. The finding that only a short segment of target DNA is deleted suggests that the nicking at (1) and (3) is concerted (see below). The phosphotyrosine bond at the 5′ end of the T-DNA is exchanged for a phosphodiester bond with the target DNA. The T-DNA is covalently integrated in e).

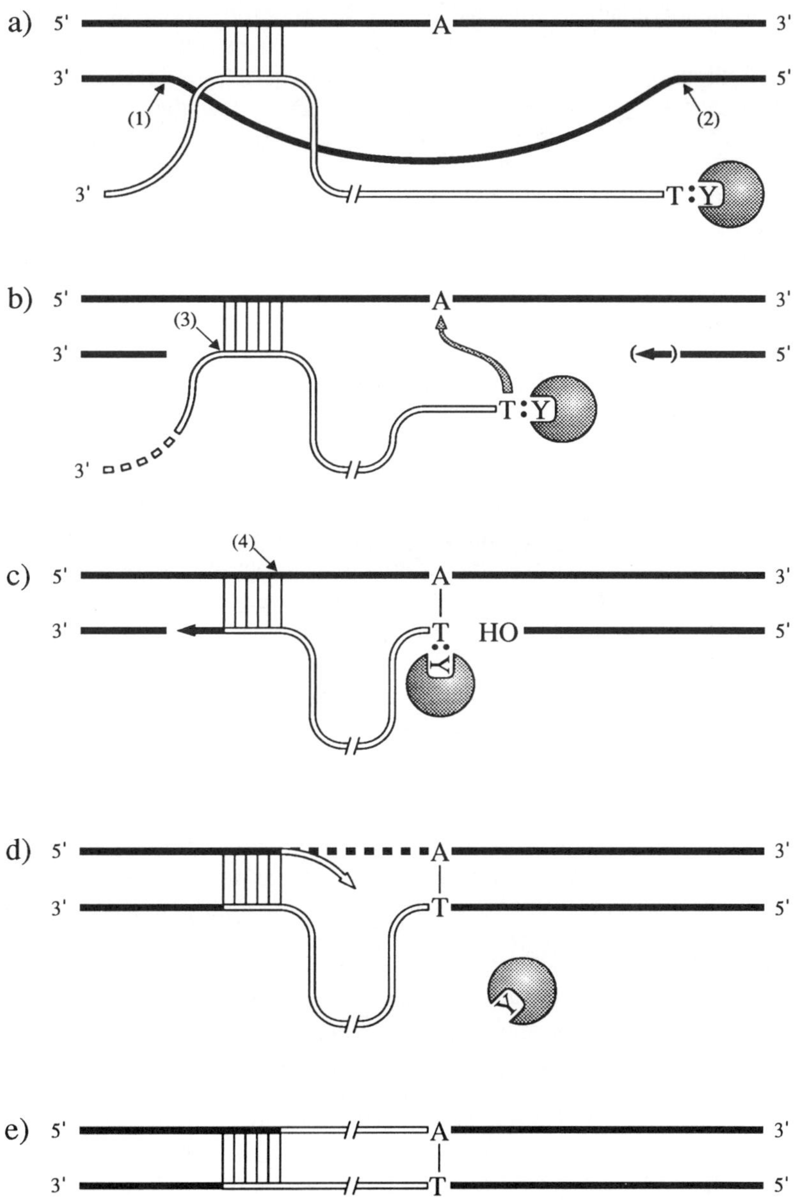

Figure 3. Model for integration of T-DNA. See text for explanation.

PROTEINS INVOLVED IN T-DNA INTEGRATION

In the absence of an *in vitro* integration assay very little can be concluded about the enzymology of T-DNA integration. However, from the data accumulated so far some clearcut predictions and some speculations seem permissible. The most interesting aspect is the interplay between bacterial and plant proteins acting directly at the level of the gene-prokaryotic/eukaryotic parasitism in its purest form.

The strongest candidate for involvement of bacterial proteins is VirD2. This protein, in combination with others, cleaves the T-DNA out of its bacterial "cage" and renders it mobile by covalently attaching to it and driving it through bacterial and plant membranes (55). The energy contained in the covalent T-DNA-VirD2 phosphotyrosine bond may well be conserved through this travel and used for the final chimeric DNA-DNA bond. A similar cleavage-ligation reaction has been observed for the replication cycle of bacteriophage ∅X174, although in this case the prokaryotic kingdom is not transversed and the viral DNA is ligated again to viral DNA (80, and refs. therein). The VirD1 protein is absolutely required for activity of VirD2 to cleave double-stranded border sequences inside the bacterial cell, as well as *in vitro* (81, and refs. therein). However, persistence of the VirD1 protein in the T-DNA complex has neither been documented nor ruled out. For the sequence-specific ligation step of a single-stranded DNA VirD2 complex to a 3′ end of a cleaved border sequence the activity of VirD1 is not required (77). If indeed VirD2 is instrumental in establishing the T-DNA-specific link to plant DNA, the independence of this process from specific plant sequences has yet to be explained. Modifications of the VirD2 protein inside the plant cell, affiliations of VirD2 with other virulence or general agrobacterial proteins, and/or connections with plant proteins remain possibilities.

Two or more other agrobacterial virulence proteins seem to have a role perhaps (only) inside the plant. Proteins VirE2 and VirF have been shown to act inside the plant cell (82, and refs. therein; 83, and refs. therein). Whereas not much else is known about the activity of VirF, the VirE2 protein clearly has an activity as non-sequence-specific single-stranded DNA binding protein. As such it may protect the T-strand from degradation, may aid in nuclear transport and/or may have a role in T-DNA integration. It may be instrumental in homology recognition (in analogy to the single-stranded DNA binding protein RecA) or could, on the contrary, prevent the bulk of the T-strand from interacting with plant DNA. Indeed, a particular mechanism should be involved to protect the interior of the T-DNA from close contacts with plant DNA and to concentrate the integrative capacity to the T-DNA ends. This idea is supported by recent results which indicate that VirE2 is not required for efficient T-DNA integration although in its absence T-DNA deletions are common (L. Rossi et al., unpublished data).

Plant proteins will have to aid in the annealing as well as the ligation step of the right T-DNA junction. The T-DNA/plant DNA/VirD2 complex suggested in steps b) and c) in Figure 3 will have to be stabilized; VirD2, by complexing with plant and other proteins, will have to take over the mode of non-sequence-specific ligation. Another activity, plant and/or bacterial in origin, may be required to keep the ends of T-DNA in proximity; in comparison to the length of a T-DNA the target deletions are very small. The complement to the 5′ end synapsis is the assimilation step of the 3′ T-DNA end. As discussed above, this step exhibits analogies to illegitimate (or nonhomologous) recombination. Just like any foreign DNA (84), T-DNA is dependent on repair enzymes for recombination into plant DNA. Particular plant mutants defective in DNA repair are therefore expected to be deficient also for T-DNA integration. Indeed, radiation-sensitive mutants of *Arabidopsis thaliana* have been isolated in which T-DNA integration but not T-DNA transfer is severely impaired (E.R. Signer, personal communication).

Very little is known about plant proteins involved in recombination in plants (85), whereas several genes involved in repair have been identified and characterized in yeast (86) and humans (87). We follow the T-DNA integration model and incorporate speculatively yeast or mammalian enzymes, where appropriate. The incisions at (1) (3′ of the melted DNA structure, Figure 3) could be performed by the equivalent of the human XPG protein (88,89) or of its homologue, the yeast Rad2 protein (90) or the murine FEN-1 protein (91). Incisions at the 5′ side of the damage, such as are postulated at position (2), (3) and (4) in Figure 3, may be achieved by a plant equivalent of the yeast protein couple Rad 1/Rad 10 (92, 93) or the human protein pair ERCC1/XPF (94). The small deletion observed in the plant genome upon T-DNA integration would fit with this model for the excision repair mechanism because the 3′ and 5′ incision in this model are thought to be coordinated and such excisions result in deletion of a similar number of bases as is observed for T-DNA integration (87). A 5′ to 3′ exonuclease such as postulated in step d) has been reported to be required for recombination in yeast (95). In addition, polymerases and helicases will be required. Proteins will be necessary for stabilization of the T-strand/plant DNA heteroduplex because the actual recombination process involves pairing of only very few bases.

An especially exciting discovery is that transcription and repair are coupled processes in bacteria, yeast and mammalian cells (96). Many proteins of the eukaryotic transcription initiation complex are also members of the nucleotide excision repair machinery. The fact that T-DNA can integrate preferentially into potentially transcriptionally active regions of the plant genome (20-22) and the notion that enzymes active in DNA repair have to perform functions in T-DNA integration are at least consistent with a transcription-repair link.

COMPLEX T-DNA ARRANGEMENTS

As discussed in the section NUMBER AND CONFORMATION OF INTEGRATED T-DNA UNITS, DNA integrates frequently as direct and indirect repeats, often in combination. This integration pattern is seen in a variety of host plants: in *Petunia,* for instance, multiple units of T-DNA inserts were recovered, mostly in inverted configuration. In addition, the complexity seemed to be linked to a low level of expression of transgenes (34). In a detailed study of a number of tomato transformants, complex arrangements of direct and indirect repeats as well as combinations thereof were described (35). Interestingly, vectors derived from nopaline-type *Agrobacterium* C58 generally seem to lead to more complex T-DNA insertions than octopine-type strain derived vectors. Complex T-DNA arrangements have been detected in *Arabidopsis* lines mutated by T-DNA tagging (37).

It is very difficult to explain the generation of the complex T-DNA arrangements in mechanistic terms. Direct DNA-mediated transformation also gives rise to multimeric arrangements of transgenes in direct and indirect configuration (97, 98). Because of this and because of the absence of detectable amounts of such structures in the agrobacterial cells, it has to be assumed that polymerization happens in the plant before or during integration. Interesting in this respect is the arrangement of one tagged *Arabidopsis* mutant; this line carries a T-DNA pair inverted over the right border, with 82 bp of target sequence duplicated in inverted orientation, flanking the T-DNA (C. Koncz, personal communication). The inversion event must have happened after initial contact of the T-DNA with the plant target sequence. However, in the absence of sequence information of other complex integration events it is difficult to speculate about mechanisms. It seems to be a general but as yet unexplained observation that T-DNA in tagged lines is frequently comprised of repeated units (37; K.A. Feldmann, personal communication; C. Koncz, personal communication).

Different T-DNAs stemming from the same (99) or even different *Agrobacterium* strains (36) have also been found integrated in a right border/right border configuration. In the first case the right border sequences of the two *Agrobacterium rhizogenes* T-DNAs were found directly linked to each other. The precise right border-right border junction can of course not be explained by ligation of single-stranded T-DNA units. Ligation of double-stranded versions is one possible explanation.

To complete the list of known complications, T-DNA-inside-T-DNA configurations have also been observed (B. Reiss, personal communication). It is not known whether the special selection pressure which was exerted through T-DNA-mediated activation of adjacent plant sequences had any influence on the formation of the "T-DNA nest."

T-DNA INTEGRATION AND PLANT METABOLISM

Wounding of plant cells seems to be a prerequisite for the transformation process by *Agrobacterium*. There may be several reasons for this. One of them is the dependence of the activation of the virulence genes on a factor produced by plants as a response to wounding. In early experiments, it was demonstrated that cell division triggered by wounding correlated with the timepoint of maximal competence for transformation of the plant tissue (100). This in turn suggests that DNA synthesis may be required for some stage of T-DNA transfer and/or integration.

It could indeed be demonstrated that the frequency of transformation is dependent on the physiological state of a tobacco suspension culture. Optimal transformation frequency correlated with the growth rate of the plant cells (101). It is likely that DNA synthesis and/or cell division are necessary for T-DNA integration. The stage of the cell cycle of protoplasts also has an influence on the integration pattern of plasmid DNA, with cells in S-phase incorporating integrated transgenes to higher copy numbers than cells in M-phase (102).

STABILITY OF INTEGRATED T-DNA

In general, once the T-DNA is integrated, it is part of the plant cell's genomic composition, and it is stably maintained and transmitted faithfully to daughter cells. Plants regenerated from such cells transmit T-DNA to their progeny as a Mendelian trait. Consistent with this is the finding that T-DNA is organized in nucleosomes with a repeat length identical to that of the bulk of nucleosomes (23, 103).

However, several conditions of transformation, tissue culture, selection or T-DNA structure, in several combinations, have led to documented rearrangements of already integrated T-DNA. a) Tissue culture mediated genetic alteration, termed somaclonal variation, whatever its exact mechanism, can be made responsible for rearrangements of T-DNA in tissue kept in culture for numerous generations (104). b) A different kind of apparent instability in fact concerns transformation at an epigenetic level (47). Cloned crown gall tumor lines, that were originally hormone-independent upon subculturing, have been found unstable in expression of the tumorous phenotype due to lack of T-DNA or lack of T-DNA expression. An explanation for the former phenomenon is expression of non-integrated T-DNA. c) An especially interesting set of observations concerns the regenerants and sexual offspring, of cloned teratoma lines (105) as well as of a shooty mutant (31). In some reported examples tissue regenerating from the original teratoma lines contained the same complement of T-DNA inserts as did the original line. However, haploid tissue derived from the line, as well as sexually produced offspring, had lost the whole set of T-DNA inserts. It was not resolved whether meiosis had actively led to the loss of T-DNA or whether meiosis had led to selection of cells out of an originally chimeric plant that did not contain the transgenes in the first place. In a different communication actual instability of T-DNA segments was documented; also in this case the presence of the T-DNA-located plant hormone genes must have been responsible for the selective advantage of plants lacking them (48). Shoots regenerated from a teratoma line by culturing in the presence of phytohormones, as well as their offspring, specifically lacked the tumor genes. However, the end-fragments of the T-DNA had remained.

More relevant for modern plant breeding and plant molecular genetics are examples of T-DNA instability in transgenic plants and their offspring in which selection for or against tumor genes could not have interfered. Two out of 22 progeny of a particular tomato transformant were reported to carry a T-DNA complement with an altered restriction pattern (10). The original T-DNA structure was shown to consist of an especially complicated array of tandem and inverted T-DNA repeats (35). In the two altered progeny plants T-DNA units were lost and other deletion events were also documented. The presence of T-DNA repeats, albeit in direct configuration, could be correlated to continued instability at the locus of *Arabidopsis thaliana* lines mutated and tagged by T-DNA inserts (Koncz, personal communication). In one line an originally dimeric T-DNA repeat suffered, upon meiosis, severe deletions on both ends of the insert; another line, containing originally a trimeric insert, underwent continuous deletions upon outcrossing.

It is presently unclear how representative these examples are. Complex T-DNA arrays, possibly in combination with complicated and unstable target sequences, may lead to especially high rates of genomic instabilities at these specific loci. However, these instabilities are probably not T-DNA specific; they may reflect a general phenomenon of plant DNA in its state of genomic flux. T-DNA, as the experimentally testable transgene, merely serves as an example for general plant genome stability/flexibility. Once more, T-DNA serves as a "window" to a particular aspect of a plant's life.

In very general terms, in the absence of specific T-DNA or target DNA derived complications, integrated copies of T-DNA are remarkably stable. Numerous field tests employing a multitude of scorable/selectable markers, testify to this statement. However, absolute stability of transgenic DNA, if not vital for survival of the plant, probably does not exist.

At the end of the scale there are plants containing *Agrobacterium rhizogenes* DNA that survived evolution for hundreds of millions of years. It was demonstrated that *Nicotiana glauca* contains genes with a remarkably high homology to their original *Agrobacterium rhizogenes* counterpart (106,107). Also in *Nicotiana tabacum Agrobacterium rhizogenes*-like genes, with a high (73%) identity in nucleic acid composition to the present day *Agrobacterium rhizogenes rolC* genes, have been detected (T. Ichikawa and F. Meins, personal communication). Given the age of *Agrobacterium* [it diverged from its relative *Rhizobium* about 250 million years ago (108)], it seems probable that a very early transformation event persisted to recent species. The degrees of selection in this process cannot easily be assessed as potential advantages for the "transgenic" sequences for the life of the affected plant species cannot be estimated. The degree of divergence of the plant-derived T-DNA sequences from the (modern) *Agrobacterium rhizogenes* T-DNA genes cannot easily be incorporated into an evolutionary clock since selective pressures and specific mutation rates of prokaryotic sequences within a eukaryotic organism cannot be estimated.

THE HOST RANGE OF T-DNA INTEGRATION

The mechanism of T-DNA integration has been examined mainly in dicotyledonous plants because they are more easily transformable. Even among the dicots sequence analysis of T-DNA/plant DNA junctions has been confined chiefly to *Nicotiana tabacum* and *Arabidopsis thaliana*. Yet it is the monocotyledons which are especially interesting for mechanistic studies because the difficulty in obtaining monocot plants transformed with *Agrobacterium* may point to differences in the quantity or quality of the interactions of pathogen and host. *Agrobacterium*-mediated transformation also has practical signficance because of their agronomic importance.

The first reported transformed monocot plant was *Asparagus* (109). Integrated T-DNA was not analyzed at the sequence level but hybridization experiments indicated normal T-DNA integration. Confirmation that even in grasses T-DNA integration follows a mechanism at least similar to that in dicotyledonous plants stems from a report on rice transformation (110). Although target sequences were not determined, junctions between T-DNA and rice DNA resemble those found for tobacco and *Arabidopsis*: T-DNA at the right junction extends to the base at which T-DNA is nicked in the bacterium, whereas the left boundaries were found distributed over the whole border region. This distribution thus conforms to that observed for dicotyledons (see Figure 2). These results suggest a uniform mechanism of T-DNA integration.

Pertinent to this discussion is a recent discovery of a low level of T-DNA transfer to yeast cells (111). In yeast the efficiency of homologous recombination is very high. Indeed, integration of introduced DNA into yeast DNA occurs predominantly by homologous recombination. Analysis of yeast transformants isolated from *Agrobacterium* - *Saccharomyces cerevisiae* mating experiments revealed that T-DNA had integrated *via* its homology to nuclear DNA. Judging from the transformation events analyzed it seems that it is the recipient organism with its specific chromosomal structure and its specific enzymology that dictates the mechanism of integration of at least the majority of events.

THE USE OF T-DNA INTEGRATION FOR GENE ISOLATION

Exogenous or endogenous elements integrating at random are ideal for creating mutations and at the same time molecularly tagging the mutated gene. As such nomadic elements plant transposons and T-DNA are used, frequently in combination. The preferential recovery of T-DNA insertions in transcribed regions of the plant genome, which may or may not be the result of the selection pressure exerted on the transformants, is an ideal condition for the establishment of T-DNA tagged mutants.

Several variations of T-DNA mutagenesis and gene tagging can be used, and more are possible. Figure 4 schematically represents gene inactivation by T-DNA insertion (A), enhancer or promoter trapping (B) and activation tagging (C). Inactivation of a gene by T-DNA insertion (reviewed in 112 and 113) usually leads to a loss-of-function phenotype and can be scored only in the homozygous state. However, this approach cannot be used for plants such as tobacco with its allotetraploid genome. The ideal model plant for insertional mutagenesis is *Arabidopsis thaliana*. In the first successful demonstrations of the application of this procedure, mutants were genetically analyzed for linkage to T-DNA and the tagged region was identified and cloned. With plant DNA adjacent to T-DNA as probe, the corresponding wild-type sequences were isolated and used to complement the original mutants by transformation (114-117). Many more interesting genes have been isolated, with this pathway.

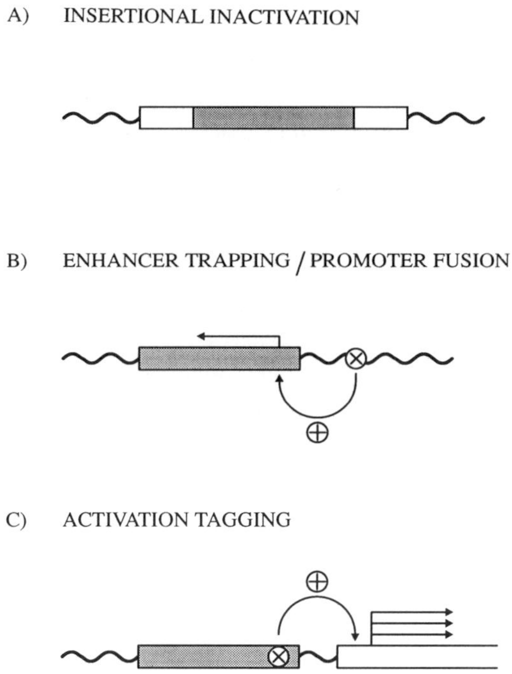

Figure 4. T-DNA tagging strategies. See text for explanation.

There are several mutant collections containing T-DNA insertions; roughly half of all *Arabidopsis* genes have been presently tagged (K.A. Feldmann, personal communication). Interestingly, not all mutants in a population mutagenized with T-DNA are actually tagged (37). The mechanism by which T-DNA may just "hit" plant DNA, mutate it and then get lost is not understood but is likely to exist. Somaclonal variation in the currently used transformation procedures is kept to a minimum and therefore should not be made responsible for the high fequency of untagged mutants.

T-DNA can also be employed for the isolation of plant promoters or enhancers with the use of a promoterless or enhancerless selectable or scorable gene in the T-DNA (route B, Figure 4). Tissue specific or developmentally regulated promoters have been isolated with such procedures (e.g., 21,22,118,119). For enhancer trapping strategies, genes located on T-DNAs, containing only weak promoters, were used to capture plant enhancers *in situ* (120,121). The pattern of transgene expression visualizes the activity of the regulatory region to which the crippled T-DNA gene had been linked; the observed degree of transgene activation indeed showed a dramatic variation from transformant to transformant, and between different tissues within a transgenic line. Again, because transformants had been selected for expression of an intact transgene, the positions of the insertions were *a priori* biased in favor of gene activity.

In the reverse approach (Figure 4C) genes adjacent to integrated T-DNA become activated because of a strong and constitutively acting enhancer located at the edge of T-DNA (122). In this activation tagging scheme, dominant phenotypes are created. Thus the ploidy level of the plant in which the experiment is conducted is not important, and mutants can be identified as primary transformants. An additional advantage is that the tagged gene is not destroyed. With this technique an auxin independent mutant was isolated by selection of regenerating transformed protoplasts on auxin-free medium (123).

SUMMARY AND PERSPECTIVES

Over the past 18 years a wealth of information on T-DNA integration has been assembled. Insertion events have been analyzed on the genetic, physiological and mechanistic level. Transgenes shipped into plant genomes by means of T-DNA vectors have been shown to overcome phylogenetic barriers - they behave as proper plant genes, with faithful regulatory expression and transmission to the progeny. Exceptions to this are teaching us features of regulation of plant gene expression which we otherwise would have taken much longer to recognize. Thus the activity of transgenes in plants represents a window to the behavior of plant genes.

We have much to learn about how T-DNA "chooses" its target locus. In the future, except for random tagging experiments, it will be desirable to insert exogenous DNA at defined target positions in the plant genome. Site-directed mutagenesis is an important part in analyzing genomes and their functions. Gene targeting in plants, although documented, is still a low efficiency procedure (see 124). By dissecting individual steps in the T-DNA integration pathway it may be possible to "convince" T-DNA to recognize larger stretches of homology than it normally would. By placing the segment homologous to the plant target area close to the left border and/or by modifying bacterial or plant proteins active in homology recognition, it may be possible to improve the still low level of gene targeting. Also, by increasing the chance of creating a double-strand DNA break close to the homology target area, as has been shown for extrachromosomal homologous recombination (125), the frequency of targeting may become high enough to be realistic.

The mechanism of T-DNA integration remains an active and fertile area of investigation. Bacterial as well as plant proteins involved in this reaction will be identified, through the use of existing or new bacterial or plant mutants. From studies of DNA repair mutants, we will learn more about the enzymology of T-DNA integration. *In vitro* experiments are already being used to dissect steps of the integration pathway. Most importantly, target DNA is not to be viewed as chemically pure nucleic acid but as "living" chromatin. It will be fascinating to deduce the rules T-DNA follows in its choice of target chromatin: possibly physical proximity to the nuclear periphery, activity due to replicative and/or transcriptional functions, availability due to a particular stage in the cell cycle or other criteria.

With exceptions, T-DNA has been reported to be stably transmitted to the offspring. This is highly important for crop improvement through transformation (with *Agrobacterium* or with direct DNA-mediated uptake). As numerous field-tests are under way, the stability of T-DNA, on the genetic as well as on the epigenetic level, is being assessed rigorously. However, as stated before, we cannot expect transgenic DNA to be more stable than plant DNA proper, on the genetic or epigenetic level.

Finally, T-DNA will continue to be used to explore the mysteries of the plant genome. T-DNA's nomadic nature allows us to use it to tag plant genes known only by phenotype but not understood at the biochemical level. What nature has evolved as a plant pathogen has become a fascinating instrument for plant biologists and a valuable tool for agriculture.

ACKNOWLEDGMENTS

We would like to thank all colleagues who contributed unpublished information: Karin Blok, Ken Feldmann, Stanton Gelvin, Paul Hooykaas, Takanari Ichikawa, Csaba

Koncz, Erich Lanka, Yasunori Machida, Fred Meins, Bernd Reiss and Luca Rossi. We appreciate the numerous valuable comments on the manuscript by Mary-Dell Chilton. We thank our group, especially Holger Puchta, for critical remarks. We are grateful to Luigina Beffa for her help with the manuscript.

REFERENCES

1 Amabile-Cuevas, C.F. and Chicurel, M.E. (1992) Cell 70, 189-199.
2 Dreiseikelmann, B. (1994) Microbiol. Rev. 58, 293-316.
3 Hooykaas, P.J.J. and Beijersbergen, A.G.M. (1994) Annu. Rev. Phytopathol. 32, 157-179.
4 Kado, C.I. (1993) in Bacterial Conjugation (D. B. Clewell, ed.) pp. 243-254, Plenum Press, New York, NY.
5 Winans, S.C. (1992) Microbiol. Rev. 56, 12-31.
6 Zambryski, P. (1992) Annu. Rev. Plant Physiol. Plant Mol. Biol. 43, 465-490.
7 Hohn, B. (1992) in Development. The Molecular Genetic Approach (V.E.A. Russo, S. Brody, D. Cove and S. Ottolenghi, eds.) pp. 206-216, Springer-Verlag, Berlin.
8 Lessl, M. and Lanka, E. (1994) Cell 77, 321-324.
9 Ambros, P.F., Matzke, A.J.M. and Matzke, M.A. (1986) EMBO J. 5, 2073-2077.
10 Chyi, Y.-S.,Jorgensen, R.A.,Goldstein, D.,Tanksley, S.D. and Loaiya-Figueroa, F. (1986) Mol. Gen. Genet. 204, 64-69.
11 Wallroth, M., Gerats, A.G.M., Rogers, S.G., Fraley, R.T. and Horsch, R.B. (1986) Mol. Gen. Genet. 202, 6-15.
12 Thomas, C.M., Jones, D.A., English, J.J., Carroll, B.J., Bennetzen, J.L., Harrison, K., Burbidge, A., Bishop, G.J. and Jones, J.D.G. (1994) Mol. Gen. Genet. 242, 573-585.
13 Saitoh, Y. and Laemmli, U.K. (1993) Cold Spring Harbor Symp. Quant. Biol. 58, 755-765.
14 Breyne, P., Van Montagu, M., Depicker, A. and Gheysen, G. (1992) Plant Cell 4, 463-471.
15 Allen, G.C., Hall, G.E., Childs, L.C., Weissinger, A.K., Spiker, S. and Thompson, W.F. (1993) Plant Cell 5, 603-613.
16 Mlynarova, L., Loonen, A., Heldens, J., Jansen, R.C., Keizer, P., Stiekema, W.J. and Nap, J.-P. (1994) Plant Cell 6, 417-426.
17 Dietz, A., Kay, V., Schlake, T., Landsmann, J. and Bode, J. (1994) Nucl. Acids Res. 22, 2744-2751.
18 Zambryski, P., Depicker, A., Kruger, K. and Goodman, H.M. (1982) J. Mol. Appl. Genet. 1, 361-370.
19 Ursic, D., Slightom, J.L. and Kemp, J.D. (1983) Mol. Gen. Genet. 190, 494-503.
20 Herman, L., Jacobs, A., Van Montagu, M. and Depicker, A. (1990) Mol. Gen. Genet. 224, 248-256.
21 Kertbundit, S., De Greve, H., Deboeck, F., Van Montagu, M. and Hernalsteens, J.-P. (1991) Proc. Nat. Acad. Sci. U.S.A. 88, 5212-5216.
22 Koncz, C., Martini, N., Mayerhofer, R., Koncz-Kalman, Z., Körber, H., Redei, G.P. and Schell, J. (1989) Proc. Nat. Acad. Sci. U.S.A. 86, 8467-8471.
23 Coates, D., Taliercio, E.W. and Gelvin, S.B. (1987) Plant Mol. Biol. 8, 159-168.
24 Pröls, F. and Meyer, P. (1992) Plant J. 2, 465-475.
25 Craigie, R. (1992) TIG 8, 187-190.

26 Mooslehner, K., Karls, U. and Harbers, K. (1990) J. Virology 64, 3056-3058.
27 Scherdin, U., Rhodes, K. and Breindl, M. (1990) J. Virology 64, 907-912.
28 Withers-Ward, E.S., Kitamura, Y., Barnes, J.P. and Coffin, J.M. (1994) Genes Dev. 8, 1473-1487.
29 Müller, H.-P. and Varmus, H.E. (1994) EMBO J. 13, 4704-4714.
30 Depicker, A., Herman, L., Jacobs, A., Schell, J. and Van Montagu, M. (1985) Mol. Gen. Genet. 201, 477-484.
31 Lemmers, M., De Beuckeleer, M., Holsters, M., Zambryski, P., Depicker, A., Hernalsteens, J.P., Van Montagu, M. and Schell, J. (1980) J. Mol. Biol. 144, 353-376.
32 Zambryski, P., Holsters, M., Kruger, K., Depicker, A., Schell, J., Van Montagu, M. and Goodman, H.M. (1980) Science 209, 1385-1391.
33 Spielmann, A. and Simpson, R.B. (1986) Mol. Gen. Genet. 205, 34-41.
34 Jones, J.D.G., Gilbert, D.E., Grady, K.L. and Jorgensen, R.A. (1987) Mol. Gen. Genet. 207, 478-485.
35 Jorgensen, R., Snyder, C. and Jones, J.D.G. (1987) Mol. Gen. Genet. 207, 471-477.
36 De Block, M. and Debrouwer, D. (1991) Theor. Appl. Genet. 82, 257-263.
37 Castle, L.A., Errampalli, D., Atherton, T.L., Franzamann, L.H., Yoon, E.S. and Meinke, D.W. (1993) Mol. Gen. Genet. 241, 504-514.
38 Grevelding, C., Fantes, V., Kemper, E., Schell, J. and Masterson, R. (1993) Plant Mol. Biol. 23, 847-860.
39 Hobbs, S.L.A., Kpodar, P. and DeLong, C.M.O. (1990) Plant Mol. Biol. 15, 851-864.
40 Hobbs, S.L.A., Warkentin, T.D. and DeLong, C.M.O. (1993) Plant Mol. Biol. 21, 17-26.
41 Yadav, N.S., Vanderleyden, J., Bennett, D.R., Barnes, W.M. and Chilton, M.-D. (1982) Proc. Nat. Acad. Sci. U.S.A. 79, 6322-6326.
42 Simpson, R.B., O'Hara, P.J., Kwok, W., Montoya, A.L., Lichtenstein, C., Gordon, M.P. and Nester, E.W. (1982) Cell 29, 1005-1014.
43 Holsters, M., Villarroel, R., Gielen, J., Seurinck, J., De Greve, H., Van Montagu, M. and Schell, J. (1983) Mol. Gen. Genet. 190, 35-41.
44 Kwok, W.W., Nester, E.W. and Gordon, M.P. (1985) Nucl. Acid Res. 13, 459-471.
45 Peerbolte, R., Leenhouts, K., Hooykaas-van Slogteren, G.M.S., Hoge, J.H.C., Wullems, G.J. and Schilperoort, R.A. (1986) Plant Mol. Biol. 7, 265-284.
46 Peerbolte, R., Leenhouts, K., Hooykaas-van Slogteren, G.M.S., Wullems, G.J. and Schilperoort, R.A. (1986) Plant Mol. Biol. 7, 285-299.
47 Van Lijsebettens, M., Inzé, D., Schell, J. and Van Montagu, M. (1986) J. Mol. Biol. 188, 129-145.
48 Yang, F. and Simpson, R.B. (1981) Proc. Nat. Acad. Sci. U.S.A. 78, 4151-4155.
49 Otten, L., De Greve, H., Hernalsteens, J.P., Van Montagu, M., Schieder, O., Straub, J. and Schell, J. (1981) Mol. Gen. Genet. 183, 209-213.
50 De Greve, H., Leemans, J., Hernalsteens, J.-P., Thia-Toong, L., De Beuckeleer, M., Willmitzer, L., Otten, L., Van Montagu, M. and Schell, J. (1982) Nature 300, 752-755.
51 Barton, K.A., Binns, A.N., Matzke, A.J.M. and Chilton, M.-D. (1983) Cell 32, 1033-1043.
52 De Block, M., Herrera-Estrella, L., Van Montagu, M., Schell, J. and Zambryski, P. (1984) EMBO J. 3, 1681-1689.
53 Horsch, R.B., Fraley, R.T., Rogers, S.G., Sanders, P.R., Lloyd, A. and Hoffmann, N. (1984) Science 223, 496-498.

54 Deroles, S.C. and Gardner, R.C. (1988) Plant Mol. Biol. 11, 365-377.
55 Citovsky, V. and Zambryski, P. (1993) Annu. Rev. Microbiol. 47, 167-197.
56 Martineau, B., Voelker, T.A. and Sanders, R.A. (1994) Plant Cell 6, 1032-1033.
57 Bakkeren, G.,Koukolíková-Nicola, Z.,Grimsley, N. and Hohn, B. (1989) Cell 57, 847-857.
58 Tinland, B., Schoumacher, F., Gloeckler, V., Bravo Angel, A.-M. and Hohn, B. (unpublished data).
59 Lee, K.Y., Lund, P., Lowe, K. and Dunsmuir, P. (1990) Plant Cell 2, 415-425.
60 Offringa, R., de Groot, M.J.A., Haagsman, H.J., Does, M.P., van den Elzen, P.J.M. and Hooykaas, P.J.J. (1990) EMBO J. 9, 3077-3084.
61 Offringa, R., Franke-Van Dijk, M.E.I., de Groot, M.J.A., van den Elzen, P.J.M. and Hooykaas, P.J.J. (1993) Proc. Nat. Acad. Sci. U.S.A. 90, 7346-7350.
62 Gheysen, G.,Van Montagu, M. and Zambryski, P. (1987) Proc. Nat. Acad. Sci. U.S.A. 84, 6169-6173.
63 Matsumoto, S.,Ito, Y.,Hosoi, T.,Takahashi, Y. and Machida, Y. (1990) Mol. Gen. Genet. 224, 309-316.
64 Ohba, T., Yoshioka, Y., Machida, C. and Machida, Y. (1995) Plant J. (in press).
65 Gheysen, G.,Villarroel, R. and Van Montagu, M. (1991) Genes Dev. 5, 287-297.
66 Mayerhofer, R., Koncz-Kalman, Z., Nawrath, C., Bakkeren, G., Crameri, A., Angelis, K., Redei, G.P., Schell, J., Hohn, B. and Koncz, C. (1991) EMBO J. 10, 697-704.
67 Murnane, J.P.,Yezzi, M.J. and Young, B.R. (1990) Nucl. Acids Res. 18, 2733-2738.
68 Roth, D. and Wilson, J. (1988) in Genetic Recombination (R. Kucherlapati and G. R. Smith, eds.) pp. 621-653, Amer. Soc. Microbiol., Washington, DC.
69 Koncz, C.,Németh, K.,Rédei, G.P. and Schell, J. (1994) in Homologous Recombination and Gene Silencing in Plants (J. Paszkowski, ed.) pp. 167-189, Kluwer Academic Publishers, Dordrecht.
70 Herrera-Estrella, A., Van Montagu, M. and Wang, K. (1990) Proc. Nat. Acad. Sci. U.S.A. 87, 9534-9537.
71 Shurvinton, C.E., Hodges, L. and Ream, W. (1992) Proc. Nat. Acad. Sci. U.S.A. 89, 11837-11841.
72 Howard, E.A., Zupan, J.R., Citovsky, V. and Zambryski, P.C. (1992) Cell 68, 109-118.
73 Tinland, B., Koukolíková-Nicola, Z.,Hall, M.N. and Hohn, B. (1993) Proc. Nat. Acad. Sci. U.S.A. 89, 7442-7446.
74 Rossi, L., Hohn, B. and Tinland, B. (1993) Mol. Gen. Genet. 239, 345-353.
75 Koukolíková-Nicola, Z., Raineri, D., Stephens, K., Ramos, C., Tinland, B., Nester, E.W. and Hohn, B. (1993) J. Bacteriol. 175, 723-731.
76 Citovsky, V., Warnick, D. and Zambryski, P. (1994) Proc. Nat. Acad. Sci. U.S.A. 91, 3210-3214.
77 Pansegrau, W., Schoumacher, F., Hohn, B. and Lanka, E. (1993) Proc. Nat. Acad. Sci. U.S.A. 90, 11538-11542.
78 Jasper, F., Koncz, C., Schell, J. and Steinbiss, H.-H. (1994) Proc. Nat. Acad. Sci. U.S.A. 91, 694-698.
79 Yusibov, V.M., Steck, T.R., Gupta, V. and Gelvin, S.B. (1994) Proc. Nat. Acad. Sci. U.S.A. 91, 2994-2998.
79a Tinland, B., Hohn, B. and Puchta, H. (1994) Proc. Nat. Acad. Sci. U.S.A. 91, 8000-80004.
80 Hanai, R. and Wang, J.C. (1993) J. Biol. Chem. 268, 23830-23836.
81 Scheiffele, P.R., Pansegrau, W. and Lanka, E., J. Biol. Chem. (in press).

82 Citovsky, V., Zupan, J., Warnick, D. and Zambryski, P. (1992) Science 256, 1802-1805.
83 Regensburg-Tuïnk, A.J.G. and Hooykaas, P.J.J. (1993) Nature 363, 69-71.
84 Köhler, F., Cardon, G., Pöhlman, M., Gill, R. and Schieder, O. (1989) Plant Mol. Biol. 12, 189-199.
85 Tissier, A.F. and Signer, E.R. (1994) in Homologous Recombination and Gene Silencing in Plants (J. Paszkowski, ed.) pp. 157-166 Kluwer Academic Press, Dordrecht.
86 Prakash, S., Sung, P. and Prakash, L. (1993) Annu. Rev. Genet. 27, 33-70.
87 Hoeijmakers, J.H.J. and Bootsma, D. (1994) Nature 371, 654-655.
88 O'Donovan, A., Scherly, D., Clarkson, S.G. and Wood, R.D. (1994) J. Biol. Chem. 269, 15965-15968.
89 O'Donovan, A., Davies, A.A., Moggs, J.G., West, S.C. and Wood, R.D. (1994) Nature 371, 432-435.
90 Habraken, Y., Sung, P., Prakash, L. and Prakash, S. (1993) Nature 366, 365-368.
91 Harrington, J.J. and Lieber, M.R. (1994) Genes Dev. 8, 1344-1355.
92 Fishman-Lobell, J. and Haber, J.E. (1992) Science 258, 480-484.
93 Bardwell, A.J., Bardwell, L., Tomkinson, A.E. and Friedberg, E.C. (1994) Science 265, 2082-2088.
94 Biggerstaff, M., Szymkowski, D.E. and Wood, R.D. (1993) EMBO J. 12, 3685-3692.
95 Huang, K.N. and Symington, L.S. (1993) Mol. Cell. Biol. 13, 3125-3134.
96 Hanawalt, P.C., Donahue, B.A. and Sweder, K.S. (1994) Current Biol. 4, 518-521.
97 Riggs, C.D. and Bates, G.W. (1986) Proc. Nat. Acad. Sci. U.S.A. 83, 5602-5606.
98 Schocher, R.J., Shillito, R.D., Saul, M.W., Paszkowski, J. and Potrykus, I. (1986) Bio/Technology 4, 1093-1096.
99 Jouanin, L., Bouchez, D., Drong, R.F., Tepfer, D. and Slightom, J.L. (1989) Plant Mol. Biol. 12, 75-85.
100 Binns, A.N. and Thomashow, M.F. (1988) Annu. Rev. Microbiol. 42, 575-606.
101 An, G. (1985) Plant Physiol. 79, 568-570.
102 Kartzke, S., Saedler, H. and Meyer, P. (1990) Plant Sci. 67, 63-72.
103 Schèfer, W., Weising, K. and Kahl, G. (1984) EMBO J. 3, 373-376.
104 Peerbolte, R., Ruigrok, P., Wullems, G. and Schilperoort, R. (1987) Plant Mol. Biol. 9, 51-57.
105 Yang, F., Montoya, A.L., Merlo, D.J., Drummond, M.H., Chilton, M.-D., Nester, E.W. and Gordon, M.P. (1980) Mol. Gen. Genet. 177, 707-714.
106 Furner, I.J., Huffman, G.A., Amasino, R.M., Garfinkel, D.J., Gordon, M. and Nester, E.W. (1986) Nature 319, 422-427.
107 Aoki, S., Kawaoka, A., Sekine, M., Ichikawa, T., Fujita, T., Shinmyo, A. and Syono, K. (1994) Mol. Gen. Genet. 243, 706-710.
108 Otten, L., Canaday, J., Gérard, J.-C., Fournier, P., Crouzet, P. and Paulus, F. (1992) Mol. Plant-Microbe Interact. 5, 279-287.
109 Bytebier, B., Deboeck, F., De Greve, H., Van Montagu, M. and Hernalsteens, J.-P. (1987) Proc. Nat. Acad. Sci. U.S.A. 84, 5345-5349.
110 Hiei, Y.,Ohta, S.,Komari, T. and Kumashiro, T. (1994) Plant J. 6, 271-282.
111 Beijersbergen, A. (1993) Ph.D. Thesis, Leiden.
112 Feldmann, K.A. (1991) Plant J. 1, 71-82.
113 Koncz, C., Németh, K., Rédei, G.P. and Schell, J. (1992) Plant Mol. Biol. 20, 963-976.
114 Marks, M.D. and Feldmann, K.A. (1989) Plant Cell 1, 1043-1050.

115 Herman, P.L. and Marks, M.D. (1989) Plant Cell 1, 1051-1055.
116 Koncz, C., Mayerhofer, R., Koncz-Kalman, Z., Nawrath, C., Reiss, B., Redei, G.P. and Schell, J. (1990) EMBO J. 9, 1337-1346.
117 Yanofsky, M.F., Ma, H., Bowman, J.L., Drews, G.N., Feldmann, K.A. and Meyerowitz, E.M. (1990) Nature 346, 35-38.
118 Teeri, T.H., Herrera-Estrella, L., Depicker, A., Van Montagu, M. and Palva, E.T. (1986) EMBO J. 5, 1755-1760.
119 Herman, L.M.F., Van Montagu, M. and Depicker, A. (1986) Mol. Cell. Biol. 6, 4486-4492.
120 Goldsbrough, A. and Bevan, M. (1991) Plant Mol. Biol. 16, 263-269.
121 Topping, J.F., Wei, W. and Lindsey, K. (1991) Development 112, 1009-1019.
122 Walden, R., Hayashi, H. and Schell, J. (1991) Plant J. 1, 281-288.
123 Hayashi, H., Czaja, I., Lubenow, H., Schell, J. and Walden, R. (1992) Science 258, 1350-1353.
124 Morton, R. and Hooykaas, P.J.J. (1995) Molecular Breeding (in press).
125 Puchta, H., Dujon, B. and Hohn, B. (1993) Nucl. Acid Res. 21, 5034-5040.
126 Tinland, B. and Hohn, B. (1993) in Virus Strategies (W. Doerfler and P. Böhm, eds.) pp. 349-359, Verlag Chemie, Weinheim.

METAL PRECIPITATION BY MARINE BACTERIA: POTENTIAL FOR BIOTECHNOLOGICAL APPLICATIONS

Bradley M. Tebo

Marine Biology Research Division and
Center for Marine Biotechnology and Biomedicine
Scripps Institution of Oceanography
University of California, San Diego
9500 Gilman Drive
La Jolla, CA 92093-0202

INTRODUCTION

Bacteria interact with metals in various ways and thereby are important agents in the geochemical cycling of metals. In the past decade or so, with the development and applications of chemical techniques to accurately measure the concentrations of metals in the environment, interest has turned to attempting to understand the role that biology plays in geochemical cycling. The picture that is beginning to emerge is that the biota is extremely important in metal cycling in diverse environments, on short time scales of hours to days rather than years. In addition, cycling is limited not only to the metals that are required as trace nutrients for biological systems, but rather most other metals cycle as well. These metal cycles are driven by diverse chemical and biological processes. In general, the biochemical mechanisms by which the biota affect geochemical cycling are very poorly understood. However, as we begin to explore the myriad possibilities, it is clear that a variety of these mechanisms may be exploited for biotechnological purposes.

Metal–microbe interactions are currently exploited in a number of different technologies including leaching, extraction, beneficiation, adsorption and ion exchange. For example, in the mining industry it has long been recognized that bacteria are important agents in the oxidation and solubilization of a variety of metal sulfide ores and that such bacteria can be exploited to leach and recover valuable metals from low grade ores. It is becoming clear that knowledge in this field has increased to a point where use of bacteria for metal-related processes is poised to expand beyond mining into a wide variety of industries.

The focus of this review is on the potential biotechnological applications of metal precipitation by marine bacteria. In the broadest definition, metal precipitation refers to the transformation of a soluble metal to an insoluble one. Metal precipitation as used in this article will refer to the formation of both metal precipitates, which are amorphous and

Genetic Engineering, Vol. 17, Edited by J.K. Setlow
Plenum Press, New York, 1995

hydrated particulate forms of metals, as well as minerals, which are more crystalline and anhydrous. With aging and dehydration metal precipitates can be transformed into minerals and thus can be thought of as precursors in the formation of some minerals (1). Biomineralization is a term frequently used in reference to biological metal precipitation; however, because true minerals are not necessarily formed, this can be misleading. In practice metal precipitation is often measured by operationally defining soluble metals as those that pass through a certain size filter (usually 0.2 to 0.4 μm). However, smaller particles (and even bacteria) and colloids that are also precipitates clearly exist. In the context of this article, the precise size range is not as important as the idea that the process (metal precipitation) or the product (the metal precipitate) have properties that make them potentially useful. These metal precipitates or minerals can exist intracellularly or extracellularly, either attached or unattached to the cell surface. Metal precipitation as defined by filtration would also include metals adsorbed to cell surfaces since such metal is no longer "in solution." This adsorption of metals to living or dead cells, which does not have to involve metabolic energy or transport processes, has been termed "biosorption" (2). Biosorption is the metal-microbe interaction that has been the most widely studied in different organisms and is perhaps the most promising approach for near term application in metal removal and recovery processes. Although aspects of biosorption will be discussed in this article, it will not be a major focus as there have been many recent articles, reviews, and even books devoted to various aspects of this topic (1-13). Rather, this article focuses on metal precipitation that occurs as a result of metabolic and/or catalytic processes, which has been largely unexploited by industry but, in the author's opinion, represents the most promising biotechnological frontier both for metal removal and recovery applications as well as the synthesis of new materials.

Although I try to focus this review on marine bacteria, the metabolic diversity of marine bacteria is largely unexplored. Since many processes in terrestrial and marine environments are likely to be similar, I frequently draw from those terrestrial processes with the belief that marine counterparts exist. One should keep in mind, however, that other processes may be unique in marine bacteria. Marine bacteria live in a high ionic medium with high concentrations of some metals. These metal ions may compete for the binding and uptake sites of nutrient metals, which in many marine environments are in extremely low concentrations. Thus, marine bacteria may have specialized mechanisms for the uptake of metal ions that they require for nutritional purposes. For example, specializations that may characterize some marine bacteria might be high affinity or high specificity metal uptake systems. In addition, in the marine environment there are regions of extreme physical conditions, such as those found in the deep sea and at deep sea hydrothermal vents, that do not occur elsewhere on earth, and in which novel metal–microbe interactions may exist. Thus, marine bacteria represent a new frontier for exploration of potential processes and biomolecules for biotechnology.

This review is divided into several sections. I begin by giving a brief overview of the metal-microbe interactions in marine organisms and at deep-sea hydrothermal vents. I then introduce the types of metal-microbe interactions that lead to the transformation of metals between their particulate, soluble, and sometimes gaseous phases, focusing on those interactions that lead to metal removal from solution *via* metal precipitation and mineralization. The potential applications of these processes for biomaterials and metal removal and recovery are then discussed. Finally, I provide a summary of the research on a unique marine metal-precipitating bacterium that appears to be well-suited for various applications and end with a brief discourse on future research needs in the area of microbial metal precipitation.

METAL–MICROBE INTERACTIONS IN MARINE ORGANISMS

Most of the research in the area of metals in marine organisms has either focused on metallothioneins, small cysteine-rich metal-binding proteins, and their role in the detoxification of heavy metals in a large variety of eukaryotic organisms (e.g., 14) and cyanobacteria (15) or on the effects of metals on phytoplankton growth and productivity (e.g., 16-19). Very few studies of metal metabolism in marine bacteria have been performed. The mechanism of iron acquisition (20-23), the effect of heavy metals on growth yields (24), the sorption of lead on marine bacteria and bacterial exopolymers (25-27) and the uptake of cadmium (28,29) have been reported for a few marine bacteria. A copper binding compound that is excreted into the surrounding medium has also been characterized from a marine *Vibrio* species (30,31). Because of the importance of Mn cycling in the marine environment, studies of Mn binding, oxidation, and reduction with marine bacteria are probably the most numerous (e.g., 32-35).

METAL–MICROBE INTERACTIONS AT HYDROTHERMAL VENTS

The microbial populations colonizing the warm and hot environments of deep-sea hydrothermal vents are of considerable interest for biotechnology. With regard to metal-microbe interactions, hydrothermal vent habitats are relatively unexplored in terms of the types of interactions and microbial diversity. The extreme temperatures and pressures found at hydrothermal vents suggest that microbes inhabiting vents will have unusual resistances and adaptations to extreme physical and chemical conditions. Temperatures at hydrothermal vents range from ambient seawater (~ 2°C) to over 350°C and pressures from about 200 to 500 atmospheres or even higher in certain cases. The chemistry of hydrothermal fluids are also characterized by a relatively low pH, high metal and sulfide concentrations as well as a lack of oxygen. Hydrothermal vents are sites of active mineral deposition and dissolution in which microorganisms play an important role. Mineral phases include metal sulfides, sulfates, oxides, hydroxides, silicates and carbonates. The discharge of hydrothermal fluids was responsible for the formation of massive sulfide ore deposits of zinc, copper, iron, silver and gold that are presently being mined.

Metal-precipitating bacteria are abundant around hydrothermal vents (36-41). Because non-specific (chemical) metal deposition is common around vents, however, it is not always possible to determine whether metal precipitation is due to activities (direct or indirect) of the microbes. Bacteria encrusted with iron and manganese precipitates are frequently found colonizing the outer surfaces of hydrothermal vent animals such as vestimentiferan tube worms and mussels and iron hydroxides in modern and ancient metal deposits have been observed to precipitate on or within microbial filamentous mats (42-44). Mn oxide deposits also form within sediments as hydrothermal fluids percolate up through the crust although the role of microbes in this deposition has not been investigated. Biogenic metal sulfide mineral formation at hydrothermal vents, while not expected to be quantitatively important, could also occur. In fact, the association of bacteria with the organic tubes of polychaete worms has been implicated in metal sulfide mineralization, specifically marcasite (FeS_2), on high temperature sulfide chimneys and thermophilic bacterial sulfate-reduction has been demonstrated in hydrothermal sediments (45). For a more detailed discussion of metal-microbe interactions at hydrothermal vents the reader is referred to a recent review on the topic (46).

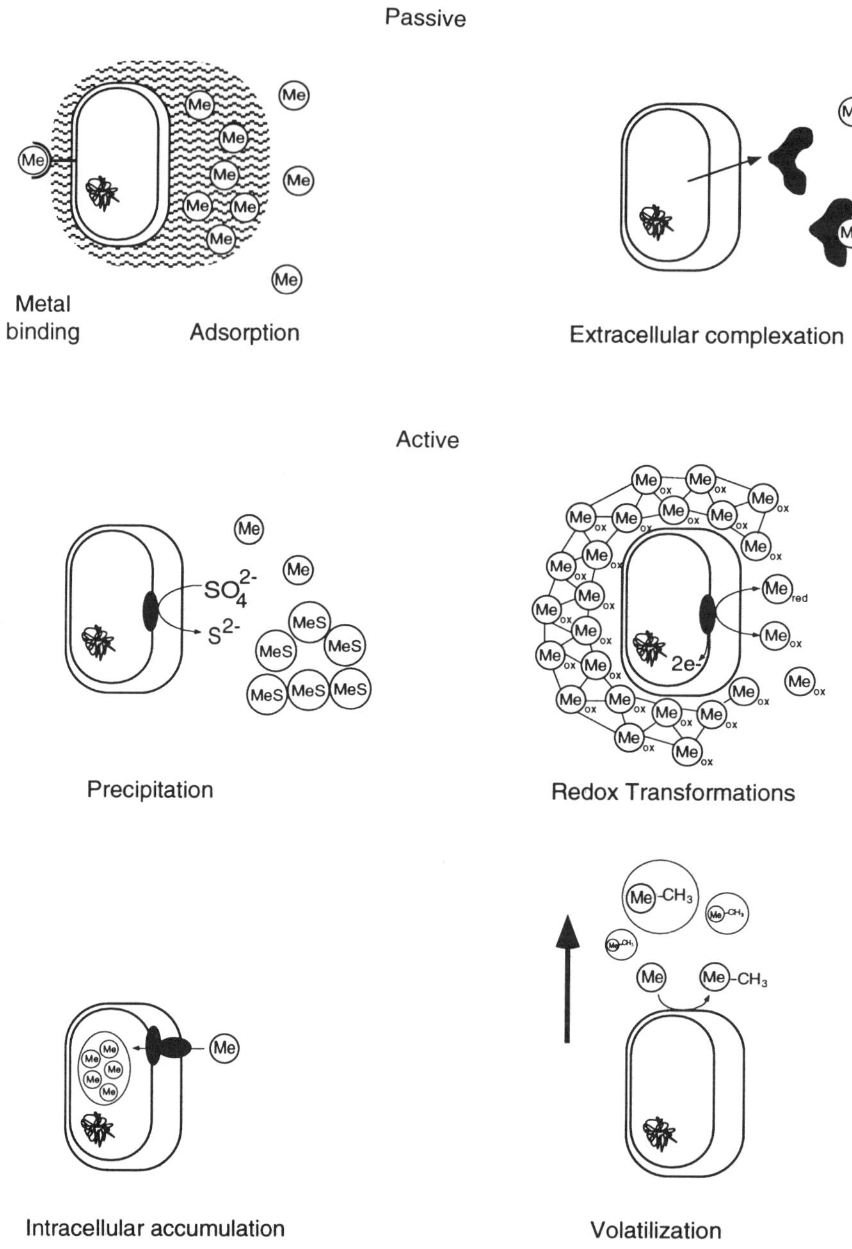

Figure 1. Microbial interactions with metals. Me, metal ion; MeS, metal sulfide; Me_{red} and Me_{ox}, reduced and oxidized metal; Me-CH_3, methylated metal.

METAL-MICROBE INTERACTIONS CAUSING METAL REMOVAL

The role of metals in microbiological systems has been a subject of much investigation during the past two or three decades. These studies have focused on the metabolic requirements of cells for metals, the toxic effects of metals on cells and the microbiological transformations of metals. While the literature on these topics is large, almost all of these studies have focused on terrestrial and freshwater organisms. In general, very little is known about the role of metals in marine bacteria.

The general classes of metal-microbe interactions that occur in terrestrial organisms are likely to occur in marine bacteria. However, because marine bacteria are subject to a completely different metal regime than most other bacteria, the details are apt to differ. Metal-microbe interactions can be broadly classified as being either passive or active (47). Passive processes do not require the direct participation of living organisms and can occur whether the microbes are alive or dead, although the processes originally depend on the growth and metabolism of the cell. Active processes are those that require metabolically active microbes; cellular metabolism and/or enzymatic activity are implicit in active processes.

Passive Removal of Metals

Passive metal-microbe interactions occur as consequences of the net negative charge of microbial cell surfaces and their exopolymers or through reactions with extracellular complexing agents that can be attached to cell surfaces or released into solution (Figure 1).

Biosorption. Microbial biomass can accumulate metal ions through processes of adsorption, ion exchange, chelation (coordination), and covalent bonding (Figure 1; 1,4,5,10,47,48). This passive sorption of metals or biosorption (2) is mediated by various functional groups that can be present on either live or dead microbial cells: carboxyl, hydroxyl, sulfhydryl, amino, imino, imidazole, sulfate and sulfonate groups are present in most biological polymers present on the surface of cells, in the cell wall and within the cytoplasm. These polymers can either be the complex polysaccharides or glycoproteins that form discrete capsules or aggregates around the cell or be due to the components of the cell wall itself.

The classic example of application of biosorption for metal precipitation or removal from solution is in the activated sludge process for wastewater treatment. Activated sludge contains a variety of microorganisms that produce copious quantities of extracellular polymers which act to remove metals both through electrostatic interactions and trapping of particles within the polymer matrix. Other systems studied for biosorption include fungi and algae, which have been studied for their ability to take up metals and radionuclides such as thorium and uranium on their cell walls (4,13).

Extracellular Complexation. Microorganisms can produce extracellular complexing agents that, once outside the cell, interact with metals in a fashion that no longer requires metabolic energy of the cell (Figure 1). Specifically, microbes can produce chemicals that have a high affinity for binding or complexing metals. The classic example is the production of siderophores by microorganisms. Siderophores are chelating agents that have very high binding constants for ferric (Fe(III)) ions and are produced by microbes to facilitate the uptake of iron into the cell. Siderophores are also capable of binding other metals such as nickel, uranium and copper (6). The cells produce specific transport proteins that recognize and bind the siderophore-Fe(III) complex and thereby take up the iron needed for cellular processes. Although this subsequent accumulation of iron is an active process, the interaction between the siderophore and the metal is not, and therefore, for the purposes of this discussion is considered a passive interaction. Immobilized siderophores have been used as complexing agents in metal removal processes (4). Marine bacteria produce siderophores that are in some cases unique (20-23,49-52). Alterobactin, the siderophore with the highest stability constant known, is of marine origin (23).

The marine bacterium *Vibrio alginolyticus* produces extracellular copper-binding proteins in response to Cu exposure (31). The proteins, whose synthesis is induced by Cu, are believed to play a role in the detoxification of Cu by altering the speciation of Cu in seawater (30). Cu-complexing agents have also been shown to be produced by cyanobacteria. Whereas in some cases these agents appear to be siderophores (53), in other cases the nature of the ligands is uncertain (54).

Active Removal of Metals

Active metal removal can occur *via* extracellular precipitation, redox reactions, intracellular accumulation, or volatilization (Figure 1; 5). With the exception of the volatilization reactions such as the methylation or reduction of mercuric ion (55), these processes often result in the formation of solid phase materials (i.e., metal precipitates and minerals). Biological precipitation and mineralization (biomineralization) is divided into those processes that are biologically-induced and those that are biologically-controlled (56). Biologically-induced metal precipitation and mineralization usually occur external to the organisms as a result of metabolic activity and products changing the environment. Biologically-induced precipitation and mineralization processes can be brought about directly *via* metal binding to a specific cell component or catalysis of a redox reaction or indirectly as a consequence of metabolism that changes the environmental conditions (e.g., pH) or produces a reactive extracellular product (e.g., hydrogen sulfide or hydrogen peroxide). Extracellularly catalyzed redox transformations and precipitation reactions are types of biologically-induced mineralization processes. Examples of these processes are the oxidation of iron and manganese (57) and the formation of metal sulfides through the activities of sulfate-reducing bacteria (47).

Biologically-controlled mineralization in microorganisms usually occurs within cells as a result of specific enzymes or matrices and often serves an important biological function. Whereas biologically-*induced* metal precipitation and mineralization processes result in metal precipitates or minerals that typically have varying size classes and no unique morphology, biologically-*controlled* mineralization processes often result in well defined minerals with a narrow size distribution (58). The classic example of biologically-controlled mineralization is the formation of magnetite within magnetosomes in magnetotactic bacteria (59). Intracellular accumulation, however, can also occur without mineral formation. Mechanisms are known by which microbes accumulate metals to high levels as soluble complexes with macromolecules. Examples of active metal accumulation by bacteria are the accumulation of silver (60,61), cadmium (29,62), germanium (63,64), and nickel (65,66).

Bacterial metal precipitation and mineralization processes may be exploited in certain biotechnological applications and it is these active processes that are the major focus of this review.

Volatilization. Some metals can exist as volatile species. The most notable is mercury (Hg) which in its elemental state (Hg^0) is volatile at ambient temperature and atmospheric pressure. Thus organisms that reduce ionic Hg bring about its volatilization (Figure 1). A much more widespread mechanism of volatilization is brought about by many different microbes that interact with elements by methylating them. Methylation can result in the formation of volatile species and has been proposed as a mechanism adopted by microorganisms for the detoxification of a variety of elements such as Hg, arsenic (As), lead (Pb), selenium (Se), tellurium (Te), cadmium (Cd) and tin (Sn). The formation of methylmercury compounds is perhaps the most well studied process because it is the methylated forms of mercury that are concentrated in the food chain. These volatilization processes are thus important elemental transformations in the environment. The converse is the demethylation of the methylated compounds, which thereby decreases the volatility and hence mobility of the elements. Although volatilization increases the mobility of elements in the environment, which is not usually desirable, there may be situations when such a process may be necessary for removing metals from the system. There also has been recent interest in combining microbially-enhanced volatilization reactions with condensation of the gases as a means for metal removal and recovery (67).

Intracellular Accumulation and Biomineralization. Microorganisms require certain metals as nutrients and therefore have mechanisms to take them up from the environment. This accumulation of metals involves active metabolism to bind and transport metal ions (Figure 1). Often the metal uptake systems are specific for certain metals enabling them to be accumulated against a large concentration gradient. In some cases the amount of metals taken up can be substantial. In addition, non-essential metals are often taken up, usually non-specifically through energy-dependent transport systems intended for other metals. These non-essential metals and even nutrient metals at higher concentrations can be toxic. Some microbes have evolved mechanisms that sequester metals intracellularly, in certain cases as a mechanism for metal detoxification.

Metal accumulation can, in specialized cases, lead to the formation of intracellular minerals, particularly of iron. In theory, intracellular mineralization is a more controlled process and may result in crystalline and amorphous materials with unique properties. This biomineralization process has been most noted in the magnetotactic bacteria which, depending on the organism, can produce either an iron oxide, magnetite (Fe_3O_4), or iron sulfides, greigite (Fe_3S_4) or pyrrhotite ($Fe_{1-x}S$) (68-70). Another form of iron mineral is the iron storage mineral, ferritin (referred to as bacterioferritin in bacteria). Ferritin has been most studied in mammals where it functions in iron sequestration or storage and thus functions in the buffering and homeostasis of iron. A variety of different bacteria produce ferritins which are related to, but distinct from, eukaryotic ferritin and function not only as iron storage molecules but also as cytochromes (71,72). The protein component of ferritin provides a lattice in which iron is deposited as a relatively inert mineral, an amorphous oxyhydroxide ferric phosphate in bacterioferritins or ferrihydrite ($5Fe_2O_3 \cdot 9H_2O$) in mammalian ferritins (73). Reconstitution of the bacterioferritin iron core *in vitro* without phosphate produces ferrihydrite (73).

Other bacteria produce intracellular polyphosphates which may complex metals and thereby function as a detoxifying agent. Cadmium, aluminum, and zinc have all been reported to be accumulated as phosphate or polyphosphate granules or bodies in prokaryotes (62,74-76). It's unclear whether these metals are simply adsorbed to the polyphosphates or whether metal phosphate minerals are actually formed.

Magnetite formation. Bacteria that were magnetotactic were discovered in the 1970s. These bacteria orient and propel themselves along magnetic field lines. The literature on these organisms has been reviewed extensively over the years (e.g., 59,70,77-80). The microbes precipitate single domain ferromagnetic minerals intracellularly which are contained within a membrane. This membrane-bound ferromagnetic mineral is called a magnetosome. Magnetotactic bacteria possess up to 1000 such magnetosomes (81) which are usually arranged in a chain parallel to the long axis of the cell and provide sufficient energy to align the cell in the earth's magnetic field. There are two possible magnetic polarities that magnetotactic bacteria could have depending on the magnetic dipole of the cell, one that allows them to swim parallel to the field direction and the other antiparallel to the field direction. Because magnetotactic bacteria in the Northern Hemisphere swim primarily to the north, while those in the Southern Hemisphere swim predominantly south, this behavior allows the bacteria to migrate downward, apparently to maintain themselves at the proper (low) oxygen tension.

Although a lot of what we know about magnetotactic bacteria comes from the study of a magnetite-producing freshwater strain, *Magnetospirillum magnetotacticum* MS-1, marine or brackish water isolates are common and at least one marine strain has been isolated and described (82,83). Also it was recently demonstrated that a freshwater sulfate-reducing bacterium could form magnetite (84), suggesting that marine counterparts are most likely abundant given the importance of sulfate reduction in marine systems. Magnetotactic bacteria can accumulate iron up to 20,000–40,000-fold over the extracellular iron concentration (85) and therefore are believed to play an important role in remnant magnetism in soils and sediments (80). Originally it was believed that magnetite would only be synthesized under microaerobic conditions, however, we now know that some magnetotactic bacteria can produce magnetosomes under completely anaerobic conditions (82,83) as well as under aerobic conditions (86,87). Magnetotactic bacteria and bacterially produced magnetites have a number of biotechnological applications in bioremediation and biomaterials (see below).

Bacterioferritin formation. Bacterioferritin is generally believed to serve as an iron storage compound for bacteria (88); however, its precise role in bacteria is uncertain. A variety of bacteria have been shown to produce cytochrome-like bacterioferritins including *Azotobacter vinelandii, Escherichia coli, Pseudomonas aeruginosa* and *Nitrobacter winogradskyi* (71,88-90). Bacterioferritins resemble ferritins of eukaryotes in that they are protein shells composed of 24 subunits surrounding a core of several thousand iron atoms. The bacterioferritin gene (*bfr*) of *E. coli* has been cloned, sequenced (71) and overproduced in *E. coli*, allowing for better characterization of the protein (90). *bfr* codes for a protein of 158 amino acids (MW 18,495) and possesses iron oxidation activity. Bacterioferritins actively sequester iron both *in vitro* and *in vivo*. The non-heme iron of bacterioferritins can be removed by reducing agents and the cores reconstituted by addition of Fe(II) aerobically (73). When bacterioferritins cores are reconstituted in the absence of phosphate, superparamagnetic crystalline ferrihydrite forms (88). Iron core formation results from a two step process in which there is a slow nucleation (oxidation) step catalyzed by the apoferritin followed by a second more rapid autocatalytic stage (72,91,92). The bacterioferritin cores can have a range of chemical composition and crystal structure suggesting that mineralization is not rigidly controlled by the cells (protein) and may be affected by various factors such as Fe concentration, the Fe/bacterioferritin ratio, pH and phosphate. Although studies of bacterioferritin are really in their infancy, this system represents an important model for the biological production of materials in defined microenvironments and may someday have useful applications for controlled production of biomaterials. In addition, because various metal ions such as zinc (Zn^{2+}), cadmium (Cd^{2+}), copper (Cu^{2+}) and manganese (Mn^{2+}) bind to ferritin (mainly the mineral component, 93), bacterioferritins may have some use for metal removal and recovery.

Precipitation Reactions. Extracellular metal precipitation occurs as a result of the production of a metabolite that combines with a metal ion, metal complex, or other metal-containing compound forming a solid phase (i.e., insoluble precipitate; Figure 1). This precipitation reaction is dependent on the solubility product (K_{sp}) of the precipitated salt: when the product of the concentrations of the dissolved species exceeds the K_{sp} of the solid then the formation of the solid is thermodynamically favored. For a dissolution–precipitation reaction of the form,

$$Me_aX_b \leftrightarrow aMe^{b+} + bX^{a-},$$
$$K_{sp} = [Me^{b+}]^a \times [X^{a-}]^b.$$

In some cases, microbes may excrete compounds that bring about changes in the environmental conditions, such as redox potential or pH, that indirectly affect metal speciation increasing one of the ion concentrations so that the K_{sp} is exceeded and thereby causing metals to precipitate.

Two of the major classes of these precipitates are metal sulfides and metal phosphates although microbial activities can also bring about the precipitation of metal carbonates, oxides and hydroxides (Table 1). In microbially-induced metal sulfide precipitation microorganisms that produce hydrogen sulfide (H_2S), such as sulfate-reducing bacteria or other dissimilatory sulfur-reducing bacteria, can precipitate large quantities of metals. In fact, the formation of black ferrous sulfide is often used as an indicator for sulfide formation and is a taxonomic trait characteristic of certain clinically important bacteria. Similarly, in metal phosphate precipitation, microbial activities are necessary for the production of orthophosphate (HPO_4^{2-}). A variety of metals may form insoluble metal sulfide or metal phosphate precipitates (Table 1), some of which may be exploited for metal-removal processes. In both metal sulfide and metal phosphate formation, the activities are enzymatically mediated *via* their involvement in the formation of the metabolic byproduct (H_2S or HPO_4^{2-}). Since sulfate reduction is an obligately anaerobic process and the phosphatase system an aerobic process, these two systems have different potential applications and can precipitate different metals.

Metal Sulfide Formation. The classic example of a metal precipitation reaction is the production of hydrogen sulfide by sulfate-reducing bacteria and the subsequent precipitation of metal sulfide. Although sulfate reduction is not confined to marine environments, because of the high levels of sulfate in seawater, it is a dominant process in anaerobic marine sediments where sulfate serves as the primary electron acceptor for the anaerobic decomposition of organic matter. Sulfate-reducing bacteria grow anaerobically on organic matter and reduce sulfate to hydrogen sulfide,

$$SO_4^{2-} + 2H^+ + 2(CH_2O) \rightarrow H_2S + 2H_2O + 2CO_2$$

where (CH_2O) represents the basic unit of organic matter. The hydrogen sulfide so formed dissociates depending on pH:

$$H_2S \xleftrightarrow{pK_a^1=7} H^+ + HS^- \xleftrightarrow{pK_a^2=13.5} 2H^+ + S^{2-} \qquad (1)$$

where the pK_as shown are the acidity constants for an ionic strength of 0.5 M (94). It is the sulfide ion (S^{2-}) which reacts with metals (Me^{2+}) forming insoluble metal sulfides,

$$Me^{2+} + S^{2-} \longleftrightarrow MeS.$$

pH is an important determinant of S^{2-} concentration (equation 1) and thus any organism living in a sulfidic environment that increases pH (e.g., through production of ammonium) will enhance metal sulfide formation. It has been well documented that sulfate-reducing bacteria are important agents in metal sulfide precipitation in the environment and are probably responsible for the formation of certain metal sulfide ores (95).

Metal Phosphate Formation. Another common metal precipitation reaction is the formation of metal phosphates. Many bacteria produce cell surface-associated enzymes known as phosphatases that cleave phosphate groups from organophosphate compounds. The organic moieties are utilized as growth substrates while most of the phosphate is released into solution as hydrogen phosphate (HPO_4^{2-}) which may combine with certain metals forming insoluble metal phosphates (96). Alkaline phosphatases are the best studied example and are synthesized under conditions of phosphate starvation, and are thus important for the uptake of phosphate as a nutrient requirement. Acid phosphatases are less well characterized and they may serve different functions for different organisms. It is an acid phosphatase that has been described to be important in metal phosphate formation (97).

Metal phosphate formation has been mostly studied by Macaskie and Dean and their colleagues, studies that have primarily focused on a species of *Citrobacter*, originally

Table 1. Solubility Products for Some Metal Precipitates

Compound	Solubility Product
Metal sulfides	
Cadmium sulfide (CdS)	$10^{-27.0}$
Cobalt sulfide (CoS)	$10^{-25.5}$
Cupric sulfide (CuS)	$10^{-36.1}$
Cuprous sulfide (Cu_2S)	$10^{-48.5}$
Ferrous sulfide (FeS)	$10^{-18.1}$
Lead sulfide (PbS)	$10^{-27.5}$
Manganese sulfide (MnS)	$10^{-13.5}$
Mercuric sulfide (HgS)	$10^{-52.7}$
Nickel sulfide (NiS)	$10^{-23.9}$
Silver sulfide (Ag_2S)	$10^{-50.1}$
Zinc sulfide (ZnS)	$10^{-24.7}$
Metal phosphates	
Ferric phosphate ($FePO_4$)	$10^{-26.4}$
Ferrous hydrogen phosphate ($FeHPO_4$)	10^{-26}
Ferrous phosphate ($Fe_3(PO_4)_2 . 8H_2O$)	10^{-36}
Strontium hydrogen phosphate ($SrHPO_4$)	10^{-27}
Uranyl phosphate (UO_2HPO_4)	10^{-48}
Metal carbonates	
Cadium carbonate ($CdCO_3$)	$10^{-13.7}$
Ferrous carbonate ($FeCO_3$)	$10^{-10.7}$
Lead carbonate ($PbCO_3$)	$10^{-13.1}$
Manganese carbonate ($MnCO_3$)	$10^{-9.3}$
Zinc carbonate ($ZnCO_3$)	10^{-10}
Metal oxides and hydroxides	
Cadmium hydroxide ($Cd(OH)_2$)	$10^{-14.4}$
Chromium hydroxide ($Cr(OH)_3$)	$10^{-29.8}$
Cupric hydroxide ($Cu(OH)_2$)	$10^{-19.4}$
Cupric oxide (CuO)	$10^{-20.4}$
Ferric hydroxide ($Fe(OH)_3$)	$10^{-38.8}$
Ferric oxyhydroxide (FeOOH)	$10^{-41.5}$
Ferrous hydroxide ($Fe(OH)_2$)	$10^{-15.1}$
Lead hydroxide ($Pb(OH)_2$)	$10^{-10.9}$
Lead oxide (PbO)	$10^{-15.3}$
Manganese hydroxide ($Mn(OH)_2$)	$10^{-12.8}$
Mercuric oxide (HgO)	$10^{-25.4}$
Zinc oxide (ZnO)	$10^{-16.9}$

Data from refs. 94, 106 and 162

isolated from a metal-polluted soil (97,98). When growing on an organo-phosphate such as glycerol 2-phosphate, *Citrobacter* induces a cell-surface associated acid phosphatase that cleaves the organic moiety from the compound, releasing HPO_4^{2-} into the medium which can combine with metals. Cadmium, lead and uranium have all been shown to be effectively removed by this *Citrobacter* sp. (99,100).

Extracellular Redox Transformations. A redox reaction is a reaction that involves the transfer of electrons from or to an element. In reduction the element accepts an electron(s) while in oxidation the element gives up an electron(s). A variety of oxidation and reduction reactions are catalyzed by living microorganisms or, in some cases, by reactive sites on their cell surfaces (101) and many elements within the periodic table are susceptible to redox reactions under the normal conditions that exist in the environment (Figure 1). Often these redox transformations bring about the precipitation of solid phases because the new species has reduced solubility; thus these redox reactions may be exploited for precipitation and recovery processes.

Some of the important redox transformations that result in metal precipitation are given in Table 2. The extracellular reduction reactions are often referred to as microbial dissimilatory metal reduction to distinguish them from reduction reactions that take place inside of cells as a part of assimilatory metabolism or metal detoxification. Reduction of elements such as chromium (Cr), selenium (Se) and uranium (U), can lead to insoluble or low solubility materials. Examples of elements that form solid phases upon oxidation are iron (Fe) and manganese (Mn) which occur as metal hydroxides, oxyhydroxides, or oxides. Both of these oxidation and reduction processes are types of biologically-induced mineralization. In general, more research has focused on metal reduction than metal oxidation reactions because of their importance in anaerobic environments where they are coupled to the oxidation of organic matter, including organic pollutants. Several reviews on dissimilatory metal reduction have been published recently (35,102-104) and therefore, rather than re-review this literature exhaustively, I will include a few examples that may have particular applications in biotechnology and for which marine bacteria may provide unique properties.

Iron- and Manganese-Depositing Microorganisms. The biomineralization of iron (Fe) and manganese (Mn) is important in the environment and is primarily due to microbial processes. A variety of different iron and manganese minerals can be formed microbially *via* both biologically-controlled and biologically-induced processes including oxides, phosphates, carbonates and sulfides. For example, the iron-oxide magnetite (Fe_3O_4) is produced intracellularly by magnetotactic bacteria (59) and extracellularly by dissimilatory iron-reducing bacteria (105).

Iron is the fourth most abundant element in the earth's crust. A variety of different iron solids can form as a result of oxidation or reduction reactions. Metallic iron (Fe^0), ferrous (Fe(II)) and ferric (Fe(III)) are the primary forms of iron in the environment. In the presence of air, the soluble form of Fe(II), Fe^{2+}, is stable only at low pH (<5). At higher pHs in the presence of oxygen Fe(II) autooxidizes to Fe(III), which, in the absence of complexing agents, is unstable and rapidly hydrolyzes forming insoluble hydroxides or oxides. Fe(III) may also combine with phosphate forming insoluble ferric phosphate ($FePO_4$). In the presence of hydrogen sulfide Fe(III) is reduced to Fe(II) and the Fe(II) can precipitate forming iron sulfides (FeS or FeS_2). Fe(II) can also precipitate as ferrous carbonate ($FeCO_3$) or ferrous phosphate ($Fe_3(PO_4)_2$). In many cases, microbial activities play an important role in the formation of these iron precipitates.

A variety of different bacteria have been implicated in the oxidation and precipitation of iron. These so-called "iron bacteria" may either possess enzymatic mechanisms for Fe(II) oxidation or may have cell surface properties that enhance the precipitation of Fe. The oxidation of Fe(II) to Fe(III) may provide energy for certain autotrophic bacteria, most notably the acidophilic iron-oxidizing bacterium, *Thiobacillus ferrooxidans*. *T. ferrooxidans* derives its energy and reducing power from the oxidation of Fe(II) and various reduced sulfur compounds (HS^- , $S_2O_3^=$, etc.) and can grow with CO_2 as its sole source of carbon. *T. ferrooxidans* is an important agent in causing acid mine drainage and has been exploited for the leaching of a variety of metal ores. Because this organism is acidophilic, the Fe(III) that it produces remains soluble at the low pH of the environment and only precipitates when pH increases. Many other organisms, however, are capable of oxidizing Fe(II) at more neutral pH.

Table 2. Some Oxidation and Reduction Reactions that Lead to Lower Solubility and Metal Precipitation

Process	Reaction	Possible Products
Oxidation Reactions		
Iron oxidation	Fe(II) → Fe(III)	Fe oxides, hydroxides and oxyhydroxides (minerals such as goethite, lepidocrocite, ferrihyrite, etc.)
Manganese oxidation	Mn(II) → Mn(III or IV)	Mn oxides, hydroxides and oxyhydroxides (minerals such as todorokite, birnessite, manganite, hausmannite, etc.)
Cobalt oxidation	Co(II) → Co(III)	Co oxides, hydroxides and oxyhydroxides
Copper oxidation	Cu(I) → Cu(II)	(ppt)
Reduction Reactions		
Gold reduction	Au(I) → Au^0	elemental Au
Silver reduction	Ag(I) → Ag^0	elemental Ag
Molybdenum reduction	Mo(VI) → Mo(V)	blue, precipitates at neutral pH
Technetium reduction	Tc(VII) → Tc(IV)	?
Chromate reduction	Cr(VI) → Cr(III)	$Cr(OH)_3$
Selenite reduction	Se(IV) → Se^0	elemental Se (red)
Uranium reduction	U(VI) → U(IV)	UO_2 (uranite)
Vanadium reduction	V(V) → V(IV) → V(III)	black and opaque colloids

Manganese is the fifth most abundant metal in the earth's crust where it exists as ferromanganese oxide deposits or as components of a variety of other rocks and minerals. In nature, the important oxidation states are II, III and IV; the latter two are the forms found in Mn oxide minerals. Mn(II) is the predominant soluble form of Mn in the environment. Although Mn(IV) is the thermodynamically favored form of Mn in the presence of oxygen, Mn(II) is quite stable at neutral pH because the energy of activation for Mn(II) oxidation is large (106). Thus, in the environment, when one sees Mn oxide deposits they are often due to Mn(II) oxidation catalyzed by microorganisms (33). As a consequence of simple redox chemistry, all organisms that oxidize Mn(II) are also able to oxidize Fe(II). This is because Mn oxides are strong oxidants that can adsorb and chemically oxidize soluble Fe(II) (107,108).

Iron- and manganese-depositing bacteria are widespread in the environment but are not phylogenetically distinct groups of microorganisms (32). Often, the activities of iron- and manganese-depositing organisms are inferred by their visual identification in environmental samples (57). The mechanisms by which these organisms oxidize Fe and Mn are diverse and may include enzymatic mechanisms, specific metal binding exopolymers, or nonspecific interactions with cell components or metabolites (32,33). Of the neutrophilic iron bacteria, only the stalked bean-shaped *Gallionella* appears to be capable of chemolithoautotrophic growth. For *Gallionella* the iron oxides accumulate on the stalk. Other iron- and manganese-depositing bacteria are recognized by their morphology such as *Metallogenium, Hyphomicrobium or Pedomicrobium, Siderocapsa,* and *Leptothrix*, which produce characteristic star shaped patterns, aggregates, capsules and sheaths, respectively. There are also a great variety of *Pseudomonad* spp. that have been isolated as Mn oxidizers (32,33,95).

Microorganisms that either produce enzymes that catalyze metal precipitation or metal-binding polymers that oxidize and precipitate metals specifically may have important applications for metal-removal processes. Unfortunately, very little is known about the biochemical mechanisms by which these organisms catalyze metal precipitation, although many are characterized by having acidic exopolymers on which the iron and/or manganese oxides are deposited.

One organism that may have some useful biotechnological applications is *Leptothrix discophora* (109). *L. discophora* is a sheathed bacterium and obligately aerobic. The most studied strain, SS–1, is a sheathless mutant that still maintains its ability to oxidize iron and manganese. SS–1 produces at least two extracellular iron- or manganese-oxidizing proteins, one that oxidizes Fe(II) and one that oxidizes Mn(II) (110-112). In sheath-forming strains of *L. discophora* the proteins are presumed to be components of the sheath. The manganese-oxidizing protein has an apparent molecular mass of 105-110 kD and a K_M for Mn^{2+} of 10 μM (110,111,113) while the iron-oxidizing protein has a molecular mass of ~150 kD. The gene coding for the manganese-oxidizing protein, *mofA*, has been cloned and sequenced; however, its sequence bears no significant homology with other proteins in the protein sequence data bases (114). The putative protein has 34 amino acids on the N-terminus with homology to the consensus signal peptide sequence, consistent with the fact that protein is transported outside the cell. It also has a region with a cluster of tryptophan residues that have been hypothesized to play a role in Mn^{2+} binding (114).

A variety of marine manganese-oxidizing bacteria have been isolated, although most of them are not very well characterized (37,38,115-118). Therefore their potential usefulness has not been fully explored. On the other hand, spore-forming *Bacillus* species are frequent manganese-oxidizing isolates (32,115,118,119). One particular strain of a spore-forming marine manganese oxidizing *Bacillus* sp., strain SG–1, has also been isolated (120-122) and studied in detail (122-132). The fact that it is the spores and not the vegetative cells that oxidize metals makes this organism a particularly promising microbe for biotechnological applications. This research on SG-1 is summarized in a separate section below.

Magnetite (Fe_3O_4) is another type of iron oxide mineral that can be formed biologically, but rather than by oxidation, it forms by reduction of Fe(III) under suitable conditions of pH and E_H (133,134). This extracellular magnetite formation results from the incomplete reduction of Fe(III) minerals and is catalyzed by dissimilatory iron- and sulfate-reducing bacteria (105,134,135). Extracellular magnetite formation has been studied in both the freshwater iron reducer *Geobacter metallireducens* strain GS–15 and in the marine sulfate-reducing bacterium *Desulfuromonas acetoxidans* (134,135). Magnetite formation in these strains is not due directly to enzymatic activity, but is due to reaction of reduced Fe(II) formed as a result of iron reduction with some of the remaining Fe(III) (105). However, while magnetite is known to form abiologically by reaction of Fe(II) with Fe(III) anaerobically, active metabolism of GS–15 was necessary to form magnetite under the culture conditions and media used (105).

Other minerals of iron and manganese besides oxides can also be formed by bacteria (57,95). Siderite ($FeCO_3$) or rhodochrosite ($MnCO_3$) will form anaerobically in cultures of *G. metallireducens* strain GS–15 and *D. acetoxidans* when grown in high bicarbonate concentrations (105,135) and vivianite ($Fe_3(PO_4)_2 \cdot 8H_2O$) can also form through the result of Fe(II) production *via* dissimilatory Fe(III) reduction (35). Bacterial sulfate reduction can also lead to the reduction of Fe(III) to Fe(II) by hydrogen sulfide. Fe(II) reacts rapidly with sulfide forming insoluble iron sulfide precipitates (iron monosulfide, FeS; pyrite, FeS_2; and greigite, Fe_3S_4).

Uranium (U) Reduction. Microbial reduction of uranium (U) leads to the transformation of soluble U(VI) to insoluble U(IV). This process is important in the biogeochemical cycling of uranium (136) and is believed to have led to the formation of uranium ore deposits. On a global scale, organic rich anaerobic marine sediments are the largest sink for dissolved uranium (136). There is a body of evidence that suggests that microbial U(VI) reduction is the driving force for this process and that it is tied to the onset of sulfate reduction. In most natural environments uranium is in the form of uranyl(VI)-carbonate complexes which upon microbial reduction is converted to insoluble uraninite (UO_2) (137).

The reduction of soluble U(VI) to insoluble U(IV) is a reaction that has been shown to be microbially mediated (138) and enzymatically catalyzed (137). Various sulfate-reducing bacteria have been shown to reduce uranium by a mechanism that involves cytochrome c_3 (139,140). It has been suggested that such microbial U(VI) reduction could be used for removing uranium from contaminated environments (141). Uranium-reducing bacteria include dissimilatory iron-reducing bacteria, e.g., *Geobacter metallireducens* and *Shewanella putrefaciens*, that can employ U(VI) as an electron acceptor for the oxidation of organic matter as well as sulfate-reducing bacteria (e.g., *Desulfovibrio* species) that don't appear to grow with U(VI) as the sole electron acceptor.

Chromium (Cr) Reduction. Cr(VI) is a highly toxic, soluble and hence mobile oxidation state of chromium in the environment which tends to associate with oxygen in the form of the oxidizing agents chromate (CrO_4^{2-}) and dichromate ($Cr_2O_7^{2-}$). Reduction of Cr(VI) to Cr(III) results in the precipitation of chromium, apparently as $Cr(OH)_3$. Cr(III) is also the most stable form of Cr and the largest reservoir of chromium on earth. Because this form is much less mobile than Cr(VI) there has been interest in microbes that can reduce Cr(VI). Although it has been suggested that organisms might employ Cr(VI) as a terminal electron acceptor for anaerobic growth (142), evidence supporting this suggestion is equivocal (103). Nevertheless, it is clear that a wide variety of organisms are capable of reducing Cr(VI) both aerobically and anaerobically (142) and that it appears, at least in some cases, to result from co-metabolism. The sulfate-reducing bacteria that reduce uranium(VI), *Desulfovibrio desulfuricans* and *D. vulgaris*, are also capable of reducing Cr(VI). Like U(VI) reduction, Cr(VI) appears to be reduced by cytochrome c_3 (103). Recently a consortium of sulfate-reducing bacteria with a high tolerance of Cr(VI) was obtained from an electroplating sludge and shown to effectively reduce Cr(VI) (143). Given the importance of sulfate reduction in marine environments, it may be that Cr(VI) reduction is a common property of marine sulfate-reducing bacteria.

APPLICATIONS OF METAL-PRECIPITATING BACTERIA

Biological applications of metal-precipitating bacteria fall into two broad classes: biomaterials, in which the bacteria are used for the production of metal precipitates or minerals with specific properties, and metal removal and recovery, in which transforming metals from a soluble to an insoluble form is the primary goal. Microbial metal precipitation has a number of industrial applications including environmental metal detoxification and clean-up, waste treatment, recovery of pollutant or precious metals, and production of catalysts or adsorbents. In this section I will touch upon a few of the applications where bacteria may make useful contributions to biotechnology. Most of these applications have been with terrestrial organisms; however, marine bacteria will find increasing applications as more organisms and their metal-precipitating properties become known.

Biomaterials

Metal precipitation by bacteria has the potential to yield useful materials by virtue of the fact that the precipitates themselves may have useful properties. Metals are known to catalyze a variety of chemical reactions and metal-containing solids can enhance reactions by providing adsorption and coordination sites for reactants. Although this area of biotechnology is relatively unexplored there are general types of materials that could be industrially (commercially) applied, such as catalysts, adsorbents, or other specific mineral forms. The only real requirements of the microbial process are that the product (mineral or precipitate) has useful properties, that its structure and properties are reproducible, and it is economical relative to other means of production.

One class of metal precipitate that has potential applications because of their material properties is the manganese oxides. Mn oxides are fairly strong oxidants that can oxidize a number of different organic and inorganic compounds nonspecifically. Mn oxides have even been found to react with recalcitrant organic compounds such as humic substances producing low molecular weight organic compounds (144). Mn oxides have highly charged surfaces so they adsorb a variety of different metals (and organics), thereby affecting the mobility and bioavailability of other metals and organics in the environment. Certain Mn oxide minerals have a number of industrially important uses including as catalysts, sorbents and electrical conductors. It is the crystalline structure of certain Mn oxides that make them suitable for industrial application. Some forms, for example, todorokites which are found in deep-sea manganese nodules, have crystalline channels (pores) within their mineral structure that allow them to absorb and release positively charged cations and the manganese within the mineral lattice can accept varied numbers of electrons. Certain structurally-disordered Mn oxides are also used as depolarizers in dry-cell batteries (145). Most of the Mn oxide used industrially, however, is synthesized chemically or electrolytically because most natural ores are relatively impure and have lower catalytic activities. Microorganisms that oxidize Mn could be used to produce oxides. For example, under certain conditions marine manganese-oxidizing bacteria form todorokites (146-148). Microbially-formed metal precipitates may have some properties that are different from those produced chemically and it may be possible to use Mn oxidizing bacteria for the controlled production of Mn oxides.

Microorganisms may be used to direct the synthesis of materials with unique properties. The design and synthesis of materials with nanometer scale dimensions is an important area of research in materials science. Materials with small particles can possess novel electronic, chemical, catalytic, optical and magnetic properties (149). Recently, bacterial cell walls have been used as a matrix for the mineralization of metal precipitates as threadlike microfibers called "bionites" (149). The mineralized fibers were produced with special strains of *B. subtilis* that were selected for their ability to grow as web cultures. From these cultures threads of uniform diameter could be drawn, treated with various metal-containing solutions, and the mineralized bionites would form. The bionite system may provide opportunities to manipulate the synthesis of new materials such as immobilized cell enzyme systems or materials with unusual magnetic or optical properties (149).

Other biominerals that are produced by certain bacteria, such as magnetite and ferritin, which, because of their magnetic properties, may also be useful materials. Bacterial magnetites have unique advantages over those produced synthetically because of the membrane that surrounds the mineral (150). The membrane prevents aggregation of the magnetic particles and permits macromolecules to be conjugated to them at higher levels than artificially coated abiotic magnetites. Bacterial magnetites have a variety of biotechnological and medical applications. Bacterial magnetites are being tested in electronics as memory devices for computers and in the production of audio-visual recording tape (150). Bacterial magnetites have been used for enzyme, antibody and DNA immobilization (150,151) and may be employed as biosensors. Bacterial magnetites have also been incorporated into red blood cells and leukocytes (151). Magnetized cells may be manipulated by applying a magnetic field. Thus magnetic control could aid in guiding immunocompetent cells or leukocytes to the sites of tumors for treatment or visualization by magnetic resonance imaging, or aid in the purification of cells cultured for the production of useful compounds, such as interferon (150). Bacterial magnetites may also have applications as projectiles for transferring DNA into cells with ballistics. The membrane surrounding the magnetite allows more DNA to be immobilized per particle and the "transformed" cells could be selected based on their magnetic properties (150).

Metal Removal and Recovery

Bacteria have broad potential applications in technologies that require the removal or immobilization of metals and their subsequent recovery. These applications fall into two types, namely the recovery of valuable metals and the remediation and treatment of wastes. These two types of applications require somewhat different microbial properties, yet the technologies for application are basically the same.

Recovery of Precious Metals. Microbial leaching of metals has become common practice in solubilizing and recovering metals from ores. This "biomining" is both cost effective and able to recover metals from low grade ores. Just as metal-solubilizing bacteria may be useful, metal-precipitating bacteria may find applications in the recovery of trace levels of precious metals, either from dilute solutions or mineral leachates. In this case the selectivity of microbes for metals is used for the removal of economically valuable metals from solutions containing low concentrations of the desirable metal or from solutions that contain high concentrations of unwanted metals. Many metals, such as cobalt and nickel, are considered strategic elements to the United States because essentially all resource needs are imported from other countries. Technologies that would be able to recover valuable metals such as these from our own resources (e.g., seawater) would reduce our dependence on foreign countries. While this idea has been around for a long time, the technologies have not been available. Biologically-based technologies, however, are relatively unexplored in this area, and given that organisms living in dilute solutions have solved their problems for obtaining essential elements, it may be that some of these organisms have unique mechanisms that can be exploited for metal recovery. With suitable organisms in hand, then engineered systems, similar to those for waste treatment, could be exploited for metal recovery (see below).

Bioremediation and Waste Treatment. The topic that has received the most attention with respect to biological metal removal is for the remediation of metal pollution and treatment of metal-containing wastes. The term bioremediation is used to describe the use of living organisms (usually microorganisms) to remove or detoxify toxic or unwanted chemicals from an environment. Unlike organic pollutants which can ultimately be degraded to CO_2 and water, metal, radionuclide and related pollutants such as metalloids, organometals, organometalloids and metallic particulates require unique approaches. Figure 2 depicts a number of important elements that occur as pollutants in different environments. These pollutants can be found in liquid and solid industrial effluents and as byproducts during mining, smelting and refining operations, in leather tanning operations, in film processing plants, and in other manufacturing processes. Radionuclides are frequently released in nuclear fuel reprocessing plants and mobilized during uranium mining. The result of the release of these pollutants is that a variety of terrestrial, aquatic, and marine environments are impacted which interferes with our recreational activities and jeopardizes our water supplies and food resources. A problem of recent concern to many coastal states is the presence of metal-contaminated sediments, especially in estuaries and bays bordered by industrial activity. Such sediments are a constant source of metal into the overlying water, a problem that is magnified when such sediments are disturbed due to natural or anthropogenic causes such as storm events or dredging activities.

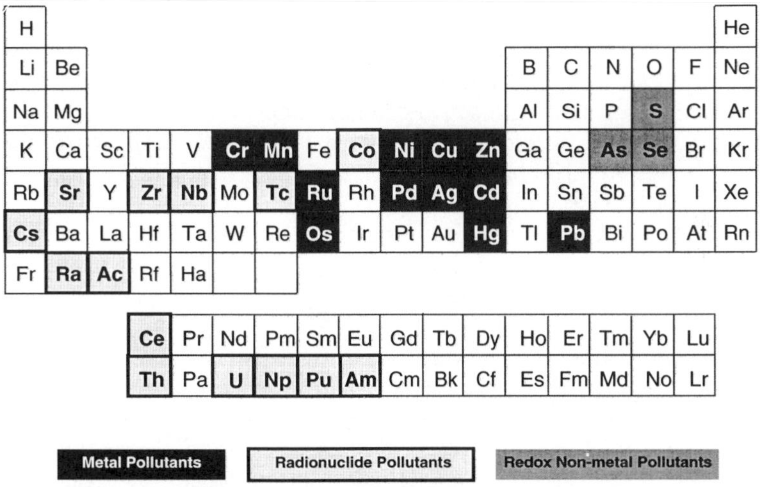

Figure 2. The periodic table of pollutants. Some of the major metals, radionuclides and redox-sensitive non-metal elements that occur as pollutants in different environments.

Metals and radionuclides or their compounds are pollutants because of their toxicity, i.e., their environmental health effects. Factors that influence metal toxicity include metal speciation and concentration, both of which affect the mobility (movement) of pollutants in the environment. In some cases, the elements can be transformed to a state that is either nontoxic or has reduced toxicity. Often these are insoluble or particulate phases. Metal precipitation by microorganisms could perhaps be used to immobilize and detoxify metal pollutants *in situ* in certain environments. However, maintaining the nontoxic metal species for extended periods of time is problematic and hence the ultimate solution for unwanted metal and radionuclide pollution is to remove them from the environment. This requires physically separating the metals from the environmental milieu. Most conventional techniques such as ion exchange or chemical precipitation, however, are often ineffective at reducing metal and radionuclide concentrations to acceptable levels.

Biological-based technologies have received great interest in relation to the treatment of wastes and for the remediation of environmental pollution. Most biological processes to date have employed sorption onto a variety of different types of biomass, including bacteria, yeast, fungi, algae and their biopolymers, to remove the metals and radionuclides from solution. Living systems exploiting active metal precipitation processes are receiving more attention recently, particularly in the decontamination of wastes where metals (and metalloids) are not at levels toxic to the desirable organism.

Biological Properties Amenable for Use in Metal Removal and Recovery Technologies. In selecting microbial processes, whether passive or active, for exploitation in metal removal and recovery processes, a number of factors are necessary for the technology to be successful (75,152,153):

1. The microbes must be able to remove large quantities of metal. Preferably this would be extracellularly so that metal recovery would be simplified and possibly permit the recycling of the organisms.
2. Metal removal must be selective enough to remove specific metals or mixtures of metals from solutions which often contain other ions or chemicals that compete with or inhibit metal removal or complex the targeted metal(s).
3. Metal removal must occur over a range of environmental conditions (e.g., pH, temperature, metal concentration).
4. The microbes must be able to do at least as well as other physicochemical procedures such as chemical precipitation, ion exchange, solvent extraction, and adsorption onto activated charcoal.
5. It must be cost effective relative to physicochemical treatments. If the recovered metal has any value it can help offset the cost of producing or reusing the organisms.
6. The microbes should be resistant to physical and chemical stresses and possibly even biological effects that might affect them in the metal-removal process.

To date, biosorption has achieved the most interest for applications in metal removal because it doesn't require live or growing microbes. Thus, one doesn't have to worry about maintaining activity or growth of cells which can be drastically affected by various environmental parameters including physical parameters (e.g., temperature), the chemical environment (e.g., pH, redox potential and the presence of toxic or competitive substances), as well as biological effects (e.g., competition for growth substrates by other microbes, attack by viruses or predation). Biosorption is also very rapid and efficient and often inexpensive, employing waste biomass from other biotechnologies (e.g., fermentation). Frequently the loaded biomass can be treated so that the valuable elements can be recovered for recycling or waste minimization. Thus biosorption has been shown to be a feasible alternative to physical and chemical treatments for removing metal and radionuclide pollutants from liquid waste (154).

Active metal-removal processes have a number of different advantages depending on the application. Frequently active metal-removal processes can reduce metal concentrations to levels lower than those achieved through biosorption and may have a higher degree of metal specificity or metal affinity. Since metals are transformed into particulate phases they are "fixed" and less susceptible to desorption that could be induced by changes in pH, chemical complexation, or competition with other metals. Because the organisms are alive, there is more versatility for the development of tailored processes. For example, there is the potential for the cells to metabolize organometallic compounds or for multiple organisms and processes to be combined into a "synergistic" process for waste treatment. Furthermore, there is the potential for living systems to be used in continuous processes for extended time periods without the need to regenerate the biomass and living systems may provide additional capabilities for bioremediation of organic pollutants and nitrate (154). Finally, and perhaps one of the greatest advantages, is that these active microbial metal-removal processes may lend themselves to genetic manipulation. Those that may lend themselves most readily to genetic improvement are those in which the activity is catalyzed by proteins that may be modified or hyperexpressed to enhance metal-removal processes.

The different active processes, i.e., redox transformations, precipitation reactions and intracellular accumulation, have different advantages depending on the application. Redox transformations allow the alteration of metal speciation and permit the process to be tailored to those metals whose solubilities vary with redox state. Metal-precipitation reactions have applications to a broad spectrum of metal pollutants and mixed wastes. The use of intracellular accumulation is somewhat more limited: since the metals are taken up within the cells, cell disruption is required in order to recover them and re-use is not possible.

Probably the most critical issue for potential application of an active microbial metal-removal process is to evaluate whether it will require growth or just the activity of the microbes. Clearly, the problem will be more complex if growth of the organisms has to be sustained (152). To sustain growth the treatment system essentially becomes either a batch or a continuous culture. In either case a variety of different factors can affect growth including the availability of the necessary nutrients (including metal nutrients) required for growth, the concentration in the medium of the target metal(s) to be removed, or the potential for components in the medium to be toxic, to complex, or to compete for uptake of either metals required for growth or the target metal(s). Some of these problems are diminished in continuous flow-through systems because the input of the various chemicals will be relatively constant and the organisms will have a chance to adapt. However, the flow rate will be limited to that which the organisms can physiologically keep up. For wastes with extremely high levels of substances that potentially could interfere in the bioprocess, a pretreatment step could be performed and the biological process used as a secondary treatment to further reduce metal concentrations. One advantage of a growing system, however, is that because the organisms are constantly growing and making new cells, the system is more self sustaining and thus potentially can have a greater capacity for metal removal.

Any treatment system will be substantially simplified if non-growing active microbes, either immobilized or in suspension, can be used. Although some of the same considerations are necessary as for treatment systems involving growth, not having to sustain growth makes this approach much more attractive for possible application. The active microbial biomass can be prepared under controlled conditions, presumably enabling the production of a product that has high capacity and specificity for target metals. Cells attached to supports can be employed in batch or column systems if they possess good mechanical properties. If the metals can be stripped or separated from the biomass such that the biomass can be re-used and the metal recovered, then the technology becomes even more feasible. Several microbial metal-removal processes that show promise for application are described in more detail below.

Applications of Metal Removal and Recovery Biotechnology. Treatment processes involving active growth of microbes for metal removal are already widespread. Constructed systems employing living microbes are used for metal removal in sewage treatment facilities, in meanders and impoundments for metal removal, and in constructed wetlands (4). Sewage treatment processes remove pollutant metals, although not always to the required levels (4). In sewage treatment facilities metals are bound mostly by the exopolymer produced by live bacteria (8,155). Although the bacteria are living, metals are removed mostly *via* passive adsorption. Primary sedimentation removes about 40 to 60% of the total metal (with the exception of nickel and manganese) and secondary treatment in activated sludge or trickling filters removes another 40 to 60% (155).

In the treatment of acid mine drainage constructed wetlands and impoundments are being employed more frequently and are very effective at removing toxic metals (4). In these "natural" systems both the plants and the microbes contribute to metal removal. The algae and plants take up and adsorb metals during growth. Upon their death, further entrapment of particulates and biosorption result in metal removal from solution and, with settling, transport to the sediments. In the sediments, the dead biomass fuels the growth of sulfate-reducing bacteria resulting in hydrogen sulfide production and hence active metal-sulfide precipitation. So long as these sediments remain anaerobic, resolubilization of the metal sulfides should not occur.

The use of these "natural" treatment processes, however, are somewhat limited in their application because they are subject to seasonal effects on growth which in many parts of the world are very limiting and there is often not sufficient contact of the water with the living biomass. Engineered treatment processes optimized for microbial metal removal and recovery (i.e., with the use of organisms specifically selected for their metal-removal abilities), perhaps operated in a fashion analogous to the activated sewage process for wastewater treatment, where part of the activated sludge from the settling tank is used as an inoculum for the aeration tank, offer approaches for treating large volumes of industrial waste. However, all of these systems, while effective for metal removal, do not provide much opportunity for metal recovery and recycling. In sewage treatment facilities the endproduct is still a waste that has to be disposed of and this can present further problems if the metal concentrations in the sewage sludge are high. The "natural" systems, because the metals are continuing to accumulate, are generally believed to have a limited lifetime and eventually will need to be restored. Thus, these approaches solve short-term problems (i.e. water treatment), but long-term solutions really require recovery and recycling of the materials.

Immobilized living biomass, where microbial cells are attached to a solid substrate which supports biofilm development or incorporated into a matrix, may have industrial applications for both metal removal and recovery (4,156,157). Immobilized biomass can be employed in a number of different bioreactor designs (2). There can be associated problems with trying to sustain growth in the presence of potentially toxic substances. However, this can be overcome by employing organisms selected for their resistances to toxic metals (158,159) or by employing enzymatically-mediated metal accumulation which permits decoupling of cell growth from metal accumulation (156). This area of biotechnology is only beginning to develop, but it shows promise because active metal-removal mechanisms can be exploited (increasing our arsenal of biological weapons against metal pollution) and these systems may be more readily tailored to specific applications with greatly improved results as compared to simple biosorption. These systems may also be ideal for use as polishing systems for the final scrubbing of the water after a primary treatment.

Active metal-precipitation processes are being studied for bioremediation of metal and radionuclide pollution. A variety of different bacteria are able to reduce chromium(VI) (chromate and dichromate) to Cr(III) which is much less toxic and soluble. Thus this process may be exploited for the remediation of contaminated waters and soils (103,160). *Bacillus subtilis* and sulfate-reducing bacteria such as *Desulfovibrio desulfuricans* and *D. vulgaris* are all capable of reducing Cr(VI). It should thus be possible to use marine sulfate-reducing bacteria for the bioremediation of Cr in marine environments.

Uranyl-carbonate complexes are common forms of U in waste waters from uranium mining and milling operations. Carbonate is often used to extract U(VI) in mining operations, in nuclear fuel processing and from ion exchange resins and biosorbents. Microbial uranium reduction has the potential to precipitate U as U(IV) from all of these sources (141). The iron- and manganese-reducing bacterium *Geobacter metallireducens* GS-15 can enzymatically reduce soluble U(VI) to insoluble U(IV) (138). The sulfate reducing bacterium *Desulfovibrio desulfuricans* is capable of reducing U(VI) in a variety of different waters including acid mine drainage and groundwaters from the DOE site in Hanford, Washington (141). Although the limits weren't tested, U(VI) were reduced to levels less than 30 nM (7 ppb) and about 11 g of U(VI) were reduced per gram of cells. Reduction occurred both at pH 4 and at pH 7.4 with live as well as cells freeze-dried and stored aerobically. U(VI) reduction in *Desulfovibrio* is mediated by cytochrome c_3 (140) and is tolerant to high U(VI) concentrations. Although biosorption is often inhibited by other ions such as Mn^{2+}, Ca^{2+}, SO_4^{2-}, Co^{2+}, Cu^{2+} and Ni^{2+}, only high levels of Cu completely inhibited U(VI) reduction by *Desulfovibrio* (141). An alternate approach to uranium removal and recovery employs uranium precipitation as uranyl phosphate rather than uranium reduction (100). A *Citrobacter* sp. when grown on organic phosphate, a condition that induces the activity of an acid phosphatase, causes a variety of different metals to be precipitated as metal phosphates. This *Citrobacter* sp. accumulates several times its own weight in uranyl phosphate (HUO_2PO_4) at the cell surface and the uranium can be removed with bicarbonate. Employing these uranium- precipitating microbes in bioreactors offers a promising approach for removing and recovering uranium from liquid wastes. An excellent review on the application of biotechnology to the treatment of radioactive wastes, including both active and passive biologically-based methods, has been published (12) and to which the reader is referred for additional potential applications of metal-precipitating bacteria.

One of the more promising biotechnological approaches to metal/radionuclide removal and recovery is biomagnetic separation (161-163). Biomagnetic separation involves the use of microorganisms to precipitate metals on the cell surface producing a coating that can adsorb a variety of metal ions and organic compounds. In the process of metal precipitation, the microbes become either paramagnetic or ferromagnetic and can then be concentrated and removed from suspension by high gradient magnetic separation (HGMS). The activities of the microbes are involved in three ways: 1) the bacteria enhance metal precipitation through their metabolic (biochemical) activities; 2) the metal precipitates that form *via* microbial action have more favorable surface properties (e.g., higher surface area) than those produced by alternate chemical means and act as good adsorbents; and 3) the microbes can be manipulated (e.g., through the use of nutrients or genetically) to enhance the magnetic properties of the precipitates, the microbes themselves, or both.

Preliminary experiments have demonstrated the feasibility of removing certain metals through the formation of metal phosphates and metal sulfides, or by co-precipitation or adsorption to such phases, in combination with recovery by HGMS (162). Metals and radionuclides that have been removed as phosphates include U, Ru, Sr, Co, Cs, Ce Zr, ^{242}Pu, ^{241}Am and ^{60}Co (162). Tests with metal sulfides produced by *Desulfovibrio* (162,164) have demonstrated that not only are metals ions that form insoluble metal sulfides (e.g., Ag, Hg, Pb, Cu, Zn, Sb, Mn, Fe, As, Ni, Sn, Al) removed from solution but other metals as well (e.g., Rh, Au, Ru, Pd, Os, Pt, Cr) (162). Apparently many of these precious metals are being collected onto the surface of FeS produced by *Desulfovibrio*. The mechanism for this appears to be due to substitutions of the metals in vacancies within the crystal structure of a highly disordered and fibrous FeS (162). Observations of this iron sulfide by electron microscopy has shown that it has a very high surface area, comparable to that of highly absorbent activated charcoal (163). This iron sulfide material was also found to adsorb metal cations with a higher capacity and affinity than activated charcoal (163). In some cases sulfate-reducing bacteria may produce a magnetic sulfide (162,165). Such organisms may have better magnetic properties and thus have an advantage for use in biomagnetic separation.

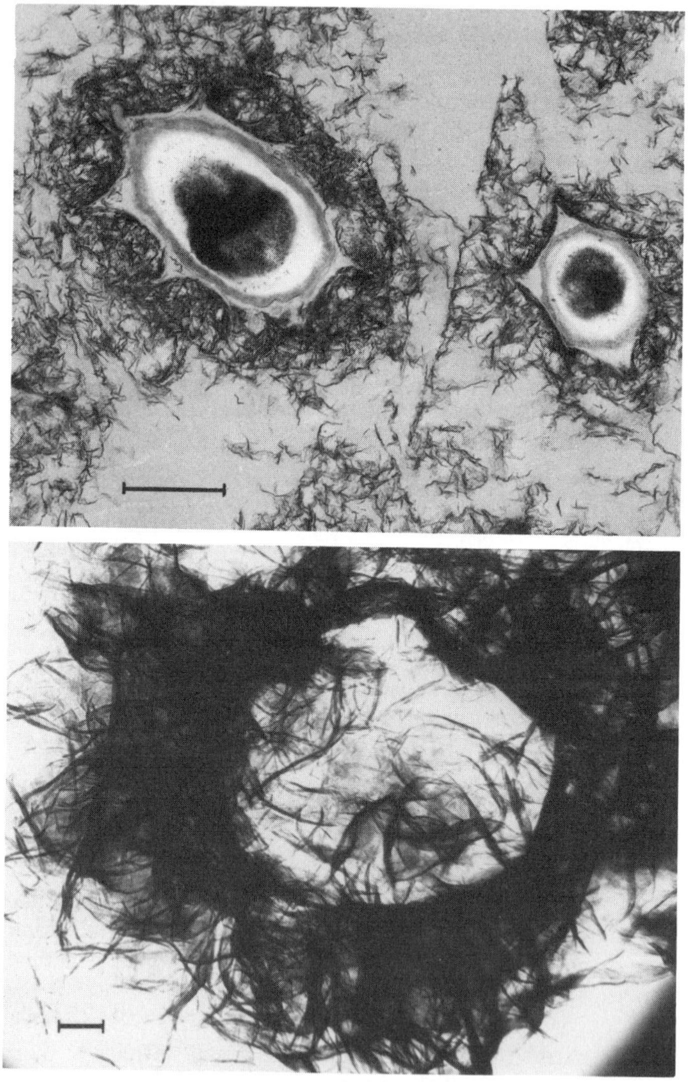

Figure 3. Top, Transmission electron micrograph of a thin section of the metal-oxidizing spores of the marine *Bacillus* sp. strain SG-1. The spores are coated with manganese oxides. Bar = 0.5 μm. Bottom: A todorokite-like manganese oxide precipitate produced by SG-1 as observed directly in the transmission electron microscope. The precipitate has been extracted to remove the organic material of the spore. Bar = 0.1 μm.

Microbially enhanced precipitation and HGMS has been successfully applied to the removal and recovery of mercury (Hg) from a contaminated industrial effluent (162). In this case several kg of *Desulfovibrio* were added to >5000 l of effluent along with some Fe(II). After 20 h incubation the effluent was processed by HGMS and in a single pass through the system, the mercury content was reduced from >2 ppm to 5 ppb. This test demonstrates the feasibility of the approach in wastewater treatment.

In estuarine and coastal marine sediments with high organic inputs, sulfate reduction usually occurs just below the sediment water interface (from millimeters to several centimeters) and results in the formation of iron sulfides which make the sediments black in appearance. Black metal sulfide rich sediments from areas around a nuclear reprocessing plant on the Irish Sea were found to concentrate the radioactivity. Eighty to ninety percent of the radioactivity could be removed by HGMS (162), suggesting that biomagnetic separation with sulfate-reducing bacteria coupled with HGMS might offer a solution to the widespread problems of metal pollution in sediments.

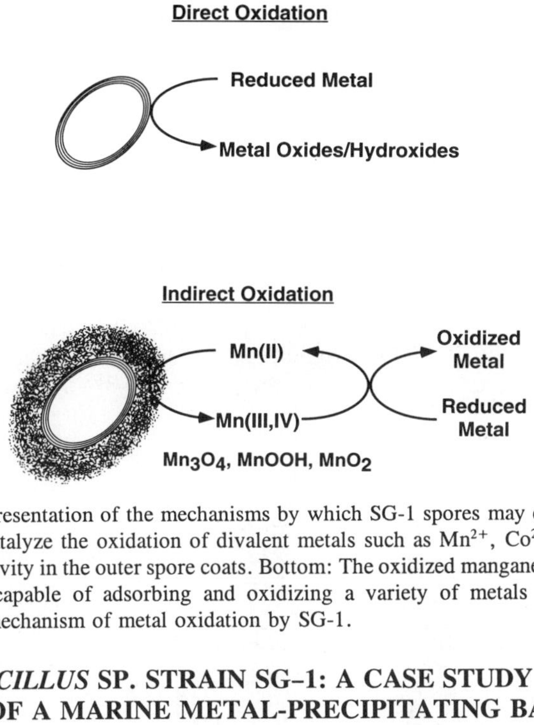

Figure 4. Schematic representation of the mechanisms by which SG-1 spores may catalyze metal oxidation. Top, The spores may catalyze the oxidation of divalent metals such as Mn^{2+}, Co^{2+} and Fe^{2+} directly as a result of the catalytic activity in the outer spore coats. Bottom: The oxidized manganese precipitates are strong oxidants that are also capable of adsorbing and oxidizing a variety of metals nonspecifically. This is considered an indirect mechanism of metal oxidation by SG-1.

THE MARINE *BACILLUS* SP. STRAIN SG–1: A CASE STUDY FOR POTENTIAL APPLICATIONS OF A MARINE METAL-PRECIPITATING BACTERIUM

One marine microorganism that has been extensively studied is the manganese(II)-oxidizing bacterium, *Bacillus* species strain SG-1. SG-1 was isolated from an enrichment culture of a nearshore marine sediment (120). In SG-1 it is the mature (dormant) spores that oxidize Mn(II) (Figure 3; 121,122,125). SG-1 catalyzes Mn oxidation under conditions where oxidation is thermodynamically favored, but kinetically slow; thus no coupling to metabolic energy is involved. As described below, SG-1 has a number of features that make it a potentially useful organism both for applications in metal removal and for production of metal oxides.

SG–1 spores are capable of binding and/or oxidizing a variety of different metals and thus may have applications for solving problems where mixed metals occur. SG-1 is capable of binding or oxidizing metals both directly on the spore surface or through the adsorption or oxidation on the Mn oxides that accumulate on the spore surface (Figure 4). Thus SG-1 spores have metal-removal capability that exploit active metal-precipitation processes as well as passive sorption. To date, we have demonstrated that SG-1 can oxidize cobalt(II) (131,166) and iron(II) (Tebo and Edwards, unpublished data) in addition to Mn(II). The spores also bind cadmium and zinc (Figure 5). From studies of the adsorption of metals onto Mn oxides, we know that a variety of other metals and radionuclides will be removed including uranium, lead, thorium, radon, cadmium, zinc, mercury, chromium, nickel and silver.

SG–1 spores have both a high affinity for metals and a high capacity for metal precipitation. SG-1 spores oxidize Mn(II) and Co(II) over a wide range of concentrations, from less than nanomolar to greater than millimolar (122,131,166). Manganese oxides can accumulate on the surface of the spores up to approximately 6 times their own weight under ideal conditions. Using radioactive tracers of Mn(II) (^{54}Mn) and Co(II) (^{57}Co), we have been able to demonstrate that SG-1 is able to bind and oxidize Mn(II) and Co(II) even at low tracer only additions of radioisotope to HEPES buffer. Both Mn(II) and Co(II) oxidation follow Michaelis-Menten kinetics. The kinetics of Co(II) oxidation as a function of Co(II) concentration are shown in Figure 6. An Eadie-Hofstee plot of the data suggests that SG-1 spores have two oxidation systems for Co(II), a high-affinity–low-rate system (K_M, 3.3×10^{-8} M; V_{max}, 1.7×10^{-15} M·spore^{-1}·h^{-1}) and a low-affinity–high-rate system (K_M, 5.2×10^{-6} M; V_{max}, 8.9×10^{-15} M·spore^{-1}·h^{-1}). The K_M for the high affinity system (33 nM) translates to an affinity on the order of 2 parts per billion (ppb) suggesting that SG-1 spores can remove metals (at least Mn and Co) to much lower levels than achieved by other chemical or biological procedures. In fact, we have demonstrated that SG-1 can remove Co(II) to levels less than parts per trillion! The radionuclide, ^{60}Co, is a significant radionuclide contaminant and product resulting from the nuclear power industry and has been identified at a number of Department of Energy sites in the United States. SG–1 spores may have useful applications for dealing with these ^{60}Co problems.

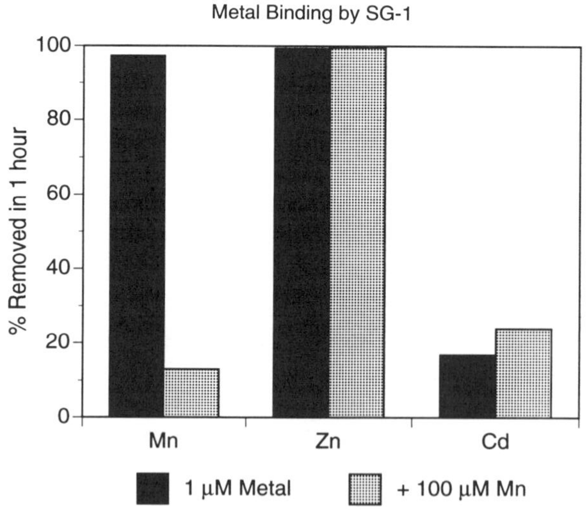

Figure 5. SG-1 is capable of binding zinc (Zn) and cadmium (Cd) in addition to manganese (Mn). The dark bars indicate the percentage of metal removal in 1 hour when the spores were incubated in solutions containing 1 μM of the indicated metal. The lighter bars indicate the percentage of metal removed when the spores were supplemented with 100 μM Mn(II) in addition to 1 μM of the indicated metal. Control experiments without spores showed no metal removal.

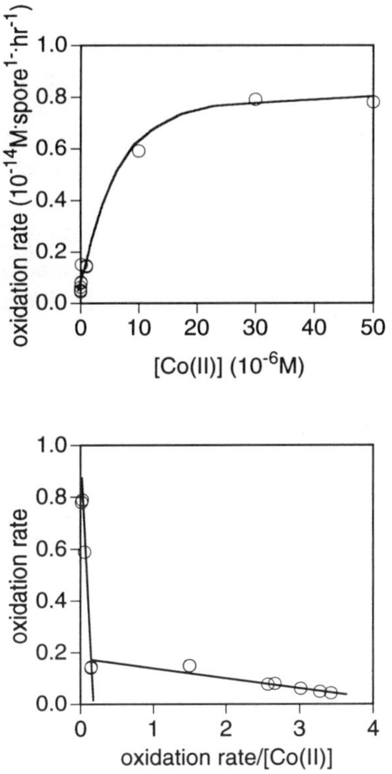

Figure 6. Kinetics of cobalt oxidation by SG-1 spores in artificial seawater. Top, Saturation of cobalt oxidation by high concentrations of Co(II) typical of Michaelis-Menten kinetics. Cobalt oxidation occurs even with tracer additions of the radioisotope ^{57}Co(II). Bottom, Eadie-Hofstee plot of the data shown in the top panel indicates that there are two cobalt oxidation sites on the spores, one with high affinity and one with low affinity. Adapted from ref. 131.

SG-1 spores have a high specificity for catalyzing metal precipitation. SG-1 catalyzes the oxidation of Mn(II) almost equally well in seawater with its abundance of potentially competing divalent cations as in freshwater or buffered distilled water. In addition, the oxidation of Mn(II) is not significantly inhibited by mixtures of other metals such as Cu, Cd, Ni and Zn.

There are a variety of other useful properties of SG-1 spores that offer advantages for applications. It is the inherent nature of bacterial spores to be resistant to physical and chemical agents and stresses. Spores are resistant to heat, UV light, toxic chemicals, detergents, denaturants, reductants and proteolytic enzymes. The fact that spores are a dormant resting stage means that growth of the organisms does not have to be maintained in order to have metal-removal activity. In fact, SG-1 spores rendered nonviable (so they don't germinate) still oxidize Mn(II) (122). Thus, if SG-1 were to be applied in the environment, one wouldn't have to worry about their uncontrolled growth, because nonviable spores could be employed. SG-1 is also active over a wide range of conditions of pH (>6), temperature (2 to 80°C), and osmotic strength (122,131,166), so the organism is environmentally versatile. Finally, the spores can be treated to remove the metal oxides and they maintain their metal-oxidizing capability (125). Thus, there is a potential for the spores to be re-cycled.

SG-1 spores may also be useful for the controlled production of manganese oxides. As described in the preceding section, Mn oxides have a number of industrial uses. We have employed SG-1 as a model system to understand the mechanisms and mineralogy of the microbially-produced Mn oxides so that the system might be manipulated for better application. The Mn mineral forms precipitated by SG-1 have been studied in controlled experiments under environmentally relevant conditions of pH and Mn concentration in both sea water and in fresh water (126,148,167). By altering the temperature and Mn concentration different Mn minerals could be formed by SG-1 and, for the most part, the phases that form are those that are thermodynamically predicted (126,127,148). In general, mixed phases of lower valence state Mn(II) or Mn(III) minerals (hausmannite, Mn_3O_4, feitknechtite, βMnOOH and manganite, γMnOOH) formed at high Mn(II) concentrations or at high temperatures (127,148). At lower Mn(II) concentrations, Mn(IV) minerals precipitated (126,148). In fresh water, the Mn(VI) minerals precipitated were characterized by a 10Å phase that collapsed to 7Å upon air drying. In sea water magnesium-rich non-collapsible 10Å Mn(IV)-manganate minerals formed (148). Thus Mn mineralization by SG-1 involves a complex interplay between the environmental conditions and catalytic macromolecules.

Finally, there is strong evidence that Mn(II) oxidation is catalyzed by a specific protein on the spore surface (125,128). This fact suggests that modern molecular approaches may be used to enhance or alter the metal-precipitating characteristics of the spores. Our group has developed a genetic system for SG-1 enabling us to create transposon mutants and complement mutants with cloned genes in SG-1 (129,130). We have discovered an operon involved in manganese oxidation that contains 7 genes (called *mnx* genes, Figure 7; 168). Sequence analysis of the genes (129) has demonstrated that one of the genes codes for a predicted 138 kD protein that is related to the family of multicopper oxidases, a fairly divergent group of proteins that use multiple copper ions to oxidize a variety of substrates (169). These predicted proteins possess signature regions for copper binding. Upstream of *mnxA* is a putative ribosome binding site and the consensus sequence for the sporulation-specific σ^K promoter. Thus, this operon would be expected to be expressed during sporulation. Expression of a β-galactosidase fusion to *mnxD* supports this prediction. Further investigation into these genes and their products, combined with chemical measurements of the surface properties and mineralogy of the metal oxides produced by the purified protein, will provide new insights into the biomineralization process and the potential utility of these biologically-produced Mn oxides. SG-1 is closely related to *B. subtilis* according to rRNA sequence analysis. Decades of genetic and biochemical research on the structure, function, and formation of bacterial spores in *B. subtilis* can most likely be directly applied to the SG-1 system.

In summary, the marine *Bacillus* sp. strain SG-1 is particularly promising for environmental applications because 1) growth does not have to be sustained, 2) viable

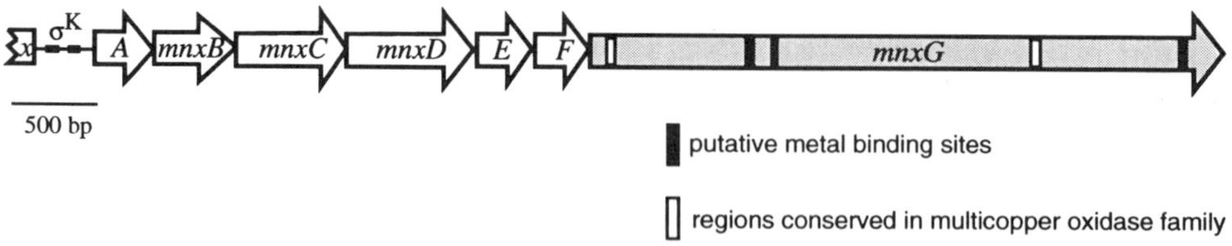

Figure 7. Map of the manganese oxidation (Mnx) operon in the marine *Bacillus* sp. strain SG-1. *mnxG* is the candidate gene coding for a manganese-oxidizing protein based on its putative metal-binding sites and other regions of similarity with proteins from the multicopper oxidase family. Heavy arrows indicate open reading frames of predicted proteins. The sporulation-specific sigma factor consensus sequence is designated by σ^K. Modified from ref. 129.

organisms do not have to be introduced into the environment, 3) spores are naturally resistant structures, 4) metal precipitation occurs over a wide range of environmentally relevant conditions (pH, temperature, metal concentration, and in both fresh and saline waters), 5) the spores have a high capacity, affinity and specificity for metals, 6) the spores may be re-used after their metal coats are stripped, and 7) the protein-mediated oxidation system can be genetically manipulated. The fact that metals are removed both through oxidative precipitation and *via* adsorption onto both the spores and their metal oxides makes them more versatile and enhances their utility for treating a variety of metal and radionuclide pollutants.

FUTURE NEEDS

In the previous sections I have not included much discussion about genetic aspects of metal precipitation. There's a good reason for this: genetic studies of metal–microbe interactions, in general, and metal precipitation, in particular, are limited, primarily because of the lack of basic knowledge of the physiology and biochemistry of these microbial interactions with metals. Certainly, our best understanding at the genetic level concerns the mechanisms of metal tolerance (resistance), such as metallothioneins in yeast and mercury resistance in bacteria. When it comes to metal precipitation, there is almost a complete lack of any genetic understanding. Genetic approaches offer promising avenues for improving metal removal processes and providing new materials.

Before genetics can make a significant impact in this field, there is a need for a better understanding at the molecular and physiological level of metal precipitation. When one considers the diversity of bacteria, few metal-precipitating organisms have been isolated and characterized. We are beginning to gain an understanding of the role of cell wall components in the passive binding and nucleation of metal precipitation and mineralization by organisms not necessarily selected for their ability to precipitate metals (e.g., 1); however, we really need to come to a similar level of understanding of the mechanisms used by bacteria that actively precipitate metals. Only then will it become more feasible to manipulate the metal-precipitating properties through physiological and genetic means.

Both physiological and genetic approaches may be used to alter functional groups to produce molecules with enhanced specificity and metal-binding capacity. For example, by growing *B. subtilis* cells under different nutrient regimes (phosphate limitation) it is possible to alter the composition, charge, porosity and morphology of the cell wall (153,170). One of the greatest advantages of the active metal precipitation process, however, is the potential to use genetics for strain improvement, because the mechanism of metal removal involves a property of the microbes rather than a product (96). Although not well studied, proteins are often involved, suggesting that a variety of molecular approaches can be taken to enhance the process, for example, overexpression or modification of the protein (171,172). Depending on the goal it may be desirable to engineer properties related to the metal-binding and precipitation process such as rate, efficiency, stability, capacity, affinity and specificity or to the mineral product, such as the form and crystal size that give minerals their electrical, optical, chemical, catalytic, physical and magnetic properties.

The ultimate goals, the design of an efficient and effective metal removal process or the controlled synthesis of minerals with unique properties, will require exploiting an organism's natural ability as well as engineering improvements. Marine bacteria are largely unexplored with regard to their natural interactions with metals. For example, metal tolerance and metal-precipitation mechanisms of hydrothermal vent bacteria or metal-acquisition mechanisms of bacteria living in the oligotrophic (nutrient poor) areas of the ocean may represent novel adaptations that marine bacteria have to cope with the metal stresses or limitation in their environment. Intracellular biomineralization has also been largely ignored in marine bacteria (perhaps with the exception of magnetotactic microbes), so this area of marine microbiology is ripe for exploration. In the next decade, with increasing interest in the health of the environment and the production of new materials, rapid advances in this new frontier of biotechnology are expected.

ACKNOWLEDGMENTS

I would like to acknowledge the dedicated work of my former graduate students for helping to lay the scientific foundation for the applied aspects of SG–1 for metal removal and mineral formation: Lori van Waasbergen, Yoon Lee and Kevin Mandernack. I also thank Margo Haygood for her critical reviews of the manuscript and Deeanne Edwards for excellent technical assistance. The University of California Toxic Substances Research and Teaching Program kindly provided the seed funding to explore the potential applications of manganese-oxidizing organisms in bioremediation. Parts of the work described herein have also been supported by grants from the National Science Foundation (MCB-9407776), the National Sea Grant College Program (National Oceanic and Atmospheric Administration, U.S. Department of Commerce grant NA36RG0537; project R/CZ-123 of the California Sea Grant College), and the California State Resources Agency. The views expressed herein are those of the author and do not necessarily reflect the views of NOAA or any of its sub-agencies. The U.S. Government is authorized to reproduce and distribute for governmental purposes.

REFERENCES

1 Beveridge, T.J. (1989) Annu. Rev. Microbiol. 43, 147-171.
2 Tsezos, M. (1990) in Microbial Mineral Recovery (Ehrlich, H.L. and Brierley, C.L., eds.) pp. 325-339, McGraw-Hill, New York, NY.
3 Eccles, H. and Hunt, S. eds., Immobilisation of Ions by Bio-sorption (1986) 266 pp., Ellis Horwood Limited.
4 Brierley, C.L., Brierley, J.A. and Davidson, M.S. (1989) in Metal Ions and Bacteria (Beveridge, T.J. and Doyle, R.J. eds.), pp. 359-382, Wiley, New York, NY.
5 Brierley, C.L. (1990) in Microbial Mineral Recovery (Ehrlich, H.L. and Brierley, C.L. eds.), pp. 303-323, McGraw-Hill, New York, NY.
6 Gadd, G.M. (1990) Experientia, 46, 834-840.
7 Gadd, G.M. (1990) in Microbial Mineral Recovery, (Ehrlich, H.L. and Brierley, C.L., eds.) pp. 249-275, McGraw-Hill, New York, NY.
8 Geesey, G. and Jang, L. (1990) in Microbial Mineral Recovery, (Ehrlich, H.L. and Brierley, C.L., eds.) pp. 223-247, McGraw-Hill, New York, NY.
9 Greene, B. and Darnall, D.W. (1990) in Microbial Mineral Recovery, (Ehrlich, H.L. and Brierley, C.L., eds.) pp. 277-302, McGraw-Hill, New York, NY.
10 McLean, R.J.C. and Beveridge, T.J. (1990) in Microbial Mineral Recovery (Ehrlich, H.L. and Brierley, C.L., eds.) pp. 185-222, McGraw-Hill, New York, NY.
11 Volesky, B., ed. (1990) Biosorption of Heavy Metals, 396 pp., CRC Press, Boca Raton, FL.
12 Macaskie, L.E. (1991) Crit. Rev. Biotech. 11, 41-112.

13 Tobin, J.M., White, C. and Gadd, G.M. (1994) J. Indust. Microbiol. 13, 126-130.
14 Hamer, D. H. (1986) Annu. Rev. Biochem. 55, 913-951.
15 Olafson, R.W. (1986) Env. Health Persp. 65, 71-75.
16 Sunda, W.G., Barber, R.T. and Huntsman, S.A. (1981) J. Marine Res. 39, 567-586.
17 Trick, C.G., Andersen, R.J., Gillam, A. and Harrison, P.J. (1983) Science 219, 306-308.
18 Hudson, R.J.M. and Morel, F.M.M. (1990) Limnol. Oceanogr. 35, 1002-1020.
19 Price, N.M. and Morel, F.M.M. (1990) Nature 344, 658-660.
20 Trick, C.G. (1989) Curr. Microbiol. 18, 375-78.
21 Buyer, J.S., de Lorenzo, V. and Neilands, J.B. (1991) Appl. Env. Microbiol. 57(8), 2246-2250.
22 Haygood, M.G., Holt, P.D. and Butler, A. (1993) Limnol. Oceanogr. 38(5), 1091-1097.
23 Reid, R.T., Live, D.H., Faulkner, D.J. and Butler, A. (1993) Nature 366(6454), 455-458.
24 Arcuri, E.J. and Ehrlich, H.L. (1977) Appl. Env. Microbiol. 33, 282-288.
25 Harvey, R.W. and Leckie, J.P. (1985) Marine Chemistry 15, 333-344.
26 Kellems, B.L. and Lion, L.W. (1989) Estuarine, Coastal and Shelf Science 28, 443-457.
27 Lion, L.W. and Rochlin, K.L. (1989) Estuarine, Coastal and Shelf Science 29, 11-22.
28 Flatau, G.N., Clement, R.L. and Gauthier, M.J. (1986) Microbios Lett. 32, 141-146.
29 Flatau, G.N., Clement, R.L. and Gauthier, M.J. (1988) Progr. Oceanogr. 21, 181-188.
30 Harwood-Sears, V. and Gordon, A.S. (1990) Appl. Env. Microbiol. 56, 1327-1332.
31 Schreiber, D.R., Millero, F.J. and Gordon, A.S. (1990) Marine Chemistry 28, 275-284.
32 Ghiorse, W.C. (1984) Annu. Rev. Microbiol. 38, 515-550.
33 Nealson, K.H., Tebo, B.M. and Rosson, R.A. (1988) Adv. Appl. Microbiol. 33, 279-318.
34 Nealson, K.H., Rosson, R.A. and Myers, C.R. (1989) in Metal Ions and Bacteria, (Beveridge, T.J. and Doyle, R.J., eds.) pp. 383-411, Wiley, New York, NY.
35 Lovley, D.R. (1991a) Microbiol. Rev. 55, 259-287.
36 Jannasch, H.W. and Wirsen, C.O. (1981) Appl. Envir. Microbiol. 41, 528-538.
37 Ehrlich, H.L. (1983) Ecological Bulletin (Stockholm) 35, 357-366.
38 Ehrlich, H.L. (1985) in Planetary Ecology (Caldwell, D.E., Brierley, J.A. and Brierley, C.L., eds.), pp. 186-194, Van Nostrand Reinhold, New York, NY.
39 Jannasch, H.W. and Mottl, M.J. (1985) Science 229, 717-725.
40 Cowen, J.P., Massoth, G.J. and Baker, E.T. (1986) Nature 322, 169-171.
41 Durand, P., Prieur, D., Jeanthon C. and Jacq, E. (1990) Comptes Rendus (Serie III) 310(7), 273-278.
42 Alt, J.C. (1988) Marine Geology 81, 227-239.
43 Juniper, S.K. and Fouquet, Y. (1988) Canad. Mineralogist 25, 859-869.
44 Karl, D.M., McMurty, G.M., Malahoff, A. and Garcia, M.O. (1988) Nature 335, 532-524.
45 Jorgensen, B.B., Zawacki, L.X. and Jannasch, H.W. (1990) Deep-Sea Research 37, 695-710.
46 Juniper, S.K. and Tebo, B.M. (1995) in The Microbiology of Deep-Sea Hydrothermal Vents (Karl, D.M., ed.) CRC Press (in press).
47 Hutchins, S.R., Davidson, M.S., Brierley, J.A. and Brierley, C.L. (1986) Annu. Rev. Microbiol. 40, 311-336.

48 Geesey, G.G. and Jang, L. (1989) in Metal Ions and Bacteria (Beveridge, T.J., Forsberg, S.W. and Doyle, R.J., eds.) pp. 325-257, Wiley, New York, NY.
49 Simpson, L. M. and Oliver, J.D. (1983) Infect. Immun. 41, 644-649.
50 Griffiths, G.L., Sigel, S.P., Payne, S. and Neilands, J.B. (1984) J. Biol. Chem. 259(1), 383-385.
51 Takahashi, A., Nakamura, H., Kameyama, T., Kurasawa, S., Naganawa, H., Okami, Y., Takeuchi, T., Umezawa, H. and Iitaka, Y. (1987) J. Antibiotics 40(12), 1671-1676.
52 Jalal, M.A.F., Hossain, M.B., van der Helm, D., Sanders-Loehr, J., Actis, L.A. and Crosa, J.H. (1989) J. Amer. Chem. Soc. 111, 292-296.
53 McKnight, D.M. and Morel, F.M.M. (1980) Limnol. Oceanogr. 25, 62-71.
54 Jardim, W.F. and Pearson, H.W. (1984) Water Research 18, 985-989.
55 Robinson, J.B. and Tuovinen, O.H. (1984) Microbiol. Rev. 48, 95-124.
56 Lowenstam, H. and Weiner, S. (1989) in On Biomineralization, Oxford University Press, New York, NY.
57 Ghiorse, W.C. and Ehrlich, H.L. (1992) Catena Supplement, 21, 75-99.
58 Frankel, R.B. (1991) in Iron Biominerals (Frankel, R.B. and Blakemore, R.P., eds.) pp. 1-6, Plenum Press, New York, NY.
59 Blakemore, R.P. (1982) Annu. Rev. Microbiol. 36, 217-238.
60 Charley, R.C. and Bull, A.T. (1979) Arch. Microbiol. 123, 239-244.
61 Pooley, F.D. (1982) Nature 296, 642-643.
62 Higham, D.P., Sadler, P.J. and Scawen, M.D. (1986) Env. Health Persp. 65, 5-11.
63 Chmielowski, J. and Klapcinska, B. (1986) Appl. Env. Microbiol. 51, 1099-1103.
64 Van Dyke, M.I., Lee, H. and Trevors, J.T. (1989) Arch. Microbiol. 152, 533-538.
65 Wood, J.M. and Wang, H.-K. (1983) Env. Sci. Tech. 17, 582A-590A.
66 Hausinger, R.P. (1987) Microbiol. Rev. 51, 22-42.
67 Gadd, G.M. (1990) in Microbiology of Extreme Environments (Edwards, C., ed.) pp. 178-210, McGraw-Hill, New York, NY.
68 Farina, M., Esquivel, D.M.S. and Lins de Barros, H.G.P. (1990) Nature 343, 256-258.
69 Mann, S., Sparks, N.H.C., Frankel, R.B., Bazylinski, D.A. and Jannasch, H.W. (1990) Nature, 343, 258-261.
70 Blakemore, R.P. and Blakemore, N.A. (1991) in Iron Biominerals, (Frankel, R.B. and Blakemore, R.P., eds.), pp. 51-67, Plenum Press, New York, NY.
71 Andrews, S.C., Harrison, P.M. and Guest, J.R. (1989) J. Bacteriol. 171, 3940-3947.
72 Harrison, P.M., Artymiuk, P.J., Ford, G.C., Lawson, D.M., Smith, J.M.A., Treffry, A. and White, J.L. (1989) in Biomineralization, Chemical and Biochemical Perspectives, (Mann, S., Webb J. and Williams, R.J.P., eds.), pp. 257-294, VCH Publishers, New York, NY.
73 Mann, S., Williams, J.M., Treffry, A. and Harrison, P.M. (1987) J. Mol. Biol. 198, 405-416.
74 Higham, D.P., Sadler, D.P. and Scawen, M.D.(1984) Science 225, 1043-1046.
75 Gadd, G.M. (1988) in Biotechnology: A Comprehensive Treatise in 8 Volumes (Rehm, H.-J., ed.) pp. 401-433, Volume 6b, VCH Verlagsgesellschaft, Weinheim.
76 Sakurai, I., Kawamure, Y., Koike, H., Inoue, Y. Kosako, Y., Nakase, T., Kondou, Y. and Sakurai, S. (1990) Appl. Env. Microbiol. 56, 2580-2583.
77 Blakemore, R.P. and Frankel, R.B. (1989) in Metal-Microbe Interactions (Poole, R.K. and Gadd, G.M., eds.), pp. 85-98, IRL Press, Oxford.
78 Mann, S. and Frankel, R.B. (1989) in Biomineralization, Chemical and Biochemical Perspectives (Mann, S., Webb, J. and Williams, R.J.P., eds.), p.389-426, VCH Publishers, New York, NY.

79 Bazylinski, D.A. and Frankel, R.B. (1992) Catena Supplement 21, 147-160.
80 Stolz, J.F. (1992) Catena Supplement 21, 133-146.
81 Spring, S., Amann, R., Ludwig, W., Schleifer, K.-H., van Gemerden, H. and Petersen, N. (1993) Appl. Env. Microbiol. 59, 2397-2403.
82 Bazylinski, D.A., Frankel, R.B. and Jannasch, H.W. (1988) Nature 334, 518.
83 Bazylinski, D.A. (1991) in Iron Biominerals, (Frankel, R.B. and Blakemore, R.P., eds.), pp. 69-76, Plenum Press, New York, NY.
84 Sakaguchi, T., Burgess, J.G. and Matsunaga, T. (1993) Science 365, 47-49.
85 Hughes, M.N. and Poole, R.K. (1989) in Metals and Micro-organisms, Chapman and Hall, London.
86 Matsunaga, T., Sakaguchi, T. and Tadokoro, F. (1991) Appl. Microbiol. Biotech. 35, 651-655.
87 Matsunaga, T., Nakamura, C., Burgess, J.G. and Sode, K. (1992) J. Bacteriol. 174, 2748-2753.
88 Smith, J.M.A., Andrews, S.C., Guest, J.R. and Harrison, P.M. (1991) in Iron Biominerals (Frankel, R.B. and Blakemore, R.P., eds.) pp. 325-337, Plenum Press, New York, NY.
89 Kurokawa, T., Fukumori, Y. and Yamanaka, T. (1989) Biochim. Biophys. Acta 976, 135-139.
90 Andrews, S.C., Smith, J.M.A., Hawkins, C., Williams, J.M., Harrison, P.M. and Guest, J.R. (1993) Eur. J. Biochem. 213, 329-338.
91 Artymiuk, P.J., Bauminger, E.R., Harrison, P.M., Lawson, M., Nowik, I., Treffry, A. and Yewdall, S.J. (1991) in Iron Biominerals, (Frankel, R.B. and Blakemore, R.P., eds.), pp. 269-294, Plenum Press, New York, NY.
92 Theil, E.C. and Sayers, D.E. (1991) in Iron Biominerals (Frankel, R.B. and Blakemore, R.P., eds.) pp. 295-305, Plenum Press, New York, NY.
93 St. Pierre, T.G., Webb, J. and Mann, S. (1989) in Biomineralization, Chemical and Biochemical Perspectives, (Mann, S., Webb, J. and Williams, R.J.P., eds.) pp. 295-344, VCH Publishers, New York, NY.
94 Morel, F.M.M. and Hering, J.G. (1993) Principles and Applications of Aquatic Chemistry, Wiley, New York, NY.
95 Ehrlich, H.L. (1990) Geomicrobiology, Marcel Dekker, New York, NY.
96 Macaskie, L.E. and Dean, A.C.R. (1989) in Biological Waste Treatment (A. Mizrahi, ed.) pp. 159-201, Alan R. Liss Inc., New York, NY.
97 Macaskie, L.E., Dean, A.C.R., Cheetham, A.K., Jakeman, R.J.B. and Skarnulis, A.J. (1987) J. Gen. Microbiol. 133, 539-544.
98 Macaskie, L.E. and Dean, A.C.R. (1982) Env. Tech. Lett. 3, 49-56.
99 Macaskie, L.E. and Dean, A.C.R. (1984) J. Gen. Microbiol. 130, 53-62.
100 Macaskie, L.E., Empson, R.M., Cheetham, A.K., Grey, C.P. and Skarnulis, A.J. (1992) Science 257, 782-784.
101 Beveridge, T.J. and Murray, R.G.E. (1980) J. Bacteriol. 141, 876-887.
102 Nealson, K.H. and Myers, C.R. (1992) Appl. Env. Microbiol. 58, 439-443.
103 Lovley, D.R. (1993) Annu. Rev. Microbiol. 47, 263-290.
104 Nealson, K.H. and Saffarini, D. (1994) Annu. Rev. Microbiol. 48, 311-343.
105 Lovley, D.R. (1991) in Iron Biominerals (Frankel, R.B. and Blakemore, R.P., eds.) pp. 151-166, Plenum Press, New York, NY.
106 Stumm, W. and Morgan, J.J. (1981) Aquatic Chemistry, Wiley, New York, NY.
107 Postma, D. (1985) Geochim. Cosmochim. Acta. 49, 1023-1033.
108 Myers, C.R. and Nealson, K.H. (1988) Geochim. Cosmochim. Acta. 52, 2727-2732.
109 Ghiorse, W.C. (1986) Biotech. Bioeng. Symp. 16, 141-148.
110 Adams, L.F. and Ghiorse, W.C. (1987) J. Bacteriol. 169, 1279-1285.

111 Boogerd, R.C. and de Vrind, J.P.M. (1987) J. Bacteriol. 169, 489-494.
112 Corstjens, P.L.A.M., de Vrind, J.P.M., Westbroek, P. and de Vrind-de Jong, E.W. (1992) Appl. Env. Microbiol. 58(2), 450-454.
113 Emerson, D. and Ghiorse, W.C. (1992) Appl. Envir. Micro. 58(12), 4001-4010.
114 Corstjens, P. (1993) Ph.D. Thesis, Leiden.
115 Nealson, K.H. (1978) in Environmental Biogeochemistry and Geomicrobiology Volume 3, Methods, Metals and Assessment (Krumbein, W.E., ed.) pp. 847-858, Ann Arbor Science Publishers Inc., Ann Arbor, MI.
116 Ghiorse, W.C. and Hirsch, P. (1982) Appl. Env. Microbiol. 43(4), 1464-1472.
117 Ehrlich, H.L. (1984) in Microbial Chemoautotrophy (Strohl, W.R. and Tuovinen, O.H. eds.), pp. 47-56, Ohio State University Press, Columbus, OH.
118 Vojak, P.W.L., Edwards, C. and Jones, M.V. (1984) Microbios. 41, 39-47.
119 Lee, Y. (1994) Ph.D. Thesis, University of California, San Diego, CA.
120 Nealson, K.H. and Ford, J. (1980) Geomicrobiol. J. 2, 21-37.
121 Nealson, K.H. and Tebo, B. (1980) Origins of Life 10, 117-126.
122 Rosson, R.A. and Nealson, K.H. (1982) J. Bacteriol. 151, 1027-1034.
123 Kepkay, P.E. and Nealson, K.H. (1982) J. Bacteriol. 151(2), 1022-1026.
124 de Vrind, J.P.M., Boogerd, F.C. and de Vrind-de Jong, E.W. (1986) J. Bacteriol. 167, 30-34.
125 de Vrind, J.P.M., de Vrind-de Jong, E.W., de Voogt, J.-W. H., Westbroek, P., Boogerd, F.C. and Rosson, R.A. (1986) Appl. Env. Microbiol. 52, 1096-1100.
126 Hastings, D. and Emerson, S. (1986) Geochim. Cosmochim. Acta, 50, 1819-1824.
127 Mann, S., Sparks, N.H.C., Scott, G.H.E. and de Vrind-de Jong, E.W. (1988) Appl. Env. Microbiol. 54, 2140-2143.
128 Tebo, B.M., Mandernack, K. and Rosson, R.A. (1988) p. 201, Annual Meeting Amer. Soc. Microbiol., Washington, D.C.
129 van Waasbergen, L.G. (1993) Ph.D. Thesis, University of California, San Diego, CA.
130 van Waasbergen, L.G., Hoch, J.A. and Tebo, B.M. (1993) J. Bacteriol. 175(23), 7594-7603.
131 Lee, Y. and Tebo, B.M. (1994) Appl. Env. Microbiol. 60, 2949-2957.
132 Mandernack, K.W. and Tebo, B.M. (unpublished data).
133 Bell, P.E., Mills, A.L. and Herman, J.S. (1987) Appl. Env. Microbiol. 53, 2610-2616.
134 Lovley, D.R., Stolz, J.F., Nord, Jr., G.L. and Phillips, E.J.P. (1987) Nature 330, 252-254.
135 Roden, E.E. and Lovley, D.R. (1993) Appl. Env. Microbiol. 59, 734-742.
136 Klinkhammer, G.P. and Palmer, M.R. (1991) Geochim. Cosmochim. Acta 55, 1799-1806.
137 Gorby, Y.A. and Lovley, D.R. (1992) Env. Sci. Tech. 26, 205-207.
138 Lovley, D.R., Phillips, E.J.P., Gorby, Y.A. and Landa, E.R. (1991) Nature 350, 413-416.
139 Lovley, D.R. and Phillips, E.J.P. (1992) Appl. Env. Microbiol. 58, 850-856.
140 Lovley, D.R., Widman, P.K., Woodward, J.C. and Phillips, E.J.P. (1993) Appl. Env. Microbiol. 59, 3572-3576.
141 Lovley, D.R. and Phillips, E.J.P. (1992) Env. Sci. Tech. 26, 2228-2234.
142 Cervantes, C. (1991) Antonie van Leeuwenhoek 59, 229-233.
143 Fude, L., Harris, B., Urrutia, M.M. and Beveridge, T.J. (1994) Appl. Env. Microbiol. 60, 1525-1531.
144 Sunda, W.G. and Kieber, D.J. (1994) Nature 367, 62-64.
145 Greene, A.C. and Madgwick, J.C. (1991) Appl. Env. Microbiol. 57, 1114-1120.

146 Takematsu, N., Kusakabe, H., Sato, Y. and Okabe, S. (1988) J. Oceanographical Society of Japan 44, 235-243.
147 Sato, Y., Okabe, S. and Takematsu, N. (1989) Geochim. Cosmochim. Acta 53, 1883-1887.
148 Mandernack, K.W., Post, J. and Tebo, B.M. (unpublished data).
149 Mendelson, N.H. (1992) Science 258, 1633-1636.
150 Matsunaga, T. (1991) Trends Biotech. 9, 91-95.
151 Matsunaga, T. (1991) in Iron Biominerals (Frankel, B. and Blakemore, R.P., eds.) pp. 79-95, Plenum Press, New York, NY.
152 Brierley, C.L., Kelly, D.P., Seal, K.J. and Best, D.J. (1985) in Biotechnology Principles and Applications (Higgins, I.J., Best, D.J. and Jones, J. eds.), pp. 163-213, Blackwell Scientific Publications, Oxford.
153 Hancock, I.C. (1986) in Trace Metal Removal from Aqueous Solution, Special Publication No. 61 (Thompson, R., ed.) pp. 25-43, The Royal Society of Chemistry, London.
154 Gadd, G.M. (1992) in Microbial Control of Pollution (Fry, J.C., Gadd, G.M., Herbert, R.A., Jones, C.W. and Watson-Craik, I.A., eds.) pp. 59-88, Cambridge University Press.
155 Sterritt, R.M. and Lester, J.N. (1986) in Immobilisation of Ions by Bio-sorption (Eccles, H. and Hunt, S., eds.) pp. 121-134, Ellis Horwood Limited, Chichester.
156 Macaskie, L.E. (1990) J. Chem. Tech. Biotech. 49, 357-379.
157 Macaskie, L.E., Clark, P.J., Gilbert, J.D. and Tolley, M.R. (1992) Biotech. Lett. 14, 525-530.
158 Silver, S. (1991) in Biodeterioration and Biodegradation 8 (Rossmoore, H.W., ed.) pp. 308-339, Elsevier, London.
159 Silver, S. (1994) Res. Microbiol. 145, 61-67.
160 Mehlhorn, R.J., Buchanon, B.B. and Leighton, T. (1994) in Emerging Technology for Bioremediation of Metals (Means, J.L. and Hinchee, R.E., eds.) pp. 26-37 Lewis Publishers, Boca Raton, FL.
161 Watson, J.H.P. and Ellwood, D.C. (1987) IEEE Transactions Magnetics, 23, 3751-3752.
162 Ellwood, D.C., Hill, M.J. and Watson, J.H.P. (1992) in Microbial Control of Pollution (Fry, J.C., Gadd, G.M., Herbert, R.A., Jones, C.W. and Watson-Craik, I.W., eds.), pp. 89-112, Cambridge University Press.
163 Watson, J.H.P., Ellwood, D.C., Deng, Q., Mikhalovsky, S. and Evans, J. (1994) in Minerals Engineering '94, Lake Tahoe, NV (in press).
164 Bahaj, A.S., Watson, J.H.P. and Ellwood, D.C. (1989) IEEE Transactions Magnetics 25, 3809-3811.
165 Frebe, A.M. and Tate, D. (1961) J. Biochem. Microbiol. Tech. Eng. 3, 29-39.
166 Tebo, B. M. and Lee, Y. (1993) in Biohydrometallurgical Technologies, I (Torma, A.E., Wey, J.E. and Lakshmanan, V.L., eds.) pp. 695-704, The Minerals, Metals, & Materials Society, Warrendale, PA.
167 Mandernack, K.W. (1992) Ph.D. Thesis, University of California, San Diego, CA.
168 van Waasbergen, L.G. and Tebo, B.M. (unpublished data).
169 Ryden, L.G. and Hunt, L.T. (1993) J. Mol. Evol. 36, 41-66.
170 Remacle, J. (1990) in Biosorption of Heavy Metals (B. Volesky, ed.) pp. 293-301, CRC Press, Boca Raton, FL.
171 Regan, L. (1993) Annu. Rev. Biophys. Biomol. Struct. 22, 257-281.
172 Timmis, K.N., Steffan, R.J. and Unterman, R. (1994) Annu. Rev. Microbiol. 48, 525-557.

INDEX